Lecture Notes in Computer Science 1947
Edited by G. Goos, J. Hartmanis and J. van Leeuwen

T0223545

Springer
Berlin
Heidelberg
New York
Barcelona
Hong Kong
London
Milan
Paris
Singapore
Tokyo

Tor Sørevik Fredrik Manne Randi Moe
Assefaw Hadish Gebremedhin (Eds.)

Applied
Parallel Computing

New Paradigms for HPC
in Industry and Academia

5th International Workshop, PARA 2000
Bergen, Norway, June 18-20, 2000
Proceedings

Springer

Series Editors

Gerhard Goos, Karlsruhe University, Germany
Juris Hartmanis, Cornell University, NY, USA
Jan van Leeuwen, Utrecht University, The Netherlands

Volume Editors

Tor Sørevik
Fredrik Manne
Assefaw Hadish Gebremedhin
University of Bergen, Department of Informatics
5020 Bergen, Norway
E-mail:{Tor.Sorevik/Fredrik.Manne/assefaw}@ii.uib.no

Randi Moe
University of Bergen, Parallab
5020 Bergen, Norway
E-mail: Randi.Moe@ii.uib.no

Cataloging-in-Publication Data applied for

Die Deutsche Bibliothek - CIP-Einheitsaufnahme

Applied parallel computing : new paradigms for HPC in industry and
academia ; 5th international workshop ; proceedings / PARA 2000,
Bergen, Norway, June 18 - 20, 2000. Tor Sørevik ... (ed.). - Berlin ;
Heidelberg ; New York ; Barcelona ; Hong Kong ; London ; Milan ;
Paris ; Singapore ; Tokyo : Springer, 2001
    (Lecture notes in computer science ; 1947)
    ISBN 3-540-41729-X

CR Subject Classification (1998): G.1-4, F.1-2, D.1-3, J.1

ISSN 0302-9743
ISBN 3-540-41729-X Springer-Verlag Berlin Heidelberg New York

Springer-Verlag Berlin Heidelberg New York
a member of BertelsmannSpringer Science+Business Media GmbH
© Springer-Verlag Berlin Heidelberg 2001
Printed in Germany

Typesetting: Camera-ready by author, data conversion by PTP Berlin, Stefan Sossna
Printed on acid-free paper   SPIN 10780911   06/3142   5 4 3 2 1 0

Preface

The papers in this volume were presented at PARA 2000, the Fifth International Workshop on Applied Parallel Computing. PARA 2000 was held in Bergen, Norway, June 18-21, 2000. The workshop was organized by Parallab and the Department of Informatics at the University of Bergen. The general theme for PARA 2000 was *New paradigms for HPC in industry and academia* focusing on:

- High-performance computing applications in academia and industry,
- The use of Java in high-performance computing,
- Grid and Meta computing,
- Directions in high-performance computing and networking,
- Education in Computational Science.

The workshop included 9 invited presentations and 39 contributed presentations. The PARA 2000 meeting began with a one-day tutorial on OpenMP programming led by Timothy Mattson. This was followed by a three-day workshop.

The first three PARA workshops were held at the Technical University of Denmark (DTU), Lyngby (1994, 1995, and 1996). Following PARA'96, an international steering committee for the PARA meetings was appointed and the committee decided that a workshop should take place every second year in one of the Nordic countries. The 1998 workshop was held at Umeå University, Sweden. One important aim of these workshops is to strengthen the ties between HPC centers, academia, and industry in the Nordic countries as well as worldwide. The University of Bergen organized the 2000 workshop and the next workshop in the year 2002 will take place at the Helsinki University of Technology, Espoo, Finland.

October 2000   Tor Sørevik
                                                                 Fredrik Manne
                                                                 Randi Moe
                                                     Assefaw Hadish Gebremedhin

Organization

PARA 2000 was organized by the Department of Informatics at the University of Bergen and by Parallab.

Organizing Committee

Conference Chair: Petter Bjørstad (University of Bergen, Norway)
Local Organization: Assefaw Hadish Gebremedhin (University of Bergen, Norway)
Fredrik Manne (University of Bergen, Norway)
Randi Moe (University of Bergen, Norway)
Synnøve Palmstrøm (University of Bergen, Norway)
Tor Sørevik (University of Bergen, Norway)
Aase Spieler (University of Bergen, Norway)

Steering Committee

Petter Bjørstad   University of Bergen (Norway)
PARA 2000 Chairman

Jack Dongarra   University of Tennessee and Oak Ridge National Laboratory (USA)

Björn Engquist   PDC, Royal Institute of Technology, Stockholm (Sweden)

Kristjan Jonasson   University of Iceland, Reykjavik (Iceland)

Kari Laasonen   Center for Scientific Computing, Espoo (Finland)

Bo Kågström   Umeå University and HPC2N (Sweden), *PARA'98 Chairman*

Risto Nieminen   Helsinki University of Technology, Espoo (Finland)

Karstein Sørli   SINTEF, Dept of Industrial Mathematics, Trondheim (Norway)

Jerzy Waśniewski   Danish Computing Centre for Research and Education (UNI•C), Lyngby (Denmark), *PARA'94-'96 Chairman*

Sponsoring Institutions

The City of Bergen
The Research Council of Norway (NFR)
Sun Microsystems AS, Norway
SGI Norge AS, Norway
The University of Bergen

Table of Contents

Speaker Information

Industrial Applications: Challenges in Modeling and Computing

Jari Järvinen[1], Juha Haataja[1], and Jari Hämäläinen[2]

[1] CSC – Scientific Computing Ltd., P.O. Box 405, FIN–02101 Espoo, Finland
{Jari.Jarvinen, Juha.Haataja}@csc.fi
[2] Valmet Corp., P.O. Box 587, FIN–40101 Jyväskylä, Finland
Jari.P.Hamalainen@valmet.com

Abstract. Complex industrial applications can be examined using modeling and computing even when theoretical and experimental approaches fail. Modeling and computing are rapidly developing tools for engineering and science, but applying these tools in practice is a challenge. This article focuses on industrial multi-physics and multi-scale problems in the production of silicon crystals and the design of paper machines.

Keywords: Modeling, computing, multi-physics, multi-scale, industry, parallel computing, numerical methods, paper machines, silicon crystals.

1 Introduction

Scientific and engineering problems can be examined using modeling and computing [5,17], but this approach poses several challenges. We focus on industrial multi-physics applications, such as the growth of silicon crystals [16] and the modeling of paper machines [12,6].

The growth of silicon crystals, shown in Fig. 1, contains coupled physical phenomena: electromagnetism, fluid dynamics, heat transfer, and structural mechanics. Thorough modeling of this application requires multi-scale modeling spanning several length and time scales [3,16].

Modeling of paper machines is a different kind of challenge. The headbox of a paper machine, presented in Fig. 2, contains flows in a complex geometry. These flows are three-dimensional, turbulent and multi-phase. A comprehensive mathematical model leads to demanding computational tasks.

Computation of a multi-physics problem is a challenge. Coupled problems lead to systems of nonlinear equations, which have to be solved iteratively. Solving these coupled systems can be a huge computational effort. We'll present some ideas on numerical solution strategies in multi-physics problems. In addition, we'll discuss how to combine computational fluid dynamics and optimization.

T. Sørevik et al. (Eds.): PARA 2000, LNCS 1947, pp. 1–16, 2001.

Fig. 1. Growth of a silicon crystal.

Fig. 2. The headbox of a paper machine.

2   Possibilities and Challenges in Industrial Applications

Industrial applications are difficult to model and compute: both the physical phenomena and the structures and geometries are typically very complex. These modeling tasks involve significant challenges for scientific computing, and high performance computing (HPC) is undoubtedly needed. Thus, HPC is an essential part of industrial research and development (R&D).

There are a multitude of modeling problems presented by the modern industry. Challenges are presented, for example, by phenomena in the human body or by the physics of miniaturized electronics components.

Microelectromechanical systems (MEMS) [20] pose a multi-physics problem, which contains electrostatistics, structural analysis, fluid dynamics and heat transfer in miniaturized geometries. In fact, modeling MEMS devices may require us to combine molecular dynamics and continuous models.

Multi-physics problems are solved increasingly often. As an example, the model may contain fluid-structure interactions. Other possibilities are free surface problems or phase-change problems. These problems are typically solved in 2D today, but in 3D in the future.

Another demanding task is the modeling the human body, which is of interest to the pharmaceutical industry. We may be interested in simulating the blood flow in 3D bifurcations, designing new drugs, or modeling how electromagnetic fields behave in living tissues.

The modeling of turbulence in fluid dynamics poses significant challenges. Today, the Reynolds stress model is applied routinely in industrial problems, and large eddy simulation (LES) is becoming feasible. However, direct numerical simulation (DNS) of the Navies-Stokes equations will be impossible for several years to come.

Another challenge are multi-phase problems. This is somewhat similar to turbulence modeling: there exist smaller and smaller phenomena arising from direct simulation, and it is less and less feasible to perform averaging. Furthermore, multi-scale problems present us with a combination of different scales (micro, meso and macro scales).

We may need to perform several simulations to solve a single problem. For example, one may be interested in finding the optimal geometry in a fluid dynamics application. This is an example of shape optimization in computational fluid dynamics (CFD). For each optimization run, one may have to solve a CFD problem tens, hundreds or thousands of times. Also, there are requirements for performing real-time interactive simulations, for example in a virtual reality environment, or a virtual learning system (Fig. 3).

3   The Three Methods of R&D

We can use three different approaches in industrial R&D: experimental, theoretical and computational (Fig. 4). All these approaches are essential and needed, and they are usually applied together to solve a problem.

Fig. 3. Industrial R&D utilizes modern computer technology: interacting with a simulation in a virtual reality environment.

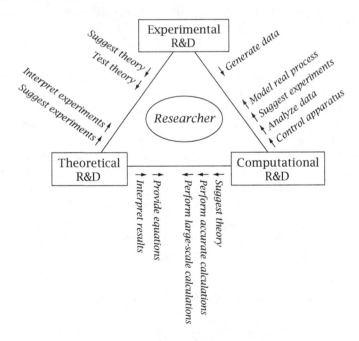

Fig. 4. The three methods of R&D.

The experimental approach is typically straightforward — basically, we measure what happens in the physical process we are interested in. Experiments are often reliable, at least to a known extent, which depends on the physical setup and the human interface of the experiment. Also, experiments are usually quantitative. However, experiments can be laborious or even impossible to carry out in complex cases. We mention the lack of commercial techniques to measure the turbulence of dense fiber-water suspensions. Experiments can also be very expensive and slow to carry out. For example, one may have to build a prototype to do the experiment.

The theoretical approach is efficient in simple cases. It is usually qualitative and suggestive — a good way to check the numerical or experimental results. However, the theoretical approach is insufficient in complex cases.

Computational modeling is repeatable, which is an asset: one does not need to build a new prototype to make a new experiment. However, there are a great many steps where mistakes can occur: mathematical modeling, selection of the numerical methods, and implementation of the methods. In addition, unanticipated hardware and software problems may occur during the computation.

4   The Challenges of Industrial HPC

Traditionally, industrial R&D has involved experiments and the internal know-how of a company. Nowadays R&D has shifted to a more innovative direction: combining the use of computation, experiments, and theory. Because all these approaches demand a different set of skills, this usually involves a collaboration of several experts.

The computational approach requires people who know how to solve mathematical models on a computer. Also, programming skills are required: real-life problems are not usually solvable using off-the-shelf software packages. Thus, R&D departments have to compete for talented people in other commercial sectors: www services, networking, etc.

The modeling and computing facilities of industrial companies are usually modest. Due to the fast development of this area, there is a continuing lack of know-how of numerical methods, programming, and computer technology (efficient workstations, clusters etc.). If one builds a computational research center from scratch, the costs easily exceed any budget. Therefore, industry is interested in co-operation with sites which have the required expertise, in addition to software and hardware resources. Usually this co-operation involves university laboratories.

However, also in the universities there is a lack of efficient training programs on HPC. At least in Finland, there may be just a single course on HPC topics, or none. Therefore, there is urgent need for teaching modeling and computing tools to the students at the university level. These courses should encourage students to use modeling techniques, numerical methods, commercial software packages, and programming languages.

5   Benefits of HPC in Industry

If everything goes according to plan, the computational approach results in high-quality products, lower production costs, and competitive advantages. Are these objectives achieved in practice? Not necessarily: research always has a possibility of failure. There is no point in researching things that are already known. Fortunately, the computational approach produces in any case knowledge, insights, and quantitative results.

We should always verify the results we get with computations. The usual way is to validate results with measurements, but also simplified theoretical models give at least qualitative results for comparison.

One should keep in mind that only long-term research produces good results. Thus, at first there has to be basic research, which produces qualitative results. After this, one may strive towards better models and quantitative results. When we have found a model, which produces good enough results, we can proceed to optimize the model. This can produce savings in costs, or improvements in the quality of products. At this step one reaches the long-term target of industrial R&D: competitive advantage.

6   Case: Czochralski Silicon Crystal Growth

6.1   Production of Crystalline Silicon

In this section we turn our attention to the electronics industry. We introduce the Czochralski crystal growth technique, which is the most commonly used method for producing the basic material, crystalline silicon, for electronics industry.

Electronics industry is the largest and the fastest growing manufacturing industry in the world. Computers and mobile phones are examples of the diverseness and extensive utilization of products manufactured by electronics industry. As another example, microelectromechanical systems are used in automotive industry to measure acceleration, vibration, inclination, pressure, and angular rate.

Crystalline silicon is a strategic material in the electronics industry. The market value of the world production is only a few billion (10^9) US dollars, but without this material, which has no proper substitute, the whole electronics industry, worth about one trillion (10^{12}) US dollars, would be very different and much smaller in volume.

6.2   How to Manufacture Silicon Crystals

Single crystal silicon is the most perfect solid material known to man, and it is available in reasonable quantities. It is manufactured mainly by two methods. The dominant one is the Czochralski method where the silicon charge is melted in a silica crucible, and the crystal is pulled upwards. The float-zone method, which is mainly applied to produce very lightly doped material for high-power applications, uses polysilicon rods as starting material. The rod is slowly passed through an induction coil.

Both methods use seed crystals. Typical crystal size is about 100 kg, and newer crystal growers are designed for crystal sizes of 300 mm in diameter and weights exceeding 300 kg (see Fig. 5).

Fig. 5. Industrially produced silicon crystals.

The Czochralski method is illustrated schematically in Fig. 6. In this method, silicon charge, made of polysilicon chunks or pellets, is first melted in a high-purity quartz crucible at above 1400°C under low-pressure argon atmosphere. After reaching the desired initial temperature and process conditions, crystal growth is initiated by dipping a small seed crystal into the melt. A cylindrical single crystal is then pulled vertically from the melt in such a way that it grows with a constant diameter except during the initial and final stages of the growth. This requires that, e.g., the pulling velocity and the heating power are carefully controlled during the growth process. The crystal rod and the crucible are usually rotated in opposite directions. Solid crystals are afterwards cut and mechanically and chemically polished to form the final products, thin silicon wafers (Fig. 7).

The commercial silicon material has evolved over a relatively short period of time, about 50 years. The material is virtually defect free. The dislocation densities are zero. Vacancy and interstitial properties, e.g., solubilities and diffusivities, are not properly known, due to very low concentrations. Contaminant concentrations are low to extremely low.

Most abundant foreign elements are intentional dopants. Electrically active dopants are added, typically between 1 ppba and 1000 ppma, to the silicon melt depending on desired electrical conductivity of the material. Oxygen, which is inherent additive in CZ material, originating from the silica crucible, is normally

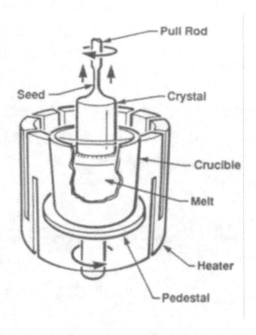

Fig. 6. A schematic configuration of Czochralski crystal growth [22].

Fig. 7. Industrially produced silicon wafers.

specified to a concentration in the vicinity of 15 ppma. Oxygen plays a role in strengthening the material against slip and provides gettering sites for deleterious impurities in high IC process temperatures. Carbon is the most abundant undesirable impurity.

Improvements in silicon wafer production are driven by the electronics industry. Material development is fairly rapid, and new smaller defect types are observed and brought under control about once a decade. It is more and more challenging and expensive to characterize the material, because the crystal sizes grow, quality requirements get more stringent, and the dominant defects get smaller. Therefore, the development costs of the growth processes skyrocket. However, at the same time, the developments in computer technology and mathematical modeling enable us to simulate the crystal growth process.

6.3 Mathematical Modeling of the Czochralski Crystal Growth

The Czochralski process is an example of multi-physics and multi-scale problem [16]. A comprehensive macro-scale model includes coupled heat transfer (conduction, convection and radiation), fluid flows (silicon melt flow, argon gas flow), magnetic field, free interfaces (phase change interface between crystal and melt, free interfaces between melt and gas and crystal and gas), mass transfer (impurities, dopants), and thermal stresses. Physical phenomena are three-dimensional, even when the geometry of the furnace is axisymmetric. This is especially the case for silicon melt flow, where the flow field in an axisymmetric crucible is three-dimensional, time-dependent and perhaps turbulent.

The growing process of single silicon crystal lasts normally 30–50 hours. Even though the growing process itself is rather slow, there are phenomena where the time scales are totally different. For instance, one can observe fluctuations in crystal properties re-occurring after every few tens of seconds. These fluctuations are caused by the constantly changing flow field of the melt. Oscillations in the melt flow reduce the homogeneity of the crystal. In the worst case, this may lead to the loss of single crystalline structure, causing loss of yield. Oscillations can partly be controlled by magnetic field. This brings new challenges to mathematical modeling [13].

Micro-scale modeling may consists of partial differential equations for vacancy and interstitial concentrations. These can be coupled with macro-scale models [3], but real use of these models requires more intensive work.

6.4 Computational Approach for the Czochralski Method

Modeling and computing offer new insight into crystal growth, and reduce dramatically the economical investments needed in experiments. However, experiments can not be totally omitted, since they offer invaluable information of the physical mechanisms. Furthermore, they form a basis for validating numerical results and provide necessary data for simulations.

Even if computer technology and numerical methods have evolved rapidly during last few years, they do not make it feasible to solve a comprehensive crystal growth model in a reasonable time. Thus, the mathematical model has to simplified or divided into parts. The subproblems have to be solved separately.

Computation of realistic silicon melt flow is no doubt the most time-consuming task. It requires the solution of three-dimensional and transient flow field coupled with the computation of magnetic field and temperature. In addition, the melt region is adjacent to two free interfaces, which have to be solved as a part of a solution. To complete the crystal growth model, the previous procedure has to couple with the global heat and mass transfer and gas flow models.

Pure melt flow has been considered in axisymmetric geometry and in 3D in [1,14,18,21]. In [25] the authors have applied massively parallel implementation in computing 3-dimensional melt flow. Global heat transfer has been studied in [2,4,19], and gas flow in [15].

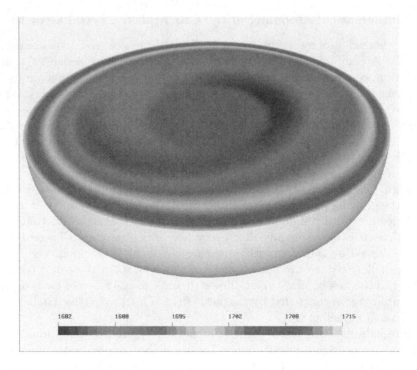

Fig. 8. Temperature distribution caused by 3D melt flow in Czochralski crystal growth.

7   Case: Modeling the Headbox of a Paper Machine

7.1   What Is a Headbox?

The headbox (Fig. 2) is the first component in the paper making process. The headbox is located at the wet end of a paper machine. The quality of produced paper is determined mostly by the fluid flow in the headbox. For example, variations in the basis weight and the fiber orientation are due to fluid dynamics at the wet end of a paper machine. These variations affect the quality of paper.

The first flow passage in the headbox is a dividing manifold called the header (Fig. 9). It is designed to distribute fiber suspension (wood fibers, filler clays and chemicals mixed in water) equally across the width of a paper machine. A turbulence generator, located after the header, dissipates fibre flocks by strong turbulent effects. Finally, the fibre suspension flows from the headbox to a forming section through a controlled outflow edge.

7.2   Turbulence and Multi-phase Phenomena

Modeling the fluid flow in the headbox contains large variety of challenges. Turbulence modeling is the most important in the turbulence generator, and in the slice channel. Turbulence determines small-scale variations of a paper sheet. The fluid velocity increases in the slice channel by a factor of ten, which changes turbulence (Reynolds stresses) in different directions. Therefore, anisotropy has to be taken into account in turbulence modeling.

Fibre suspension consists of wood fibers, filler clays and chemicals mixed in water. Small particles follows turbulence eddies but larger ones dampen turbulence. Hence, interactions of fibres and turbulence should be modeled. Even if turbulence could be modeled, interactions between time-averaged velocities of water and solids need an Eulerian multi-phase model.

The manifold tube bundle and the turbulence generator consist of hundreds of small tubes. Today, computing capacity is not enough to model those tubes in detail simultaneously with the whole headbox. Thus we are using homogenization methods to replace the tube bundles with effective boundary conditions, which take into account the avarage fluid flow through the tube bundle [6,11]. The larger the available computing capacity, the more details can be included in the computational domain, and thus the geometry of the headbox can be modeled more accurately.

7.3   Coupled Problems

Control of turbulence leads to coupled problems. The intensity of turbulence can be handled by using flexible plates inside the slice channel. The longer the vanes are, the higher turbulence intensity is introduced to the outflow from the headbox. From a modeling point of view, this leads to a coupled system where we have to solve interactions of bending of the vanes and fluid flow. Fluid flow from the headbox is a free jet and its direction and thickness is not a-priori known. This leads to a free boundary problem (Fig. 11).

7.4   Several CFD Simulations in One Problem

When CFD is coupled together with optimization or an interactive virtual reality (VR) environment, a CFD problem has to be solved in seconds. Otherwise the whole simulation would take too long.

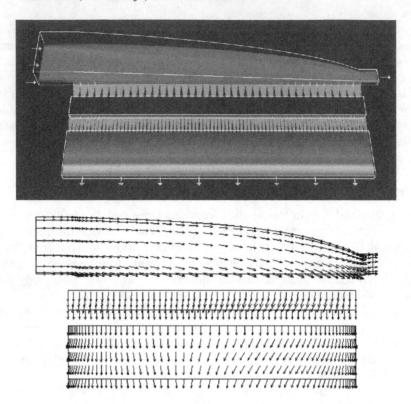

Fig. 9. Velocity magnitude (above) and velocity vectors (below) in the paper machine headbox. The shape optimization problem is to find optimal tapering of the header. The inlet of the header is on the left hand side and the main outlet is through the manifold tube bank.

Our optimization problem is to find the optimal shape of the dividing manifold such that the outlet flow rate distribution from the headbox is optimal resulting in optimal paper quality [8,7,9,23]. One example is given in Fig. 10. Three different shapes of the header (two of them are not optimized) are in the upper figure and the resulting flowrate profiles out from the header are in the lower figure. If only the tapering of the header is altered, we cannot affect on the flow rate profile on the left hand edge (see Fig. 4) because there is always a shortage of mass flow due to a boundary layer flow in the inlet tube of the header. We are also using optimization tools in order to study control problems of the headbox [10].

Today, virtual reality can be used to post-process the CFD results. In future it will be possible to run a CFD simulation in a VR environment so that the user can, for example, modify the boundary conditions and geometry during the simulation [24].

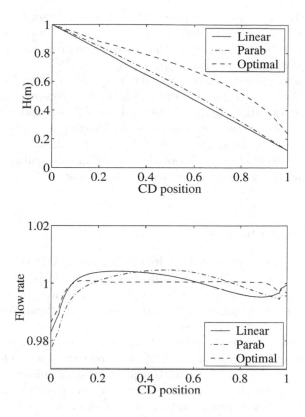

Fig. 10. Tapering of the header and related outlet mass distribution.

Fig. 11. Simulated velocity magnitude near the oulet edge of the vane headbox. A model consists of $k - \varepsilon$ turbulence model coupled with structural analysis model of the vanes. Also the shape of the free jet is modeled by using solution methods for free boundary problems.

8   The Future of HPC

Due to the increasing use on computational R&D in solving industrial problems, we see an increasing demand for HPC in industry. The successful use of computational R&D typically involves collaboration between universities, research laboratories and industry. There will be common research projects, where both basic research and new applications are developed. There is also an increasing need for high-quality courses on mathematical modeling, numerical methods, programming, and computer technology. To be able to develop current skills further, there should be good domestic and international contacts, enabling the exchange of researchers in the fast developing areas of HPC.

Efficient use of existing computer capacity is needed to solve the potentially difficult numerical models of industrial research. Thus, a company may want to build a cluster of workstations to distribute the load of these applications. However, this requires a lot of expertise, especially in setting up a parallel computing environment and making the applications scale well on a cluster. An alternative may be to use the existing computational capacity in universities or national HPC centers. However, all these approaches require systematic and organized HPC courses in universities.

9   Concluding Remarks

Modeling and computing will strengthen its role in industrial R&D in the future. The possibilities of solving difficult real-life problems will expand significantly, especially in multi-physics and multi-scale applications. However, to tackle these problems, scientists and engineers need to develop new mathematical and computational tools for industrial purposes.

In this article, we have discussed several non-trivial modeling problems which arise from real-life industrial problems. CSC is co-operating with Okmetic Ltd. and other Finnish companies to model the production of silicon material and MEMS devices. The modeling of paper machines is an active research topic at Valmet and co-operating research institutes. For example, all the headers of current paper machines manufactured by Valmet are designed by shape optimization.

These examples show that there exist demanding scientific problems to be solved in the real world applications. We believe there is need for increasing co-operation between application experts, theoretical and applied mathematicians, computer scientists, and programmers. Solving a real-life industrial problem has to be done in stages, combining different approaches, and this can take a lot of effort and time. However, the resulting competitive advantage can far outweigh the costs of research.

References

1. A. Anselmo, V. Prasad, J. Koziol, and K.P. Gupta. *J. Crystal Growth*, 134:116–139, 1993.
2. R.A. Brown, T. Kinney, P. Sackinger, and D. Bornside. *J. Crystal Growth*, 97:99, 1989.
3. E. Dornberger, D. Gräf, M. Suhren, U. Lambert, P. Wagner, F. Dupret, and W. von Ammon. Influence of boron concentration on the oxidation-induced stacking fault ring in Czochralski silicon crystals. *Journal of Crystal Growth*, 180:343–352, 1997.
4. F. Dupret, P. Nicodeme, Y. Ryckmans, P. Wouters, and M.J. Crochet. *Int. J. Heat Mass Transfer*, 33:1849, 1990.

5. Juha Haataja. Using Genetic Algorithms for Optimization: Technology Transfer in Action. In K. Miettinen, M. M. Mäkelä, P. Neittaanmäki, and J. Periaux, editors, *Evolutionary Algorithms in Engineering and Computer Science*, pages 3–22. Wiley, 1999.

6. J. Hämäläinen. *Mathematical Modelling and Simulation of Fluid Flows in the Headbox of a Paper Machine*. PhD thesis, University of Jyväskylä, 1993.

7. J. Hämäläinen, R.A.E. Mäkinen, and P. Tarvainen. Optimal Design of Paper Machine Headboxes. *International Journal of Numerical Methods in Fluids*. accepted 1999.

8. J. Hämäläinen, T. Malkamäki, and J. Toivanen. Genetic Algorithms in Shape Optimization of a Paper Machine Headbox. In K. Miettinen, M. M. Mäkelä, P. Neittaanmäki, and J. Periaux, editors, *Evolutionary Algorithms in Engineering and Computer Science*, pages 435–443. Wiley, 1999.

9. J. Hämäläinen and P. Tarvainen. CFD Coupled with Shape Optimization — A New Promising Tool is Paper Machine R&D. In L. Arkeryd, J. Bergh, P. Brenner, R. Pettersson, and B.G. Teubner, editors, *Progress in Industrial Mathematics at ECMI 98*, pages 171–178. 1999.

10. J. Hämäläinen and P. Tarvainen. CFD-based Shape and Control Optimization Applied to a Paper Machine Headbox. In *86th annual meeting of Pulp and Paper Technical Association of Canada*, pages A99–A102, 2000.

11. J. Hämäläinen and T. Tiihonen. Flow Simulation with Homogenized Outflow Boundary Conditions. In Morgan et al., editor, *Finite Elements in Fluids, New Trends and Applications*, pages 537–545. 1993.

12. E. Heikkola, R. A. E. Mäkinen, T. Rossi, and J. Toivanen. Shape optimization of a paper machine headbox. In Juha Haataja, editor, *CSC Report on Scientific Computing 1997–1998*. CSC – Scientific Computing Ltd., 1999. Www address http://www.csc.fi/reports/cr97-98/.

13. J. Heikonen, J. Järvinen, J. Ruokolainen, V. Savolainen, and O. Anttila. Simulation of Large-Scale Axisymmetric Silicon Melt Flow in Magnetic Czochralski Growth. In *to be presented in the proceedings of 3rd International Workshop on Modeling in Crystal Growth, NY, USA*, 2000.

14. J. Järvinen, J. Ruokolainen, V. Savolainen, and O. Anttila. A Stabilized Finite Element Analysis for Czochralski Silicon Melt Flow. In *to be published in the Proceedings of Mathematics in Applications*. John Wiley & Sons, 1999.

15. J. Järvinen, J. Ruokolainen, V. Savolainen, and O. Anttila. Simulation of Argon Gas Flow in a Czochralski Crystal Growth Geometry. In *to be presented in the proceedings of 3rd International Workshop on Modeling in Crystal Growth, NY, USA*, 2000.

16. Jari Järvinen. *Mathematical Modeling and Numerical Simulation of Czochralski Silicon Crystal Growth*. PhD thesis, Department of Mathematics, University of Jyväskylä, 1996. CSC Research Reports R03/96.

17. Jari Järvinen, Jouni Malinen, Juha Ruokolainen, and Martti Verho. ELMER: A multi-physical tool for academic and industrial research. In Juha Haataja, editor, *CSC Report on Scientific Computing 1997–1998*. CSC – Scientific Computing Ltd., 1999. Www address http://www.csc.fi/reports/cr97-98/.

18. K. Kakimoto, M. Watanabe, M. Eguchi, and T. Hibiya. *J. Crystal Growth*, 139:197–205, 1994.

19. H.-J. Leister, A. Seidl, and G. Mueller. In *Proc. Int. Symp. on Highly Advanced Computing, Chiba, 1994*, page 92. Jpn. Soc. Mech. Eng., Tokyo 1994.

20. R.M. Nieminen, J. Järvinen, and P. Råback. Modeling of Microelectromechanical Systems (MEMS) — Theoretical and Computational Challenges. In *Proceedings of 12th Nordic Seminar on Computational Mechanics, 22-23 October, 1999, Espoo, Finland.*

21. Y. Ryckmans, P. Nicodème, and F. Dupret. *J. Crystal Growth*, 99:702–706, 1990.

22. P.A. Sackinger and R.A. Brown. *International Journal for Numerical Methods in Fluids*, 9:453–492, 1989.

23. P. Tarvainen, R. Mäkinen, and J. Hämäläinen. Shape Optimization for Laminar and Turbulent Flows with Applications to Geometry Design of Paper Machine Headboxes. In M. Hafez and J.C. Heinrich, editors, *Proceedings of 10th Int. Conference on Finite Elements in Fluids*, pages 536–541, 1998.

24. The home page of a VISIT project supported by EC. Www address http://www.uni-stuttgart.de/visit/.

25. Q. Xiao and J.J. Derby. *J. Crystal Growth*, 152:169–181, 1995.

Data Grids for Next Generation Problems in Science and Engineering

Carl Kesselman

Information Sciences Institute, USC
4676 Admiralty Way, Suite 1001
Marina del Rey, CA 90292-6695
USA
carl@isi.edu

Abstract. One of the hallmarks of current science experiments is the large volume of data that must be managed. For example, next generation particle physics experiments will generate petabytes of data per year. These data sets form important community resources, so not only must be manage to store and process this data, but we must be able to share it as well. In this talk, I will introduce the concept of data grids, a distributed grid infrastructure for sharing and managing large data sets. Using several examples, I will present a data grid architecture and illustrate how it can be used to address a number of current applications.

T. Sørevik et al. (Eds.): PARA 2000, LNCS 1947, p. 17, 2001.

High-Performance Computers: Yesterday, Today, and Tomorrow

Erik Hagersten

Department of Computer Systems, Uppsala University, Sweden eh@it.uu.se

Abstract. Innovations in parallel computer architecture, as well as the development of many new languages, was originally driven by the frontiers of high-performance technical computing. Much of that work took place in small start-ups companies. While parallel computer technology did receive a lot of attention in the early 90's, it never became a widely accepted technique for providing computing to the masses. Only a hand-full machines were manufactured each year.

In the middle of the 90's, parallel computer technology suddenly got a boost from the new application area of parallel commercial computing. Overnight, the rocket science arena of parallel computing became the main-stream technology for many established computer manufacturers. Meanwhile, many, if not all, of the start-up struggled to stay alive.

Today, the start-ups have all disappeared and commercial computing dominates the server marked with around 85-90% of the total server market revenue. While the established computer manufacturers have brought stability and availability to the high-performance technical computing customers, innovation focus is no longer geared only towards the needs of that market segment.

In this talk I will analyze the past, current and future of high-performance computers and try to identify the good news and bad news it brings to the technical computing community.

T. Sørevik et al. (Eds.): PARA 2000, LNCS 1947, p. 18, 2001.
© Springer-Verlag Berlin Heidelberg 2001

The Evolution of OpenMP

Timothy Mattson

Intel Corp, Parallel Algorithms Laboratory, USA
tgmattso@ichips.intel.com

Abstract. Over the years, hundreds of parallel programming Application Programming Interfaces (API) have come and gone. Most grow a small user base and then fade away. A few API's, however, are "winners" and catch on with a large and committed user base. We believe that OpenMP is one of these "winners". Unlike other parallel programming APIs, OpenMP benefits from an organization that makes sure it evolves to meet programmer's changing requirements. This organization is called the OpenMP Architecture Review Board, or the ARB.

In this talk, I will briefly describe the ARB, its members, and how others can join. I will then talk about the current ARB project: OpenMP version 2.0 for Fortran. I will close with my own ideas about where OpenMP needs to go over the next few years.

T. Sørevik et al. (Eds.): PARA 2000, LNCS 1947, p. 19, 2001.
© Springer-Verlag Berlin Heidelberg 2001

JavaGrande – High Performance Computing with Java

Michael Philippsen, Ronald F. Boisvert, Valdimir S. Getov, Roldan Pozo,
José Moreira, Dennis Gannon, and Geoffrey C. Fox

Abstract. The JavaGrande Forum is a group of users, researchers, and interested parties from industry. The Forum members are either trying to use Java for resource-intensive applications or are trying to improve the Java platform, making it possible to create large-sized applications that run quickly and efficiently in Java.

In order to improve its floating point arithmetic, the Forum has suggested to add the keywords `strictfp` and `fastfp` to the Java programming language It has has worked on complex numbers, multidimensional arrays, fast object serializations, and a high-speed remote method invocation (RMI). Results about the new field of research that has been started by the JavaGrande Forum have been recognized both internationally and within Sun Microsystems.

1 JavaGrande Forum

Inspired by coffee house jargon, the buzzword *Grande* applications became commonplace.[1] Grande applications can be found in the fields of scientific, engineering or commercial computing and distinguish themselves, due to their complex data processing or through their complex input/output demands. Typical application classes include simulation and data analysis. In order to cope with the processing needs, a Grande application needs high performance computing, if necessary even parallelism or distributed computing.

The *JavaGrande Forum* [17] is a union of those users, researchers, and company representatives, who either try to create Grande applications with the help of the Java programming language, or those who try to improve and extend the Java programming environment, in order to enable efficient Grande applications with Java. The JavaGrande Forum was founded in March 1998, in a "Birds-of-a-Feather" session in Palo Alto. Since then the Forum organizes regular meetings, which are open to all interested parties, as are the web site [17] and the mailing list [18]. The scientific coordinator is Geoffrey C. Fox (Syracuse); and the main contact person within Sun Microsystems is Sia Zadeh.

---

[1] Grande, as found in the phrase "Rio Grande" is the prevalent designation for the term large or big in many languages. In American English, the term Grande has established itself as describing size within the coffee house scene.

T. Sørevik et al. (Eds.): PARA 2000, LNCS 1947, pp. 20–36, 2001.

1.1   Goals of the JavaGrande Forum

The most important goals of the JavaGrande Forum are the following:

- Evaluation of the applicability of the Java programming language and the run-time environment for Grande applications.
- Bringing together the "JavaGrande Community", compilation of an agreed-upon list of requirements and a focussing of interests for interacting with Sun Microsystems.
- Creation of prototypical implementations that have community-wide consensus, interfaces (APIs) and recommendations for improvements, in order to make Java and its run-time environment utilizable for Grande applications.

1.2   Members of the JavaGrande Forum

The participants in the JavaGrande Forum are primarily American companies, research institutions, and laboratories. More recently, JavaGrande activities can be seen in Europe, as well [19].

In addition to Sun Microsystems, IBM and Intel, as well as Least Square, MathWorks, NAG, MPI Software Technologies and Visual Numerics are participants. Cooperation with hardware dealers is crucial, especially in reference to questions dealing with high-speed numerical computing. Academic circles are represented by a number of universities, such as Chicago, Syracuse, Berkeley, Houston, Karlsruhe, Tennessee, Chapel Hill, Edinburgh, Westminster and Santa Barbara, to name a few. The American Institute for Standardization (NIST), Sandia Labs and ICASE participate, as well.

The members are organized into two working groups. Section 3 presents the results of the numerical computing working group. Section 4 dedicates itself to the working group parallelism and distribution.

Former results and activities organized be the JavaGrande Forum have been well-received by Sun Microsystems. Gosling, Lindhom, Joy and Steele have studied the work of the JavaGrande Forum. Impressive public relations work was done by panelist Bill Joy, who praised the work of the JavaGrande Forum in front of an audience of almost 21,000 at the JavaOne 1999 Conference.

1.3   Scientific Contributions of the JavaGrande Forum

The Forum organizes scientific conferences and workshops or is represented on panels; see Table 1. The most important annual event is the ACM Java Grande Conference. A large portion of the scientific contributions of the "JavaGrande Community" can be found in the conference journals (Table 1) and in some issues of *Concurrency – Practice & Experience* [9,10,11,12]. In addition, the JavaGrande Forum publishes working reports at regular intervals [34,35].

The scientific work is important for bringing the "JavaGrande Community" together and for creating cohesiveness. This make it possible to achieve consensus about ideas and to focus interests, thus making it easier to achieve the goals.

Table 1. Events organized by the JavaGrande Forum

Workshop, Syracuse, December 1996, [9]
PLDI Workshop, Las Vegas, June 1997, [10]
Java Grande Conference, Palo Alto, February 1998, [11]
EuroPar Workshop, Southampton, September 1998, [8]
Supercomputing, Exhibit and Panel, November 1998, [34]
IEEE Frontiers 1999 Conference, Annapolis, February 1999
HPCN, Amsterdam, April 1999, [33]
IPPS/SPDP, San Juan, April 1999, [7]
SIAM Meeting, Atlanta, May 1999
IFIP Working Group 2.5 Meeting, May 1999
Mannheim Supercomputing Conference, June 1999
ACM Java Grande Conference, San Francisco, June 1999, [12]
JavaOne, Exhibit and Panel, San Francisco, June 1999, [35]
ICS'99 Workshop, Rhodos, June 1999
Supercomputing, Exhibit and Panel, September 1999
ISCOPE, JavaGrande Day, December 1999, [32]
IPPS, Cancun, May 2000
ACM Java Grande Conference, San Francisco, June 2000

2   Why Java for High-Performance Computing?

In addition to the usual reasons for using Java that also apply to "normal" applications, such as portability, the existence of development environments, the (alleged) productivity-enhancing language design (with automatic garbage collection and thread support), and the existence of an extensive standard library; there are other important arguments for Grande applications.

The JavaGrande Forum designates Java as an universal language, which enables the mastering of the entire spectrum of tasks necessary for dealing with Grande applications. Java offers a standardized infrastructure for distributed applications, even in heterogeneous environments. Since the graphical interface is well-integrated (although there are a few portability problems) visualized can often be implemented as well. In addition, Java can also be used as glue code, in order to couple (interconnect) existing high-performance applications, to link computations that have been realized in other programming languages to one another, and as a universal in-between layer to function between computations and I/O. Another aspect that is no less important for the scientific and enginee-ring fields, is that Java is being taught and learned – in contrast to Fortran for which a serious shortage of new recruits is expected.

2.1   Run-Time Performance and Memory Consumption

A commonplace and persistent rumor is that Java programs run very slowly. The first available version of Java (1.0 beta) interpreted ByteCode and was, in fact, extremely slow. Since then, much has changed: almost no Java program is executed on a purely interpretive basis anymore. Instead, so-called Just-In-Time-Compilers of different coinage are being used to improve the execution times. The following comparisons give a good impression of the current state.

1. Comparative Experiment. Prechelt [30] conducted an experiment at the University of Karlsruhe, in which the same task was solved by 38 different people who were supposed to provide a reliable implementation. The experiment was not designed as a speed contest. As a result 24 Java programs, 11 C++ programs, and 5 C implementations were submitted. Amongst the 38 people taking part, all were computer science students at the post-graduate level with an average of 8 years and 100 KLOC of programming experience. The group included excellent to relatively poor programmers.

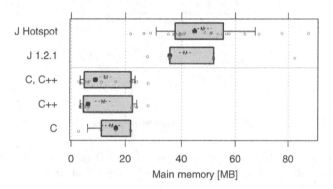

Fig. 1. Box plots of the main memory requirements of the programs; measured on Solaris 7 including data structures, program code, libraries, process administration and JVM. The individual measurements are symbolized by small circles. The M shows the mean, the black dot the median. Inside of a box, the innermost 50% of the measuremnts can be found, the H-line contains the innermost 80% of the measurements.

On average, Java (Version 1.2.1) needs four to five times as much main memory as C/C++, see Figure 1. For all languages, the interpersonal differences between the experiment participants were quite radical. The observed run-times varied from a matter of seconds to 30–40 minutes, see Figure 2. However, the medians of Java and C++ were similar. The variability within a language was much higher than the difference between the languages. The fastest five Java programs were five times faster than the median of the C++ programs; although it must be said that these programs were three times slower than the five fastest C++ programs. In this experiment, the C programs were clearly the fastest, a fact that could be explained by the fact that only few (and presumably the best) programmers selected C.

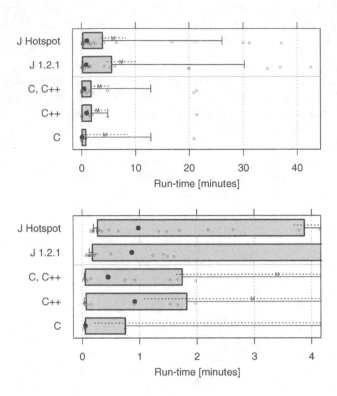

Fig. 2. Run-times of the different programs measured on Solaris 7. Enlarged in the second graph. The legend is the same as that of the previous figure.

In conclusion, it can be said that C++ showed almost no advantage over Java in terms of run-time. The ability differences of the programmers were greater than the differences in the programming languages. C++ still requires less main memory than Java.

2. Benchmark for Scientific Applications. Pozo and Miller combined five medium-sized numerical kernels in SciMark 2.0 [29] (compex valued fast Fourier transformation, Gauss-Seidel-relaxation, Monte-Carlo-integration of e^{-x^2}, multiplication of sparsely populated matrixes, and the LU-factorization of dense matrixes with pivoting). For each of these kernels, a Java and a C version are available. In addition, there is a version that fits in the cache, as well as one that does not fit.

The antiquated JVM 1.1.5 of the Netscape-Browser reaches approximately 0,7 MFlops on an Intel Celeron 366 under Linux and is, as such 135 times slower than a C-implementation on that platform. Java 1.1.8 is a huge improvement:

using the same processor (running on OS/2) about 76 MFlops have currently been achieved – only 35% less than with C.

As a whole, the speed of Java is better than its reputation suggests and further optimizations are expected. In addition, there are papers that either describe how to optimize a Java application, or that discuss which rules to follow, so that a fast Java application is created from the outset, e.g. [24,31]. Given these facts we feel that it is by no means out-of-place to seriously consider Java as a programming language for the central components of Grande applications. The results of both working groups of the JavaGrande Forum, which will be discussed below, provide support for this idea.

3 Numerical Computing Working Group

3.1 Goals of the Working Group

The Numerical Computing Working Group, that is supported by the IFIP Working Group 2.5 (International Federation for Information Processing), set its goal as the evaluation of the applicability of Java for numerical computing. Building upon that, the group aims to provide recommendations reached on a consensual basis and to implement them (at least prototypically), in order to eradicate deficiencies of the Java programming language and its run-time environment. The following sections discuss individual problems and results.

3.2 Improvement of the Floating Point Arithmetic

For scientific computing, it is important to reach acceptable speeds and precision on most types of processors. For Grande applications, it is additionally important to achieve a very high level of floating point performance on *some* processors. Numericians have learned to deal with architecture-specific Rounding errors since the beginning of floating point arithmetic, so that an exact reproducibility of bits is *seldom* of great importance – in contrast to Java's design goal of exact reproducibility of results. A too imprecise computation in the mathematical library (trigonometric functions, for example) is unacceptable, as is observable in many Java implementations up to this point in time.

Floating Point Performance. In order to achieve exact reproducibility, Java forbids common optimizations, such as making use to the associativity of operators, since these optimizations might cause differently rounded results. Further prohibitory measures affect processor features.

1. Prohibition to use the 80 Bit "double extended format". Processors of the x86-family have 80 bit wide registers that use the double extended format as defined in the IEEE Standard 754. To implement Java's bitwise reproducibility, every (temporary) result needs to be rounded to a more imprecise number format. Even if the precision control bit is set to enforce a rounding in the registers after each step, still 15 bits instead of the standard 11 bits are used to represent the exponents. In the x86-family the exponent can only be rounded slowly by transferring it from the register to the main memory and re-loading it. Many JVM implementations ignore proper rounding for exactly this reason. In addition, the two-phase rounding (immediately after the operation and during the interaction with the main memory) leads in many cases to deviations from the specification on the order of 10^{-324}. A correction of this type of mistake is extremely time-intensive and also is not undertaken in most implementations. On the basis of both of these problems on Intel processors, Java's floating point arithmetic is either wrong or two to ten times slower than possible.

2. Prohibition to use for "FMA – Fused Multiply Add" machine commands. Processors of the PowerPC-family offer a machine command, that multiplies two floating point numbers and adds a further number to them. In addition of being faster that two single operations, only one rounding is needed. Java's language definition prohibits the use of this machine command and thus sacrifices 55% of potential performance in experiments, see Figure 3. Without all optimizations, only 3,8 MFlop will be achieved. If all of the common optimizations are carried out (including the elimination of redundant boundary checks), then IBM's native Java compiler achieves 62% of the performance of an equivalent Fortran program (83,4 MFlop). If the FMA machine command could also be used – which the Fortran compiler does routinely – almost 97% of the Fortran performance could be achieved.

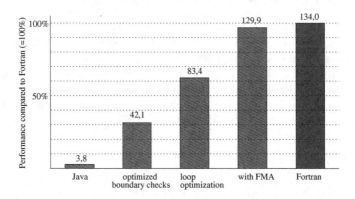

Fig. 3. Cholesky Factoring on a 67 MHz Power-2-Processor. Information given in MFlops and relative to optimized Fortran.

Keyword strictfp in Java 2. As suggested by the the JavaGrande Forum, the keyword strictfp was included in the version 1.2 of the Java programming language. Only when this attribute is used on classes or methods, will the bit-wise reproducibility of results be guaranteed. In the standard case without the keyword, Java can use the 15 bit long exponents for anonymous `double` variables to avoid the expensive memory copy and re-loading operations. In the standard case without the keyword, different numerical results may occur not only between processors with and without extended exponent lengths, but also between different JVM implementations on one platform. The reason for this is that the use of the extended exponents for anonymous variables is optional.

Reproducibility of `Math`-Functions. In addition to the strict prohibitions that limit floating point performance, Java's floating point arithmetic suffers from yet another problem regarding precision. According to the Java specification (version 1.1.x), the `java.lang.Math` library must be implemented by porting the `fdlibm` library.[2] Instead of this, most manufacturers use the faster machine commands offered by their hardware that often return slightly incorrect results, though. In order to face up to this development, Sun offers two wrapper methods for the `fdlibm` functions realized in C, which should make it simpler to achieve `fdlibm` results on all platforms. The first complete implementation of `fdlibm` in Java is due to John Brophy of Visual Numerics, making it possible to forswear the wrapper methods in the future [1].

An ideal math library would deliver foating point numbers, that at the most deviate 0,5 ulp (unit in the last place) from the actual result,[3] this means they have been as well-rounded off as possible. The `fdlibm` library itself only delivers a precision of 1 ulp, for which reason Sun decided to change the specification in Java 1.3, so that the results of the math library would be allowed a fault tolerance level of 1 ulp. Abraham Ziv, IBM Haifa, created a math library (in ANSI C), that rounds correctly, i.e., with a fault level 0,5 ulp at the most [37].

Recommendation of the keyword `fastfp`. The JavaGrande Forum is currently working on a JSR (Java Specification Request [21]) that recommends integrating a further modifier, `fastfp`, to the Java programming language. In classes and methods, that are labeled with this modifier, the use of the FMA machine command will be allowed. In addition, the math library should be extended with an additional FMA method, whose use forces the FMA command to execute. Currently, the sensibility of allowing associativity optimizations via this modifier is being examined.

3.3 Efficient Complex Arithmetic

A requirement for the use of Java for scientific computing is the efficient and comfortable support of complex numbers. With Java's expressiveness however,

---

[2] The `fdlibm` library is the free math library distributed by Sun. `Fdlibm` is considerably more stable, to a greater degree correct, and much more easily portable than the `libm` libraries available on most platforms.

[3] For numbers between 2^k and 2^{k+1}, 1 ulp $= 2^{k-52}$.

complex numbers can only be realized in the form of a `Complex` class, whose objects contain two `double` values. Complex valued arithmetic must then be expressed by means of complicated method calls, as in the following code fragment.

```
Complex a = new Complex(5,2);
Complex b = a.plus(a);
```

This has three disadvantages: first, arithmetic expressions are difficult to read without operator overloading. Secondly, complex arithmetic is slower than Java's arithmetic on primitive types, since it takes longer to create an object and more memory space is necessary than for a variable of a primitive type. In addition, temporary helper objects need to be created for almost every method call. In contrast, primitive arithmetics can use the stack to pass results. IBM analyzed the performance of class-based complex arithmetic. Using a Jacobi-relaxation, an implementation based on a class `Complex` achieved only 2% of the implementation that used two `double` instead. Thirdly, class-based complex numbers invariably cannot be fully integrated in the system of primitive types. They are not integrated into the type relationships that exist between primitive types, so that for example, the assignment of a primitive `double` value to a `Complex` object does not result in any automatic type cast. Equality tests between complex objects refer to object identities rather than to value equality. In addition to this, an explicit constructor call is necessary for a class-based solution, where a literal would be sufficient to represent a constant value.

Since scientific computing only makes up an small portion of total Java use, it is improbable that the Java Virtual Machine (JVM) or the ByteCode format will be extended to include a new primitive type `complex`, although this would probably be the best way of introducing complex numbers in Java. In addition, it is not known, if (and if so, when) Java will be extended by general operator overloading and value classes. And since even after such an extension value classes can still not be seamlessly integrated into the system of existing primitive types, the JavaGrande Forum regards that the following twin-track way to be more sensible (see Figure 4).

Class `java.lang.Complex`. The JavaGrande Forum has defined and prototypically implemented a class `java.lang.Complex` that is similar in style to Java's other numerical classes. In addition,method based complex arithmetic operations are provided. This class will be submitted to Sun in the form of a JSR.

IBM has built the semantics of this `Complex` class permanently into its native Java compiler; internally a complex type is used and the usual optimizations are carried out on it [36]. In particular, most of the method and constructor calls that prevail in the Java code using this class are replaced by stack operations. Hence, at least for the platforms that are supported, class `Complex` is supported efficiently.

Primitive Data Type `complex`. As a further JSR, the JavaGrande Forum suggests a primitive type `complex` with corresponding infix operations. In order

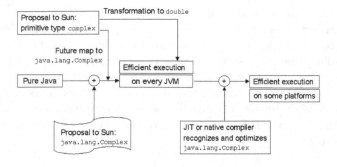

Fig. 4. Introduction of complex numbers and arithmetics

that both the ByteCode format and the existing JVM implementation do not have to undergo changes, the language extension are transformed back to normal Java in a pre-processor step. Currently it is being discussed whether it is necessary to introduce a further primitive data type `imaginary` in conjunction with the primitive data type `complex`, as it has been done in C99 [23,2].

Figure 4 shows two alternative transformations. First, the primitive data type `complex` is mapped to a pair of `double` values. Secondly, it is mapped to the `Complex` class to make use of the above mentioned compiler support. The compiler *cj* developed at the University of Karlsruhe both formally describes and prototypically realizes the transformation of the primitive type `complex` [13,14].

3.4 Efficient Multi-dimensional Arrays

In the same way as efficient and comfortable complex arithmetic must be made available, numerical computing without efficient (i.e., optimizable) and comfortable multi-dimensional arrays is unthinkable.

Java offers multi-dimensional arrays only as arrays of one-dimensional arrays. This causes several problems for optimization. Especially since index vectors cannot be mapped to memory locations. One problem is that several rows of such a multi-dimensional array could be aliases to a shared one-dimensional array. Another problem is that the rows could have different lengths. Moreover, each field access to a multi-dimensional array not only requires a pointer indirection but also causes several bound checks at run-time. By means of dataflow analysis and code cloning plus guards the optimizer can only reduce the amount of boundary checks. The optimizer can seldom avoid all run-time checks.

For this reason the JavaGrande Forum recommends a class for multi-dimensional arrays (once again in the form of a JSR), which would be mapped to one-dimensional Java arrays with a specific roll-out scheme. The compiler can then use common optimizations; algorithms can make use of the roll-out scheme to improve their cache locality.

As with the class-based complex numbers, this multi-dimensional array class requires awkward access methods instead of elegant []-notation. For this reason, a twin track solution is appropriate in this case, as well. On the one hand, IBM will build permanent support for multi-dimensional arrays into their native Java compiler. At the same time, an extended array access syntax will be set up, allowing elegant access to multi-dimensional arrays. This syntactical extension, which is quite difficult due to the necessary interaction with regular one-dimensional Java arrays, and the corresponding pre-processor will be recommended to Sun in JSR form.

3.5   More Fundamental Issues

Lightweight classes and operator overloading are the general solutions for both complex numbers, as well as for multi-dimensional arrays. Lightweight classes have value semantics, their instantiated variables cannot be changed after object creation. Thus, the problem of equality semantics does not come up at all. In addition, lightweight objects can often be allocated on the stack and by passed by copy. If the programmer could overload basic operators like +, -, *, /, [], a pre-processor would no longer be necessary.

Why does the JavaGrande Forum not try to introduce lightweight classes and operator overloading? The answer is quite pragmatic. The JavaGrande Forum hopes the above-mentioned JSRs are light enough to withstand the formal process of language alterations. The community of normal Java users remains almost completely unaffected and quite possibly will not notice the changes at all. Very few of today's Java users are even aware of the existence, much more the importance, of the `strictfp` keywords. The smaller the number of the people affected, the less-damaging the endorsement of JSR will be.

Value classes and operator overloading demand a greater change to the language as a whole and (supposedly) impact the ByteCode format and thus, the JVM. For this reason and due to the almost religious character of operator overloading, the outcome of such efforts remains open to speculation, while the above recommendations have better prospects.

4   Parallelism and Distribution Working Group

4.1   Goal of the Working Group

The Parallelism and Distribution Working Group of the JavaGrande Forum evaluates the applicability of Java for parallel and distributed computing. Actions based on consensus are formulated and carried out, in order to get rid of inadequacies in the programming language or the run-time system. The results that have been achieved will be presented in the following sections. Further work in the field of parallel programming environments and "Computing Portals" have not yet been consolidated and will not be covered in this article.

4.2   Faster Remote Method Invocation

Good latency times and high band widths are essential for distributed and parallel programs. However, the remote method invocation (RMI) of common Java distributions is too slow for high performance applications, since RMI was developed for wide area networks, builds upon the slow object serialization, and does not support any high speed networks. With regular Java, a remote method invocation takes milliseconds – concrete times depend on the number and the types of arguments. A third of that time is needed for the RMI itself, a third for the serialization of the arguments (their transformation into a machine-independent byte representation), and a third for the data transfer (TCP/IP-Ethernet).

In order to achieve a fast remote method invocation, work must be done at all levels. This means that one needs a fast RMI implementation, a fast serialization, and the possibility of using communication hardware that does not employ TCP/IP protocols.

Within the framework of the JavaParty Project at the University of Karlsruhe [20], all three of these requirements were attacked to create the fastest (pure) Java implementation of a remote method invocation. On a cluster of DEC-Alpha computers connected by Myrinet, called ParaStation, currently a remote method invocation takes about 80 μs although it is completely implemented in Java.[4] Figure 5 shows, that for benchmark programs 96% of the time can be saved, if the UKA serialization, the high-speed RMI (KaRMI), and the faster communication hardware is used. The central ideas of the optimization will be highlighted in the next sections.

UKA Serialization. The UKA serialization [15] can be used instead of the official serialization (and as a supplement to it) and saves 76%–96% of the serialization time. It is based on the following main points:

— Explicit serialization routines ("marshalling routines") are faster than those used by classical RMI that automatically derive a byte representation with the help of type introspection.
— A good deal of the costs of the serialization are needed for the time-consuming encoding of the type information that is necessary for persistent object storage. For the purposes of communication, especially in work station clusters with common file systems, a reduced form of the type encoding is sufficient and faster. The JavaGrande Forum has convinced Sun Microsystems to make the method of type encoding plugable in one of the next versions.
— Copied objects need to be transferred again for each call in RMI. RMI does not differentiate between type encoding and useful data, meaning that the type information is transferred redundantly.
— Sun has announced (without concretely naming a version) it will pick up on the idea of a separate reset of type information and user data.

---

[4] Of course, the connection of the card driver was not realized in Java.

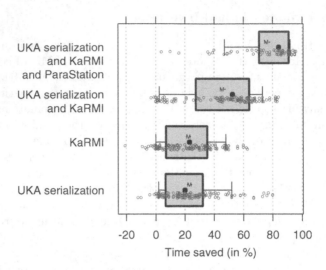

Fig. 5. The bottom three box plots each show 2·64 measured results for diverse benchmarks (64 points represent Ethernet on PC, 64 stand for FastEthernet on Alpha). The first (bottom-most) box plot shows the run-time improvement that was achieved with regular RMI and the UKA serialization. The second box plot shows the improvement that KaRMI achieves when used with Java's regular serialization. The third box plot shows the combined effect. The top line demonstrates what happens, if Myrinet cards are used in addition to the UKA serialization and KaRMI (64 measured results).

- The official serialization uses several layers of streams that all possess their own buffers. This causes frequent copying operations and results in unacceptable performance. The UKA serialization only needs one buffer, which the byte representation can be directly written in.
- Sun remains steadfast about layering for reasons of clearer object-oriented design, is though, at least improving the implementation of the layers.

KaRMI. A substitute implementation of RMI, called KaRMI, was also created at the University of Karlsruhe. KaRMI [26,28] can be used instead of the official RMI and gets rid of the following deficiencies, as well as some others found in official RMI:

- KaRMI supports non-TCP/IP networks. Based upon the work of the Java-Grande Forum, Suns plans to make this possible (still being outlined) for the official RMI-Version, as well.
- KaRMI possesses clearer layering, which will make it easier to employ other protocol semantics (i.e. Multicast) and other network hardware (i.e. Myrinet-Cards).
- In RMI, objects can be connected to fixed port numbers. Therefore, a certain detail of the network layer is passed to the application. Since tis is in conflict

with the guidelines of modular design, KaRMI only supports use of explicit
port numbers when the underlying network offers them.
– The distributed garbage collection of the official RMI was created for wide
  area networks. Although there are optimized garbage collectors for tightly
  coupled clusters and for other platforms [27], the official RMI sees no alternative garbage collector as being necessary, in contrast to KaRMI.

4.3   Message Passing in Java

While Java's mechanisms for parallel and distributed programming are based on
the client/server paradigm, MPI is a symmetrical, message-based parallel calculation model. In order to ensure that the MPI-based solutions are not lost during
the transition to Java, a sub-group of the JavaGrande Forum is working on providing an MPI binding for Java. The availability of such an interface would make
MPI-based Grande applications possible. Members of the JavaGrande Forum
have made the following recommendations for that reason:

– mpiJava: a collection of wrapper classes that reach back to the C++ binding
  of MPI through the native Java interface (JNI) [4,16].
– JavaMPI: automatically created JNI wrappers for a program with a C bin-
  ding to MPI [25].
– MPIJ: an implementation of MPI interfaces in Java, that is based on the
  C++ binding and demonstrates relatively good performance results [22].

At the moment, the sub-group is working on the unification of previous pro-
totypes [3]. One of the major issues that has arisen, is how the mechanisms of
Java can be made useful in association with MPI. Under investigation is whether
the types used in MPI could be extended with Java objects, which could then
be sent in serialized form [5]. In addition, it is being studied if and how Java's
thread model can be utilized to extend the process-based approach of MPI.

4.4   Benchmarks

The JavaGrande Forum has begun a benchmark initiative. The intentions are to
make convincing arguments for Grande applications and to uncover the weaknes-
ses in the Java platform. The responsibility for this initiative is being carried by
the EPCC (Edinburgh) [6]. Currently a stable collection of non-parallel benchmarks exists in three categories:

– Basic operations are being timed (such as arithmetic expressions, object
  generation, method calls, loop bodies, etc.)
– Computational kernels: similar to the example of SciMark, numerical kernels
  are being observed. The IDEA-encryption algorithm is also in the collection.
– Applications: The collection is made up of an Alpha-Beta search with
  pruning, a Computational-Fluid-Dynamics application, a Monte-Carlo-
  simulation, and a 3D ray-tracing.

Thread benchmarks for all three categories are being worked on. For these purposes, the basic operations are being timed (create, join, barrier, synchronized methods); some of the applications (Monte Carlo and Ray-tracer) are being implemented in parallel. In addition, for quantitative language comparisons it is intended to provide equivalent implementations in C/C++.

5 Conclusion

Contributions of the JavaGrande Forum are the keywords `strictfp` and `fastfp` for improved floating point arithmetic, work in the field of complex numbers, the high-speed serialization, the fast RMI, and finally the benchmark initiatives.

Due to the cooperation with Sun Microsystems, due to the creation of a new branch of research, and due to the focussing of interests of the "JavaGrande Community", the future holds the hope that the requirements of high performance computing will be made a reality in Java.

Acknowledgements. This compilation of the activities and results of the Java-Grande Forum used collected presentation documents from Geoffrey Fox, Dennis Gannon, Roldan Pozo and Bernhard Haumacher as a source of information. Lutz Prechelt made Figures 1 and 2 available. A word of thanks to Sun Microsystems, especially to Sia Zadeh, for financial and other support.

References

1. John Brophy. Implementation of `java.lang.complex`. http://www.vni.com/corner/garage/grande/.
2. C9x proposal. http://anubis.dkuug.dk/jtc1/sc22/wg14/ and ftp://ftp.dmk.com/DMK/sc22wg14/c9x/complex/.
3. B. Carpenter, V. Getov, G. Judd, T. Skjellum, and G. Fox. MPI for Java: Draft API specification. Technical Report JGF-TR-003, JavaGande Forum, November 1998.
4. Bryan Carpenter, Yuh-Jye Chang, Geoffrey Fox, Donald Leskiw, and Xiaoming Li. Experiments with "HPJava". *Concurrency: Practice and Experience*, 9(6):579–619, June 1997.
5. Bryan Carpenter, Geoffrey Fox, Sung Hoon Ko, and Sang Lim. Object serialization for marshalling data in a Java interface for MPI. In *ACM 1999 Java Grande Conference*, pages 66–71, San Francisco, June 12–14, 1999.
6. Java Grande Benchmarks. http://www.epcc.ed.ac.uk/javagrande
7. J. Rolim et al., editor. *Parallel and Distributed Processing*. Number 1586 in Lecture Notes in Computer Science. Springer Verlag, 1999.
8. *Proc. Workshop on Java for High Performance Network Computing at EuroPar'98*. Southampton, September 2–3, 1998.
9. Geoffrey C. Fox, editor. *Concurrency: Practice and Experience*, volume 9(6). John Wiley & Sons, June 1997.
10. Geoffrey C. Fox, editor. *Concurrency: Practice and Experience*, volume 9(11). John Wiley & Sons, November 1997.

11. Geoffrey C. Fox, editor. *Concurrency: Practice and Experience*, volume 10(11–13). John Wiley & Sons, September–November 1998.
12. Geoffrey C. Fox, editor. *Concurrency: Practice and Experience*, to appear. John Wiley & Sons, 2000.
13. Edwin Günthner and Michael Philippsen. Komplexe Zahlen für Java. In *JIT'99, Java-Informations-Tage*, pages 253–266, Düsseldorf, September 20–21, 1999. Springer Verlag.
14. Edwin Günthner and Michael Philippsen. Complex numbers for Java. *Concurrency: Practice and Experience*, to appear, 2000.
15. Bernhard Haumacher and Michael Philippsen. More efficient object serialization. In *Parallel and Distributed Processing*, number 1586 in Lecture Notes in Computer Science, Puerto Rico, April 12, 1999. Springer Verlag.
16. http://www.npac.syr.edu/projects/pcrc/HPJava/.
17. Java Grande Forum. http://www.javagrande.org.
18. Java Grande Forum, mailinglist. All Members: javagrandeforum@npac.syr.edu, Subscribe: gcf@syracuse.edu.
19. Java Grande Forum, Europe. http://www.irisa.fr/EuroTools/Sigs/Java.html.
20. JavaParty. http://wwwipd.ira.uka.de/JavaParty/.
21. The Java community process manual. http://java.sun.com/aboutJava/community process/java_community_process.html.
22. Glenn Judd, Mark Clement, Quinn Snell, and Vladimir Getov. Design issues for efficient implementation of MPI in Java. In *ACM 1999 Java Grande Conference*, pages 58–65, San Francisco, June 12–14, 1999.
23. William Kahan and J. W. Thomas. Augmenting a programming language with complex arithmetics. Technical Report No. 91/667, University of California at Berkeley, Department of Computer Science, December 1991.
24. Reinhard Klemm. Practical guideline for boosting Java server performance. In *ACM 1999 Java Grande Conference*, pages 25–34, San Francisco, June 12–14, 1999.
25. S. Mintchev and V. Getov. Towards portable message passing in Java: Binding MPI. In M. Bubak, J. Dongarra, and J. Wańiewski, editors, *Recent Advances in PVM and MPI*, Lecture Notes in Computer Science, pages 135–142. Springer Verlag, 1997.
26. Christian Nester, Michael Philippsen, and Bernhard Haumacher. Ein effizientes RMI für Java. In *JIT'99, Java-Informations-Tage*, pages 135–148, Düsseldorf, September 20–21, 1999. Springer Verlag.
27. Michael Philippsen. Cooperating distributed garbage collectors for clusters and beyond. *Concurrency: Practice and Experience*, to appear, 2000.
28. Michael Philippsen, Bernhard Haumacher, and Christian Nester. More efficient serialization and RMI for Java. *Concurrency: Practice and Experience*, to appear, 2000.
29. Roldan Pozo and Bruce Miller. SciMark 2.0. http://math.nist.gov/scimark/.
30. Lutz Prechelt. Comparing Java vs. C/C++ efficiency differences to inter-personal differences. *Communications of the ACM*, 42(10):109–112, October 1999.
31. Mark Roulo. Accelerate your Java apps! *Java World*, September 1998.
32. R. R. Oldehoeft S. Matsuoka and M. Tholburn, editors. *Proc. ISCOPE'99, 3rd International Symposium on Computing in Object-Oriented Parallel Environments*. Number 1732 in Lecture Notes in Computer Science. Springer Verlag, 1999.
33. P. Sloot, M. Bubak, A. Hoekstra, and B. Hertzberger, editors. *Proc. 7th Intl. Conf. on High Performance Computing and Networking, HPCN Europe 1999*. Number 1593 in Lecture Notes in Computer Science. Springer Verlag, 1999.

34. George K. Thiruvathukal, Fabian Breg, Ronald Boisvert, Joseph Darcy, Geoffrey C. Fox, Dennis Gannon, Siamak Hassanzadeh, Jose Moreira, Michael Philippsen, Roldan Pozo, and Marc Snir (editors). Java Grande Forum Report: Making Java work for high-end computing. In *Supercomputing'98: International Conference on High Performance Computing and Communications*, Orlando, Florida, November 7–13, 1998.
35. George K. Thiruvathukal, Fabian Breg, Ronald Boisvert, Joseph Darcy, Geoffrey C. Fox, Dennis Gannon, Siamak Hassanzadeh, Jose Moreira, Michael Philippsen, Roldan Pozo, and Marc Snir (editors). Iterim Java Grande Forum Report. In *ACM Java Grande Conference'99*, San Francisco, June 14–17, 1999.
36. Peng Wu, Sam Midkiff, José Moreira, and Manish Gupta. Efficient support for complex numbers in Java. In *ACM 1999 Java Grande Conference*, pages 109–118, San Francisco, June 12–14, 1999.
37. Abraham Ziv. Fast evaluation of elementary mathematical functions with correctly rounded last bit. *ACM Transactions on Mathematical Software*, (17):410–423, 1991.

Ocean and Climate Prediction on Parallel Super Computers

Geir Evensen

Nansen Environmental and Remote Sensing Center, Norway
`Geir.Evensen@nrsc.no`

Abstract. The availability of new powerful super computers has lead to new activities in high resolution ocean prediction and data assimilation. This paper will present results from three application areas within ocean and climate modeling.

First, data assimilation systems developed for operational oceanography will be described where information from satellite remote sensing data are integrated with numerical ocean circulation models using sophisticated assimilation techniques. These systems form a backbone for operational ocean forecasting which will become a growing industry in the years to come.

The second example discusses the use of high resolution ocean circulation models applied in met-ocean studies for supporting offshore oil industry operating in deep waters down to 2000 m. These applications require huge computer resources but provide the offshore industry with environmental information leading to optimal design criteria for rig selection and safer operation.

Finally, in climate research, coupled ice, ocean and atmospheric models covering the entire globe are used to hindcast the climate in the past and to predict the evolution of the climate system in the future. Such computations requires the models to be run for long time periods in coupled mode and requires extensive super-computer resources.

T. Sørevik et al. (Eds.): PARA 2000, LNCS 1947, p. 37, 2001.
© Springer-Verlag Berlin Heidelberg 2001

LAWRA
Linear Algebra with Recursive Algorithms [*]

Bjarne S. Andersen[1], Fred Gustavson[2], Alexander Karaivanov[1],
Minka Marinova[1], Jerzy Waśniewski[1][**], and Plamen Yalamov[3]

[1] Danish Computing Center for Research and Education (UNI•C), Technical
University of Denmark, Building 304, DK-2800 Lyngby, Denmark, emails:
`bjarne.stig.andersen@uni-c.dk`, `alexander.karaivanov@uni-c.dk`,
`minka.marinova@uni-c.dk` and `jerzy.wasniewski@uni-c.dk`, respectively.
[2] IBM T.J. Watson Research Center, P.P. Box 218, Yorktown Heights, NY 10598,
USA, email: `gustav@watson.ibm.com`
[3] Center of Applied Mathematics and Informatics, University of Rousse, 7017
Rousse, Bulgaria, email: `yalamov@ami.ru.acad.bg`

Abstract. Recursion leads to automatic variable blocking for dense linear algebra algorithms. The recursion transforms LAPACK level-2 algorithms into level3 codes. For this and other reasons recursion usually speeds up the algorithms.
Recursion provides a new, easy and very successful way of programming numerical linear algebra algorithms. Several algorithms for matrix factorization have been implemented and tested. Some of these algorithms are already candidates for the LAPACK library.
Recursion has also been successfully applied to the BLAS (Basic Linear Algebra Subprograms). The ATLAS system (Automatically Tuned Linear Algebra Software) uses a recursive coding of the BLAS.
The Cholesky factorization algorithm for positive definite matrices, LU factorization for general matrices, and LDL^T factorization for symmetric indefinite matrices using recursion are formulated in this paper. Performance graphs of our packed Cholesky and LDL^T algorithms are presented here.

1 Introduction

This work was started by Gustavson and Toledo described in [8,16]. These papers describe the application of recursion to the numerical dense linear algebra algorithms. Recursion leads to automatic variable blocking for the dense linear-algebra algorithms. This leads to modifications of the LAPACK [2] algorithms. LAPACK's level-2 version routines are transformed into level-3 codes by using recursion.

---

[*] This research was partially supported by the UNI•C collaboration with the IBM T.J. Watson Research Center at Yorktown Heights. The second and fifth authors was also supported by the Danish Natural Science Research Council through a grant for the EPOS project (Efficient Parallel Algorithms for Optimization and Simulation).
[**] Invited speaker

T. Sørevik et al. (Eds.): PARA 2000, LNCS 1947, pp. 38–51, 2001.
© Springer-Verlag Berlin Heidelberg 2001

Fortran 90 allows recursion (see [15]). The programs are very concise and the recursion part is automatic as it is handled by the compiler. The intermediate subroutines obey the Fortran 90 standard too (see [5]).

Section 2 shows the recursive Cholesky factorization algorithm. Section 3 formulates the recursive algorithm of Gaussian elimination without pivoting and LU factorization with partial pivoting. Section 4 explains two recursive BLAS: RTRSM and RSYRK. Section 5 demonstrates the factorization algorithm using the pivoting strategy introduced by Bunch-Kaufman for symmetric indefinite matrices (see [6, pp. 161–170]).

2   Cholesky Factorization

We would like to compute the solution to a system of linear equations $AX = B$, where A is real symmetric or complex Hermitian and, in either case, positive definite matrix, X and B are rectangular matrices or vectors. The Cholesky decomposition can be used to factor A, $A = L L^T$ or $A = U^T U$, where U is an upper triangular matrix and L is a lower triangular $(L = U^T)$. The factored form of A is then used to solve the system of equations $AX = B$.

A recursive algorithm of Cholesky factorization is described in detail in [18, 8]. Here we give the final recursive algorithms for the lower triangular and upper triangular cases. We assume that A is n by n.

Recursive Algorithm 21 *Cholesky recursive algorithm if lower triangular part of A is given (rcholesky):*

> *Do recursion*
> - *if $n > 1$ then*
>   - *$L_{11} :=$ rcholesky of A_{11}*
>   - *$L_{21}L_{11}^T = A_{21}$ →   **RTRSM***
>   - *$\hat{A}_{22} := A_{22} - L_{21}L_{21}^T$ →   **RSYRK***
>   - *$L_{22} :=$ rcholesky of $\hat{A}_{22}$*
> - *otherwise*
>   - *$L := \sqrt{A}$*
> *End recursion*

The matrices A_{11}, A_{12}, A_{21}, A_{22}, L_{11}, L_{21}, L_{22}, U_{11}, U_{12} and U_{22} are submatrices of A, L and U respectively.

$$A = \begin{pmatrix} A_{11} & A_{12} \\ A_{21} & A_{22} \end{pmatrix}, \quad L = \begin{pmatrix} L_{11} & \\ L_{21} & L_{22} \end{pmatrix} \text{ and } U = \begin{pmatrix} U_{11} & U_{12} \\ & U_{22} \end{pmatrix}$$

The sizes of the submatrices are: for A_{11}, L_{11} and U_{11} is $h \times h$, for A_{21} and L_{21} is $(n - h) \times h$, for A_{12} and U_{12} is $h \times (n - h)$, and for A_{22}, L_{22} and U_{22} is $(n - h) \times (n - h)$, where $h = n/2$. Matrices L_{11}, L_{22}, U_{11} and U_{22} are lower and upper triangular respectively. Matrices A_{11}, A_{12}, A_{21}, A_{22}, L_{21} and U_{12} are rectangular.

Recursive Algorithm 22 *Cholesky recursive algorithm if upper triangular part of A is given (rcholesky):*

> *Do recursion*
> - *if $n > 1$ then*
>   - $U_{11} := $ *rcholesky of A_{11}*
>   - $U_{11}^T U_{12} = A_{12} \ \rightarrow$ **RTRSM**
>   - $\hat{A}_{22} := A_{22} - U_{12}^T U_{12} \ \rightarrow$ **RSYRK**
>   - $U_{22} := $ *rcholesky of $\hat{A}_{22}$*
> - *otherwise*
>   - $U := \sqrt{A}$
> *End recursion*

The RTRSM and RSYRK are recursive BLAS of _TRSM and _SYRK respectively. _TRSM solves a triangular system of equations. _SYRK performs the symmetric rank k operations (see Sections 4.1 and 4.2).

The listing of the Recursive Cholesky Factorization Subroutine is attached in the Appendix A.

Lower triangular case   Upper triangular case

$$
\begin{pmatrix}
1 & & & & & & \\
2 & 9 & & & & & \\
3 & 10 & 17 & & & & \\
4 & 11 & 18 & 25 & & & \\
5 & 12 & 19 & 26 & 33 & & \\
6 & 13 & 20 & 27 & 34 & 41 & \\
7 & 14 & 21 & 28 & 35 & 42 & 49
\end{pmatrix}
\qquad
\begin{pmatrix}
1 & 8 & 15 & 22 & 29 & 36 & 43 \\
 & 9 & 16 & 23 & 30 & 37 & 44 \\
 & & 17 & 24 & 31 & 38 & 45 \\
 & & & 25 & 32 & 39 & 46 \\
 & & & & 33 & 40 & 47 \\
 & & & & & 41 & 48 \\
 & & & & & & 49
\end{pmatrix}
$$

Fig. 1. The mapping of 7×7 matrix for the LAPACK Cholesky Algorithm using the full storage

2.1   Full and Packed Storage Data Format

The Cholesky factorization algorithm can be programmed either in "full storage" or "packed storage". For example the LAPACK subroutine POTRF works on full storage, while the routine PPTRF is programmed for packed storage. Here we are interested only in full storage and packed storage holding data that represents dense symmetric positive definite matrices. We will compare our recursive algorithms to the LAPACK POTRF and PPTRF subroutines.

The POTRF subroutine uses the Cholesky algorithm in full storage. A storage for the full array A must be declared even if only $n \times (n + 1)/2$ elements of array A are needed and, hence $n \times (n-1)/2$ elements are not touched. The PPTRF subroutine uses the Cholesky algorithm on packed storage. It only needs $n \times (n+1)/2$ memory words. Moreover the POTRF subroutine works fast while the PPTRF subroutine works slow. Why? The routine POTRF is constructed with BLAS level 3, while the PPTRF uses the BLAS level 2. These LAPACK data structures are illustrated by the figs. 1 and 2 respectively.

Lower triangular case   Upper triangular case

$$
\begin{pmatrix}
1 & & & & & & \\
2 & 8 & & & & & \\
3 & 9 & 14 & & & & \\
4 & 10 & 15 & 19 & & & \\
5 & 11 & 16 & 20 & 23 & & \\
6 & 12 & 17 & 21 & 24 & 26 & \\
7 & 13 & 18 & 22 & 25 & 27 & 28
\end{pmatrix}
\qquad
\begin{pmatrix}
1 & 2 & 4 & 7 & 11 & 16 & 22 \\
 & 3 & 5 & 8 & 12 & 17 & 23 \\
 & & 6 & 9 & 13 & 18 & 24 \\
 & & & 10 & 14 & 19 & 25 \\
 & & & & 15 & 20 & 26 \\
 & & & & & 21 & 27 \\
 & & & & & & 28
\end{pmatrix}
$$

Fig. 2. The mapping of 7×7 matrix for the LAPACK Cholesky Algorithm using the packed storage

2.2   The Recursive Storage Data Format

We introduce a new storage data format, the recursive storage data format, using the recursive algorithms 21 and 22. Like packed data format this recursive storage data format requires $n \times (n+1)/2$ storage for the upper or lower part of the matrix. The recursive storage data format is illustrated in fig. 3. A buffer of the size $p \times (p-1)/2$, where $p = \lceil n/2 \rceil$ (integer division), is needed to convert from the LAPACK packed storage data format to the recursive packed storage data format and back. No buffer is needed if data is given in recursive format.

We can apply the BLAS level 3 using the recursive packed storage data format. The performance of the recursive Cholesky algorithm with the recursive packed data format reaches the performance of the LAPACK POTRF algorithm. A graphs with the performance between different storages is presented in fig. 4.

The Figure 4 has two subfigures. The upper subfigure shows comparison curves for Cholesky factorization. The lower subfigure show comparison curves of forward and backward substitutions. The captions describe details of the performance figures. Each subfigure presents six curves. From the bottom: The first two curves represent LAPACK PPTRF performance results for upper and lower case respectively. The third and fourth curves give the performance of the recursive algorithms, L and U, respectively. The conversion time from LAPACK packed data format to the recursive packed data format and back is included here. The fifth and sixth curves give performance of the L and U variants for the LAPACK POTRF algorithm. The IBM SMP optimized ESSL DGEMM was used by the last four algorithms.

Lower triangular case   Upper triangular case

$$
\left(
\begin{array}{ccc|cc|cc}
1 & & & & & & \\
\hline
2 & 4 & & & & & \\
3 & 5 & 6 & & & & \\
\hline
7 & 11 & 15 & 19 & & & \\
8 & 12 & 16 & 20 & 21 & & \\
\hline
9 & 13 & 17 & 22 & 24 & 26 & \\
10 & 14 & 18 & 23 & 25 & 27 & 28
\end{array}
\right)
\qquad
\left(
\begin{array}{c|cc|cccc}
1 & 2 & 3 & 7 & 10 & 13 & 16 \\
\hline
 & 4 & 5 & 8 & 11 & 14 & 17 \\
 & & 6 & 9 & 12 & 15 & 18 \\
\hline
 & & & 19 & 20 & 22 & 24 \\
 & & & & 21 & 23 & 25 \\
\hline
 & & & & & 26 & 27 \\
 & & & & & & 28
\end{array}
\right)
$$

Fig. 3. The mapping of 7×7 matrix for the LAPACK Cholesky Algorithm using the packed recursive storage

The same good results were obtained on other parallel supercomputers, for example on Compaq α DS–20 and SGI Origin 2000.

The paper [1] gives comparison performance results on various computers for the Cholesky factorization and solution.

3 *LU* Factorization

We would like to compute the solution to a system of linear equations $AX = B$, where A is a real or complex matrix, and X and B are rectangular matrices or vectors. Gaussian elimination with row interchanges is used to factor A as $LU = PA$, where P is a permutation matrix, L is a unit lower triangular matrix, and U is an upper triangular matrix. The factored form of A is then used to solve the system of equations $AX = B$.

The recursive algorithm of the Gauss LU factorization is described in detail in [3,8]. We give two recursive algorithms here. They are listed in figs. 31 and 32.

Recursive Algorithm 31 *Recursive LU factorization without pivoting (rgausslu):*

> *Do recursion*
> - *if $min(m, n) > 1$ then*
>   - $(L_1, U_1) =$ *rgausslu of* A_1
>   - $L_{11}U_{12} = A_{12} \rightarrow$ **RTRSM**
>   - $\hat{A}_{22} = A_{22} - L_{21}U_{12} \rightarrow$ **_GEMM**
>   - $(L_{22}, U_{22}) =$ *rgausslu of* $\hat{A}_{22}$
> - *otherwise*
>   - $L_1 := A_1/a_{11}$ *and* $U_1 = a_{11}$
> *End recursion*
> - *if $n > m$ then*
>   - $L U_3 = A_3 \rightarrow$ **RTRSM**

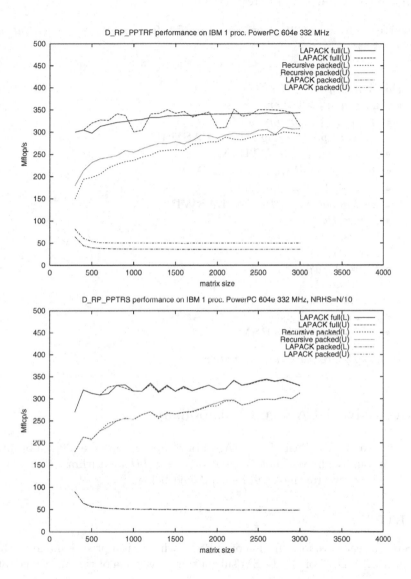

Fig. 4. The performance graphs between a full, packed and recursive packed storage data formats of the Cholesky factorization and solution algorithms run on the IBM 1 Processor PowerPC 604e @ 332 MHz computer, using the double precision arithmetic. The recursive results include the time consumed by converting from packed to recursive packed storage and vice versa. All routines call the optimized BLAS for the PowerPC architecture.

The matrices A_1, A_2, A_3, A_{12}, A_{22}, L_1, L_{11}, L_{21}, L_{22}, U_1, U_3, U_{12} and U_{22} are submatrices of A, L and U respectively, $a_{11} \in A_1$.

Recursive Algorithm 32 *Recursive LU=PA factorization with partial pivoting (rgausslu):*

> *Do recursion*
> - *if* $\min(m, n) > 1$ *then*
>   - $(P_1, L_1, U_1) = $ *rgausslu of* A_1
>   - *Forward pivot* A_2 *by* P $\rightarrow$ **_LASWP**
>   - $L_{11}U_{12} = A_{12}$ $\rightarrow$ **RTRSM**
>   - $\hat{A}_{22} := A_{22} - L_{21}U_{12}$ $\rightarrow$ **_GEMM**
>   - $(P_2, L_{22}, U_{22}) = $ *rgausslu of* $\hat{A}_{22}$
>   - *Back pivot* A_1 *by* P_2 $\rightarrow$ **_LASWP**
>   - $P = P_2 P_1$
> - *otherwise*
>   - *pivot* A_1
>   - $L_1 := A_1/a_{11}$ *and* $U_1 := a_{11}$
> *End recursion*
> - *if* $n > m$ *then*
>   - *Forward pivot* A_3 *by* P $\rightarrow$ **_LASWP**
>   - $LU_3 = A_3$ $\rightarrow$ **RTRSM**

where P_1 *and* P_2 *are permutation matrices.*

4   Recursive BLAS and Parallelism

Two recursive BLAS (Basic Linear Algebra Subprograms, see [20]) subroutines are used in our recursive Cholesky and recursive LU algorithms: RTRSM and RSYRK. These two routines will be explained below.

4.1   RTRSM

RTRSM is a recursive formulation of _TRSM, where _ is a precision and arithmetic indicator: S, D, C or Z. _TRSM subroutine solves one of the matrix equation

$$AX = \alpha B, A^T X = \alpha B, X A = \alpha B, or X A^T = \alpha B,$$

where α is a scalar. X and B are $(m \times n)$ rectangular matrices. A is a unit, or non-unit, upper or lower $(m \times m)$ triangular matrix. The matrix X is overwritten on B. We have 16 different triangular equations because A and A^T can be either upper or lower triangular matrices, and the diagonal is normal or unit.

We will introduce the recursive formulation only for one case $AX = \alpha B$, where A is lower triangular. The other cases will be similar.

The matrices A, B, and X can be partitioned into smaller submatrices, thus

$$\begin{pmatrix} A_{11} & \\ A_{21} & A_{22} \end{pmatrix} \begin{pmatrix} X_{11} & X_{12} \\ X_{21} & X_{22} \end{pmatrix} = \alpha \begin{pmatrix} B_{11} & B_{12} \\ B_{21} & B_{22} \end{pmatrix}$$

The matrices $A_{11} = A(1:h, 1:h)$, $A_{21} = A(h+1:m, 1:h)$, $A_{22} = A(h+1:m, h+1:m)$, $B_{11} = B(1:h, 1:p)$, $B_{12} = B(1:h, p+1:n)$, $B_{21} = B(h+1:m, 1:p)$, $B_{22} = B(h+1:m, p+1:n)$, $X_{11} = X(1:h, 1:p)$, $X_{12} = X(1:h, p+1:n)$, $X_{21} = X(h+1:m, 1:p)$ and $X_{22} = X(h+1:m, p+1:n)$ are submatrices of A, B and X respectively, and $h = m/2$ and $p = n/2$.

Multiplying the matrix A by X gives:

$$\begin{pmatrix} A_{11}X_{11} & A_{11}X_{12} \\ A_{21}X_{11} + A_{22}X_{21} & A_{21}X_{12} + A_{22}X_{22} \end{pmatrix} = \begin{pmatrix} \alpha B_{11} & \alpha B_{12} \\ \alpha B_{21} & \alpha B_{22} \end{pmatrix}$$

We have got two independent groups of triangular systems:

$$A_{11}X_{11} = \alpha B_{11} \qquad A_{11}X_{12} = \alpha B_{12}$$
$$A_{22}X_{21} = \alpha B_{21} - A_{21}X_{11} \qquad A_{22}X_{22} = \alpha B_{22} - A_{21}X_{12}$$

We could do a double recursion on m and n; i. e. on h and p. However, we do not do the recursion on p. This results in the following algorithm:

Recursive Algorithm 41 *Recursive algorithm for the $AX = B$ operation (one group only), where A is a lower triangular matrix (rtrsm):*

> *Do recursion*
> - *if $m > 1$ then*
>   - $A_{11}X_1 = \alpha B_1 \;\rightarrow\;$ **RTRSM**
>   - $\hat{B}_2 := \alpha B_2 - A_{21}X_1 \;\rightarrow\;$ **_GEMM**
>   - $A_{22}X_2 = \alpha \hat{B}_2 \;\rightarrow\;$ **RTRSM**
> - *otherwise*
>   - $a_{11}X_1 = \alpha B_1$
> *End recursion*

4.2 RSYRK

RSYRK is a recursive formulation of _SYRK, where _ is a precision and arithmetic indicator: S, D, C or Z. _SYRK performs one of the symmetric rank k operations:

$$C := \alpha AA^T + \beta C \text{ or } C := \alpha A^T A + \beta C$$

where α and β are scalars. A is a rectangular matrix $(m \times n)$. C is a square symmetric matrix.

We will introduce the recursive formulation only for one of the four cases of _SYRK:

$$C := \alpha AA^T + \beta C.$$

The matrices A and C can be partitioned into smaller submatrices:

$$\begin{pmatrix} C_{11} \\ C_{21} \; C_{22} \end{pmatrix} = \beta \begin{pmatrix} C_{11} \\ C_{21} \; C_{22} \end{pmatrix} + \alpha \begin{pmatrix} A_{11} \; A_{12} \\ A_{21} \; A_{22} \end{pmatrix} \begin{pmatrix} A_{11} \; A_{12} \\ A_{21} \; A_{22} \end{pmatrix}^T .$$

The matrices $A_{11} = A(1:h, 1:p)$, $A_{12} = A(1:h, p+1:n)$, $A_{21} = A(h+1:m, 1:p)$, $A_{22} = A(h+1:m, p+1:n)$, $C_{11} = C(1:h, 1:h)$, $C_{21} = C(h+1:m, 1:h)$, $C_{22} = C(h+1:m, h+1:m)$ are submatrices of A and C respectively, and $h = m/2$ and $p = n/2$. The recursion could be done again on two variables, h and p, but we do recursion on h only.

In terms of the partitioning we have three independent formulas:

$$C_{11} = \beta C_{11} + \alpha A_{11} A_{11}^T + \alpha A_{12} A_{12}^T$$
$$C_{21} = \beta C_{21} + \alpha A_{21} A_{11}^T + \alpha A_{22} A_{12}^T$$
$$C_{22} = \beta C_{22} + \alpha A_{21} A_{21}^T + \alpha A_{22} A_{22}^T$$

These three computations can be done in parallel. We now formulate a recursive algorithm as follows:

Recursive Algorithm 42 *Recursive algorithm for the $C := \alpha A A^T + \beta C$ symmetric rank k operations (rsyrk):*

> *Do recursion*
> - *if $m \geq 1$ then*
>   *Perform computation C_{11}:*
>   - *$\hat{C}_{11} := \beta C_{11} + \alpha A_{11} A_{11}^T \rightarrow$ **RSYRK***
>   - *$C_{11} := \hat{C}_{11} + \alpha A_{12} A_{12}^T \rightarrow$ **RSYRK***
>   *Perform computation C_{21}:*
>   - *$\hat{C}_{21} := \beta C_{21} + \alpha A_{21} A_{11}^T \rightarrow$ **_GEMM***
>   - *$C_{21} := \hat{C}_{21} + \alpha A_{22} A_{12}^T \rightarrow$ **_GEMM***
>   *Perform computation C_{22}:*
>   - *$\hat{C}_{22} := \beta C_{22} + \alpha A_{21} A_{21}^T \rightarrow$ **RSYRK***
>   - *$C_{22} := \hat{C}_{22} + \alpha A_{22} A_{22}^T \rightarrow$ **RSYRK***
> *End recursion*

4.3   A Fast _GEMM Algorithm

The _ of _GEMM is a precision and arithmetic indicator: S, D, C or Z. _GEMM subroutine does the following operations

$$C := \alpha AB + \beta C, C := \alpha AB^T + \beta C, C := \alpha A^T B + \beta C,$$

$$C := \alpha A^T B^T + \beta C, \text{ or } C := \alpha AB^C + \beta C, C := \alpha A^C B + \beta C$$

$$\text{and } C := \alpha A^C B^C + \beta C,$$

where α and β are scalars. A, B and C are rectangular matrices. A^T, B^T, A^C and B^C are transpose and conjugate matrices respectively.

The GEMM operation is very well documented and explained in [9,10,20]. We can see that work is done by _GEMM for both our BLAS RTRSM Section 4.1 and RSYRK Section 4.2. The speed of our computation depends very much from the speed of a good _GEMM. Good _GEMM implementations are usually developed by computer manufacturers. The model implementation of _GEMM can be obtained from netlib [20]; it works correctly but slowly. However, an excellent set of high performance BLAS, called _GEMM based BLAS was developed by Bo Kågström at the University of Umeå in Sweden, see for example [14,10]. A key idea behind _GEMM based BLAS was to cast all BLAS algorithms in terms of the simple BLAS _GEMM. Recently, the Innovative Computing Laboratory at University of Tennessee in Knoxville developed a system called ATLAS which can produce a fast _GEMM program.

ATLAS. Automatically Tuned Linear Algebra Software (ATLAS) [19]. ATLAS is an approach for the automatic generation and optimization of numerical software for processors with deep memory hierarchies and pipelined functional units. The production of such software for machines ranging from desktop workstations to embedded processors can be a tedious and time consuming task. ATLAS has been designed to automate much of this process. So, having a fast _GEMM means our RTRSM and RSYRK routines will be fast. The ATLAS GEMM is often better than _GEMM developed by the computer manufacture. What is important, the ATLAS software is available to everybody, free of charge. Every personal computer can have a good GEMM.

5 LDL^T Factorization for Symmetric Indefinite Matrices

It is well-known that the Cholesky factorization can fail for symmetric indefinite matrices. In this case some pivoting strategy can be applied (e. g. the Bunch-Kaufman pivoting [6, §4.4]). The algorithm is formulated in [4,6,13,17].

As a result we get

$$PAP^T = LDL^T,$$

where L is unit lower triangular, D is block diagonal with 1×1, or 2×2 blocks, and P is a permutation matrix.

Now let us look at a recursive formulation of this algorithm. This is given below. The recursion is done on the second dimension of matrix A, i. e. the algorithm works on full columns like in the LU factorization.

The LAWRA project on LDL^T is still going on. We have developed several perturbation approach algorithms for solving linear systems of equations, where the matrix is symmetric and indefinite. The results are presented in two papers, [7,12].

Recursive Algorithm 51 *Recursive Symmetric Indefinite Factorization (RSIF) of $A_{1:m,1:n}$, $m = n$:*

$$k = 1$$
if *(n = 1)*
  Define the pivot: 1×1, or 2×2.
  Apply interchanges if necessary
  $k = k + 1$, *or* $k = k + 2$
  If the pivot is 2×2: FLAG=1
else
  $n1 = n/2$
  $n2 = n - n1$
  RSIF *of* $A_{:,k:k+n1-1}$
  if *(FLAG = 1)*
    $n1 = n1 + 1$
    $n2 = n - n1$
  end
  update $A_{:,k:k+n2-1}$
  RSIF *of* $A_{:,k:k+n1-1}$
end

We have also obtained very good results for the Bunch-Kaufman factorization, where the matrix A is given in packed data format with about 5% extra space appended. We call it a blocked version. The recursion is not applied. Our packed blocked algorithm works faster than the LAPACK full storage algorithm. The algorithm is described and the comparison performance results for the factorization and solution are presented in [11].

References

1. B. Andersen, F. Gustavson and J. Wasniewski: "A recursive formulation of Cholesky factorization of a matrix in packed storage", University of Tennessee, Knoxville, TN, Computer Science Dept. Technical Report CS-00-441, May 2000, also LAPACK Working Note number 146 (lawn146.ps), and submitted to the Transaction of Mathematical Software (TOMS) of the ACM.
2. E. Anderson, Z. Bai, C. H. Bischof, S. Blackford, J. Demmel, J. J. Dongarra, J. Du Croz, A. Greenbaum, S. Hammarling, A. McKenney, and D. C. Sorensen. *LAPACK Users' Guide Release 3.0.* SIAM, Philadelphia, 1999.
3. B.S. Andersen, F. Gustavson, J. Waśniewski, and P. Yalamov: Recursive formulation of some dense linear algebra algorithms, in *Proceedings of the 9th SIAM Conference on Parallel Processing for Scientific Computing, PPSC99*, B. Hendrickson, K.A. Yelick, C.H. Bischof, I.S. Duff, A.S. Edelman, G.A. Geist, M.T. Heath, M.A. Heroux, C. Koelbel, R.S. Schreiber, R.F. Sincovec, and M.F. Wheeler (Eds.), San Antonio, TX, USA, March 24-27, 1999, SIAM, Scientific Computing, CDROM.
4. J.W. Demmel, *Applied Numerical Linear Algebra.* SIAM, Philadelphia, 1997.

5. J. Dongarra and J. Waśniewski, High Performance Linear Algebra Package – LA-PACK90, in Advances in Randomized Parallel Computing, Kluwer Academic Publishers, Combinatorial Optimization Series, P.M. Pardalos and S. Rajasekaran (Eds.), 1999 and available as the LAPACK Working Note (Lawn) Number 134: http://www.netlib.org/lapack/lawns/lawn134.ps

6. G.H. Golub and C.F. Van Loan, *Matrix Computations (third edition)*. Johns Hopkins University Press, Baltimore, MD, 1996.

7. A. Gupta, F. Gustavson, A. Karaivanov, J. Waśniewski, and P. Yalamov, *Experience with a Recursive Perturbation Based Algorithm for Symmetric Indefinite Linear Systems*, in: Proc. EuroPar'99, Toulouse, France, 1999.

8. F. Gustavson, *Recursive Leads to Automatic Variable Blocking for Dense Linear-Algebra Algorithms*. IBM Journal of Research and Development, Volume 41, Number 6, November 1997.

9. F. Gustavson, A. Henriksson, I. Jonsson, B. Kågström and P. Ling, Recursive Blocked Data Formats and BLAS' for Dense Linear Algebra Algorithms, in *Proceedings of the 4th International Workshop, Applied Parallel Computing, Large Scale Scientific and Industrial Problems, PARA'98*, B. Kågström, J. Dongarra, E. Elmroth, and J. Waśniewski (Eds.), Umeå, Sweden, June 1998, Springer, Lecture Notes in Computer Science Number 1541, pp. 195–206.

10. F. Gustavson, A. Henriksson, I. Jonsson, B. Kågström and P. Ling, Superscalar GEMM-based Level 3 BLAS – The On-going Evolution of Portable and High-Performance Library, in *Proceedings of the 4th International Workshop, Applied Parallel Computing, Large Scale Scientific and Industrial Problems, PARA'98*, B. Kågström, J. Dongarra, E. Elmroth, and J. Waśniewski (Eds.), Umeå, Sweden, June 1998, Springer, Lecture Notes in Computer Science Number 1541, pp. 207–215.

11. F. Gustavson, A. Karaivanov, M. Marinova, J. Waśniewski, and Plamen Yalamov, *A Fast Minimal Storage Symmetric Indefinite Solver*, in PARA2000 Proceedings (this proceedings), Bergen, Norway, June, 2000.

12. F. Gustavson, A. Karaivanov, J. Waśnicwski, and P. Yalamov, *A columnwise recursive perturbation based algorithm for symmetric indefinite linear systems*, in: Proc. PDPTA'99, Las Vegas, 1999.

13. N.J. Higham, *Accuracy and Stability of Numerical Algorithms*. SIAM, Philadelphia, 1996.

14. B. Kågström, P. Ling, and C. Van Loan, *GEMM-based level 3 BLAS: High-performance model implementations and performance evaluation benchmark*. ACM Trans. Math. Software, 1997.

15. S. Metcalf and J. Reid. *Fortran 90/95 Explained*. Oxford, New York, Tokyo, Oxford University Press, 1996.

16. S. Toledo, *Locality of Reference in LU Decomposition with Partial Pivoting*. SIAM Journal on Matrix Analysis and Applications, Vol. 18, No. 4, 1997.

17. L.N. Trefethen and D. Bau, III. *Numerical Linear Algebra*. SIAM, Philadelphia, 1997.

18. J. Waśniewski, B.S. Andersen, and F. Gustavson. Recursive Formulation of Cholesky Algorithm in Fortran 90, in *Proceedings of the 4th International Workshop, Applied Parallel Computing, Large Scale Scientific and Industrial Problems, PARA'98*, B. Kågström, J. Dongarra, E. Elmroth, and J. Waśniewski (Eds.), Umeå, Sweden, June 1998, Springer, Lecture Notes in Computer Science Number 1541, pp. 574–578.

19. R.C. Whaley and J. Dongarra, Automatically Tuned Linear Algebra Software (ATLAS), in *Ongoing Projects, The Innovative Computing Laboratory, Distributed Network Computing, Numerical Linear Algebra, Software Repositories, and Performance Evaluation,* http://www.netlib.org/atlas/, Knoxville, Tennessee, USA, 1999.
20. BLAS (Basic Linear Algebra Subprograms), in *Ongoing Projects, The Innovative Computing Laboratory, Distributed Network Computing, Numerical Linear Algebra, Software Repositories, and Performance Evaluation,* http://www.netlib.org/blas/, Knoxville, Tennessee, USA, 1999.

A   Recursive Cholesky Factorization Subroutine

```
RECURSIVE SUBROUTINE RPOTRF( A, UPLO, INFO )
  USE LA_PRECISION, ONLY: WP => DP
  USE LA_AUXMOD, ONLY: ERINFO, LSAME
  USE F90_RCF, ONLY: RCF => RPOTRF, RTRSM, RSYRK
  IMPLICIT NONE
  CHARACTER(LEN=1), OPTIONAL, INTENT(IN) :: UPLO
  INTEGER, OPTIONAL, INTENT(OUT) :: INFO
  REAL(WP), INTENT(INOUT) :: A(:,:)
  CHARACTER(LEN=*), PARAMETER :: SRNAME = 'RPOTRF'
  REAL(WP), PARAMETER :: ONE = 1.0_WP
  CHARACTER(LEN=1) :: LUPLO; INTEGER :: N, P, LINFO
  INTEGER, SAVE :: IC = 0, NMAX = 0
    N = SIZE(A,1); LINFO = 0; IF( NMAX == 0 )NMAX = N
    IF( PRESENT(UPLO) )THEN; LUPLO = UPLO
      ELSE; LUPLO = 'U'; ENDIF
    IF( N < 0 .OR. N / = SIZE(A,2) )THEN; LINFO = -1
    ELSE IF( .NOT. (LSAME(LUPLO,'U').OR.LSAME(LUPLO,'L')) )THEN
        LINFO = -2
    ELSE IF (N == 1) THEN; IC - IC + 1
      IF( A(1,1) > 0.0_WP )THEN; A(1,1) = SQRT(A(1,1))
        ELSE; LINFO = IC; ENDIF
    ELSE IF( N > 0 )THEN; P=N/2
      IF( LSAME(LUPLO,'L') )THEN
        CALL RCF( A(1:P,1:P), LUPLO, LINFO )
        IF( LINFO == 0 )THEN
          CALL RTRSM( A(1:P,1:P), A(P+1:N,1:P), UPLO=LUPLO, &
            SIDE='R', TRANSA='T' )
          CALL RSYRK( A(P+1:N,1:P), A(P+1:N,P+1:N), ALPHA=-ONE, &
            UPLOC=LUPLO )
          IF( LINFO == 0 )CALL RCF( A(P+1:N,P+1:N), LUPLO, LINFO )
        ENDIF
      ELSE
        CALL RCF( A(1:P,1:P), LUPLO, LINFO )
        IF( LINFO == 0 )THEN
          CALL RTRSM( A(1:P,1:P), A(1:P,P+1:N), TRANSA='T' )
          CALL RSYRK( A(1:P,P+1:N), A(P+1:N,P+1:N), ALPHA=-ONE, &
            TRANSA='T' )
          IF( LINFO == 0 ) CALL RCF( A(P+1:N,P+1:N), LUPLO, LINFO )
        ENDIF
      ENDIF
    ENDIF
    IF( NMAX == N )THEN; NMAX = 0; IC = 0; END IF
    CALL ERINFO( LINFO, SRNAME, INFO )
END SUBROUTINE RPOTRF
```

Solving CFD Problems with Open Source Parallel Libraries

Bill Gropp

Argonne National Laboratory, USA
gropp@mcs.anl.gov

Abstract. Parallel computers are often viewed as the only way to provide enough computing power to solve complex problems in scientific
simulation. Unfortunately, the complexity of developing and validating
a parallel program has limited the use of massive parallelism to "heroic"
programmers. One solution to this problem is to build better, more pro-
grammable parallel computers or more effective parallel languages. Unfortunately, scientific computing no longer drives computing, even for
parallel computers. As a result, the systems that many researchers use
do not provide good support for massively parallel programming, and
help is unlikely to come from computer system vendors. At the same
time, numerical algorithms are increasingly sophisticated, making it difficult for any one group to develop or maintain a state-of-the-art code.
The solution to these problems is the development of community-based
standards and software for scientific computing. Open source software in
particular has been an effective method for developing and distributing
software, in large part because it makes it easy for any interested party
to participate. Open source software such as the PETSc or Parasol libraries also makes it possible to tackle large scale scientific simulation
problems. This point is illustrated by the success of PETSc in achieving
over 220 GFlops on a fully unstructured CFD application, running on
over 6000 processors.

T. Sørevik et al. (Eds.): PARA 2000, LNCS 1947, p. 52, 2001.

High-Performance Library Software for QR Factorization

Erik Elmroth[1] and Fred Gustavson[2]

[1] Department of Computing Science and HPC2N, Umeå University, SE–901 87
Umeå, Sweden.
elmroth@cs.umu.se
[2] IBM T.J. Watson Research Center, P.O. Box 218, Yorktown Heights, NY 10598,
U.S.A.
gustav@watson.ibm.com

Abstract. In [5,6], we presented algorithm RGEQR3, a purely recursive formulation of the QR factorization. Using recursion leads us to a natural way to choose the k-way aggregating Householder transform of Schreiber and Van Loan [10]. RGEQR3 is a performance critical subroutine for the main (hybrid recursive) routine RGEQRF for QR factorization of a general $m \times n$ matrix. This contribution presents a new version of RGEQRF and its accompanying SMP parallel counterpart, implemented for a future release of the IBM ESSL library. It represents a robust high-performance piece of library software for QR factorization on uniprocessor and multiprocessor systems. The implementation builds on previous results [5,6]. In particular, the new version is optimized in a number of ways to improve the performance; e.g., for small matrices and matrices with a very small number of columns. This is partly done by including mini blocking in the otherwise pure recursive RGEQR3. We describe the salient features of this implementation. Our serial implementation outperforms the corresponding LAPACK routine by 10-65% for square matrices and 10-100% on tall and thin matrices on the IBM POWER2 and POWER3 nodes. The tests covered matrix sizes which varied from very small to very large. The SMP parallel implementation shows close to perfect speedup on a 4-processor PPC604e node.

Keywords: Serial and parallel library software, QR factorization, recursion, register blocking, unrolling, SMP systems, dynamic load balancing.

1 Introduction

A program's ability to efficiently utilize the memory hierarchy is the key to high performance on uniprocessors as well as on SMP systems. In dense linear algebra this is traditionally accomplished by formulating algorithms so that they are rich in Level 3 (i.e., matrix-matrix) operations. This is typically done by expressing matrices as a collection of submatrices (block matrices) and performing the computations as block operations on these submatrices. The block size is typically a fixed parameter that is adjusted for the specific memory hierarchy of the computer.

T. Sørevik et al. (Eds.): PARA 2000, LNCS 1947, pp. 53–63, 2001.
© Springer-Verlag Berlin Heidelberg 2001

It has recently been shown that recursion can be used to perform an automatic variable blocking that have a number of advantages over fixed blocking [5,6,7, 11]. The variable blocking leads to improved performance. It also leads to novel and/or more general algorithms that are less sensitive to parameter settings.

However, for the pure recursive QR factorization presented in [5,6] there are significant additional costs for creating and performing the updates. This problem was overcome by developing a hybrid recursive algorithm to be used for $m \times n$ matrices where n is large. In general, the hybrid recursive implementation outperforms the corresponding LAPACK algorithm, sometimes significantly.

However, as pointed out in [5,6], there are situations where the cost of the recursive overhead cannot be overcome by replacing Level 2 operations by Level 3 operations. Typically, this is the case for matrices too small to gain from matrix-matrix operations. This specific problem can be overcome by pruning the recursion tree; i.e., by stopping the recursion when n is small (e.g, $= 4$) instead of $n = 1$.

In this contribution, we present robust library software for QR factorization that is also designed to give high performance for cases where pure recursion provides too much overhead. In Section 2.1, we present the mini blocking, or register blocking, in some more detail. Sections 2.2–2.3 is devoted to concise descriptions of our modifications of the routines RGEQR3 and RGEQRF presented in [5, 6]. Technical descriptions of the new register blocking routines are presented in Sections 2.4–2.6. Performance results for our new set of uniprocessor routines are given in Section 3. In Section 4, we outline the modifications of our new parallel version of RGEQRF. Finally, in Section 5 we give some concluding remarks. The software described will be part of the IBM ESSL library version 3.2.

2   Recursive QR Factorization Using Mini Blocking

2.1   Introductory Remarks

In [5,6], we applied a pure recursive algorithm for QR factorization of an $m \times n$ matrix, using n as the recursion variable. In this algorithm a divide and conquer approach is performed until the matrix becomes a single column; i.e., $n = 1$ is used as the stopping criteria. At this point, a Householder transformation $Q = I - \tau vv^T$ is produced from this single column. Through the recursion, Householder transformations are recursively combined into a k-way aggregated transformation by computing the compact WY representation for Q.

However, there are significant additional costs for creating and performing the updates using the compact WY representation which prohibits the efficient use of the recursion for large n. This problem is overcome by using a hybrid recursive algorithm called RGEQRF [5,6] that uses the pure recursive algorithm for fixed size block columns.

The performance obtained on the PPC604e processor indicated that RGEQRF and the pure recursive RGEQR3 was all that was required. However, on "scientific computing machines" such as the IBM POWER2 the performance gains over traditional block algorithms were less significant, and for small matrices the LAPACK algorithm DGEQRF sometimes outperforms the hybrid recursive algorithm [5]. This contribution focuses on correcting the performance deficiency just described. Our approach here had been followed successfully for both Cholesky factorization and Gaussian elimination with partial pivoting [7, 9].

The new approach for recursion is old and very important for RISC type architectures, especially those with limited memory bandwidth and deep memory hierarchies. The approach is called mini blocking where the blocking parameter is small or tiny. In fact, we only concern ourselves with the tiny case and apply a special type of this blocking called register blocking or loop unrolling [1,8].

There is another way to view this approach. We call it pruning the recursion tree. Let mb be the mini blocking parameter and assume without loss of generality that mb divides $n = nb$, where n is the column dimension of the $m \times n$ matrix we wish to QR factor. In Figure 1, we give our modified QR factorization algorithm obtained by introducing the mini blocking parameter mb. Briefly, the new recursion tree has d leaves and $d - 1$ interior nodes, where $d = \lceil n/mb \rceil$. Each of the d leaves now becomes a single (m, mb) QR factorization problem consisting of mb leaves and mb - 1 interior nodes. In terms of the pure recursion problem ($mb = 1$) we have pruned the tree at nodes where (m, n) has $n \leq mb$.

The d recursive subproblems of the pure recursive QR factorization problem execute at greatly reduced Mflops/s rate on processors like IBM POWER2 and POWER3. By pruning the pure recursion tree, we replace each recursive subproblem by a direct computation which we now show executes at a much higher Mflops/s rate. Overall then, performance improves.

2.2   Adding a Blocking Parameter mb to RGEQR3

Figure 1 presents the new RGEQR3 algorithm, including the mini blocking parameter mb, for avoiding overhead from recursion for problems to small to gain from Level 3 operations. For further details about RGEQR3 without the mini blocking parameter or, equivalently set $mb = 1$; see [5,6]. In Figure 1, the if clause computation is now a single call to a subroutine named DQRA4. We have set $mb = 4$.

2.3   The Hybrid Recursive Algorithm RGEQRF

In [5,6], we only stated algorithm RGEQRF when $m \geq n$. Now we state it for arbitrary m and n in Figure 2. Note that this general algorithm is a trivial change to RGEQRF in [5,6]; i.e., the variable $k = \min(m, n)$ is used in the outer do loop instead of the variable n. The algorithm is included here for completeness of the description of this new ESSL QR factorization software.

2.4   The Mini Blocking Routine DQR24

It is fairly easy to modify a Level 2 QR factorization routine, like the LAPACK algorithm DGEQR2, to utilize a mini blocking. In this new ESSL internal rou-

RGEQR3 $A(1\!:\!m,1\!:\!n) = (Y,R,T)$! Note, Y and R replaces A, $m \geq n$
if $(n \leq mb)$ then
    compute Householder transform $Q = I - YTY^T$ such that
    $Q^T A = R$;   Return (Y,R,T)
else
    $n_1 = (\lfloor \lceil n/mb \rceil /2 \rfloor) * mb$ and $j_1 = n_1 + 1$
    call RGEQR3 $A(1\!:\!m,1\!:\!n_1) = (Y_1,R_1,T_1)$ where $Q_1 = I - Y_1 T_1 Y_1^T$
    compute $A(1\!:\!m,j_1\!:\!n) \leftarrow Q_1^T A(1\!:\!m,j_1\!:\!n)$
    call RGEQR3 $A(j_1\!:\!m,j_1\!:\!n) = (Y_2,R_2,T_2)$ where $Q_2 = I - Y_2 T_2 Y_2^T$
    compute $T_3 \leftarrow T(1\!:\!n_1,j_1\!:\!n) = -T_1(Y_1^T Y_2)T_2$.
    set $Y \leftarrow (Y_1,Y_2)$! Y is m by n unit lower trapezoidal
    return (Y,R,T), where
$$R \leftarrow \begin{pmatrix} R_1 & A(1\!:\!n_1,j_1\!:\!n) \\ 0 & R_2 \end{pmatrix} \text{ and } T \leftarrow \begin{pmatrix} T_1 & T_3 \\ 0 & T_2 \end{pmatrix}.$$
endif

Fig. 1. Recursive QR factorization routine RGEQR3 with mini blocking.

RGEQRF$(m,n,A,lda,\tau,work)$
$k = \min(m,n)$
do $j = 1, k, nb$! nb is the block size
    $jb = \min(n - j + 1, nb)$
    call RGEQR3 $A(j:m,j+jb-1) = (Y,R,T)$! T is stored in $work$
    if $(j + jb$ .LE. $n)$ then
        compute $(I - YT^T Y^T)A(j\!:\!m,j+jb\!:\!n)$ using $work$ and T via
        a call to DQTC
    endif
enddo

Fig. 2. Hybrid algorithm RGEQRF.

DQR24$(m,n,A,lda,tau,work,ldt,lwork)$
$k = \min(m,n)$
do $j = 0, k - 4, 4$
    ! Generate elementary reflectors $H(j:j+3)$ and $T(4,4)$ to produce
    ! block reflector $H(j:j+3) = I - Y * T * Y^T$ by calling
    DQRA4$(m - j, A(j,j), lda, tau(j), work(ldt * j), ldt)$
    ! Apply $H(j:j+3)^T$ to $A(j:m-1,j+4:n-1)$ from the left by calling
    DQRU4$(m\!-\!j, n\!-\!j - 4, A(j,j), work(ldt*j), ldt, A(j,j+4), ldt, work(ldt*(j+4)))$
enddo
if $(k \geq j)$ then
    if $(m \geq n)$ then
        DQRA4$(m - j, k - j, A(j,j), tau(j), work(ldt * j), ldt)$
    else ! $m < n$ and $m = k$
        Apply fixup code for cases when n not a multiple of 4
    endif
endif

Fig. 3. Mini blocking Level 2 routine DQR24.

tine, we used $mb = 4$. The algorithm was named DQR24 and it had exactly the same functionality as DGEQR2. In Figure 3 we give algorithm DQR24.

Note that DQR24 calls subroutines DQRA4 and DQRU4. In DGEQR2, the equivalent routines are DLARFG and DLARF and the corresponding value of $mb = 1$. In our ESSL implementation we chose to either call DQR24 or RGEQRF to QR factor the matrix A. Note that RGEQRF does $\lceil k/nb \rceil$ block steps in the outer do loop on j. Let $km = \max(m, n)$. For a given j the crossover criterion used was $(km - j) * (k - j)$ versus a tunable constant.

2.5   Details on DQRA4

DQRA4 does the QR factorization of $A(0 : m-1, 0 : 3)$ as a single computation. If we look at the recursion tree of an $m \times 4$ matrix A in Figure 4 we see that it calls DLARFG once at each of the four leafs to compute a Householder transformation, DQTC three times and makes three T_3 computations. In the figure, the (m, k) notation at each node represents a factorization of an $m \times k$ matrix. The DQTC (m, n, k) notation represents the update of an $m \times n$ matrix with respect to k Householder transformations. $T(a : b, c : d)$ means compute a T_3 submatrix of that size. For both DQTC and the T_3 computations, the calls are made with parameters $(m, 1, 1)$, $(m, 2, 2)$, and $(m - 2, 1, 1)$ and matrix sizes of T_3 equal to 1×1, 1×1, and 2×2. Recall that DQTC consists of seven subroutine calls, consisting of three calls to DTRMM, two calls to DGEMM, one call to DCOPY, and one call to an matrix subtract routine. For a single T_3 computation, there are five subroutine calls, consisting of three calls to DTRMM, one call to DGEMM, and one call to a matrix transpose routine. Summing over

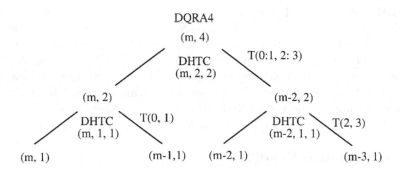

Fig. 4. Recursion tree for DQRA4 on a $m \times 4$ matrix.

the six calls, three to DQTC and three for the T_3 computations, there are 36 subroutine calls, of which 18 are to DTRMM and 9 to DGEMM. Some of the matrix sizes are tiny; e.g., 1×1, and 2×2. Note that ESSL's DTRMM and DGEMM monitors the matrix sizes and if necessary will call Level 2 BLAS such as DGEMV, DGER, and DTRMV. The point we make here is that one has a lot of subroutine calls that have argument checking (DTRMM has 11 arguments and DGEMM has 13 arguments). Additionally, we have not included the costs associated with traversing the recursion tree consisting of three interior nodes and four leaves.

The single routine DQRA4 obviates much of the book keeping described above. Look at the recursion tree again. We see that the two DQTC calls of size $(m, 1, 1)$ and $(m - 2, 1, 1)$ can be replaced by two calls to DDOT and DAXPY and some scalar floating point operations. Similarly, the DQTC call of size $(m, 2, 2)$ consists of a single call to $(2, 2, m)$ DGEMM and a single call to $(2, 2, m)$ DGEMM plus some scalar floating point operations. (Here, (m, n, k) DGEMM refers to a matrix multiplication that performs a rank-k update of an $m \times n$ matrix.) Likewise, two of the three T_3 computations are DDOT calls plus scalar floating point operations while the $T(0 : 1, 2 : 3)$ T_3 computation is a call to $(2, 2, m - 2)$ DGEMM plus scalar floating point operations. In all, there are three calls to DGEMM, three to DDOT, and three to DAXPY. Now, additionally we chose not to call DGEMM but to call DGEMM kernel routines. Hence, DQRA4 has minimal subroutine overhead plus it calls DGEMM kernels which has no argument error checking. Performance of DQRA4 is significantly better than DGEQR2 for arbitrary m and $n = 4$. The impact of using this routine for subproblems when factorizing matrices with n large is presented in Section 3.

2.6   DQRU4

This routine takes the output of DQRA4 which is $Q(0 : 3) = I - YTY^T$ and applies Q^T to an m by n matrix C. Hence, it has the same functionality as DQTC when the blocking parameter nb is 4. DQTC has seven calls (see Section 2.5). Now, DQRU4 has three calls. The first call is to DSCAL to zero out a 4 by n matrix W (work array). The next call computes $W = Y^T C$ by calling a DGEMM kernel of matrix size $(m, n, k) = (4, m, n)$. Now $W = T^T W$ is computed by inline code. T, of course, is LDT by 4 upper triangular. Lastly, a DGEMM kernel of matrix size $(m, n, 4)$ is called to compute $C = C - YW$. Note that Y is lower trapezoidal so the upper triangular part of A $(= Y$ here) must be saved and replaced with ones and zeros. This amounts to saving and restoring ten floating numbers. Finally, we note that RGEQR3 takes advantage of the extra speed of DQRU4; i.e., when $nb = 4$ RGEQR3 calls DQRU4 instead of DQTC.

3   Performance Results

Performance results for the serial RGEQRF are presented for a 160 MHz POWER2 processor and a 200 MHz POWER3 processor. The results for RGEQRF are compared to the DGEQRF which is the corresponding routine in LAPACK. Both routines are providing the same functionality. Values for tuning parameters have been set individually for each routine on each computer system.

Figures 5 and 6 shows the performance in Mflops/s for the new version of the hybrid recursive algorithm RGEQRF and the LAPACK algorithm DGEQRF

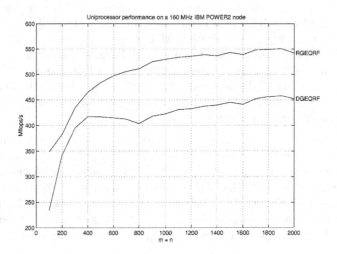

Fig. 5. Uniprocessor performance results in *Mflops* for the hybrid recursive algorithm RGEQRF and DGEQRF of LAPACK on the 160 MHz POWER2 processor.

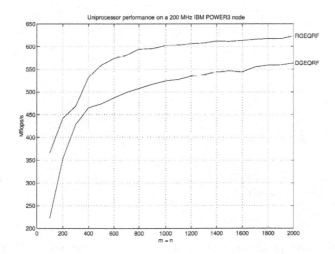

Fig. 6. Uniprocessor performance results in *Mflops* for the hybrid recursive algorithm RGEQRF and DGEQRF of LAPACK on the 200 MHz POWER3 processor.

on square matrices. For $m = n = 100, 200, \ldots, 2000$, RGEQRF outperforms DGEQRF for all tests on both machines, by at least 10%, and most often by around 20%. At best, the difference is 114 Mflops/s (49%) on the POWER2 and and 143 Mflops/s (65%) on the POWER3. In [5], DGEQRF outperformed RGEQRF for $m = n \leq 300$ on the 120 MHz POWER2. Here, we clearly see the positive change of RGEQRF with the new mini blocking incorporated, especially in the critical case for small matrices. For $m = n = 100$, the old RGEQRF was outperformed by DGEQRF by more than 10% on the 120 MHz POWER2 [5], and here the new RGEQRF with mini blocking included is 49% better on the 160MHz version of the same processor.

Performance results for tall and thin matrices are presented in Figures 7 and 8. The figures shows the performance ratio of RGEQRF to DGEQRF. Notably, RGEQRF perform better for all cases. RGEQRF without mini blocking was up to 30% slower than DGEQRF for some cases with $m = 200$. With the mini blocking, this performance deficiency is cured. For $m \geq 800$, RGEQRF outperforms DGEQRF by 30–70% for all values of n tested on POWER2. For the POWER3, the performance difference is up to almost a factor of 2.

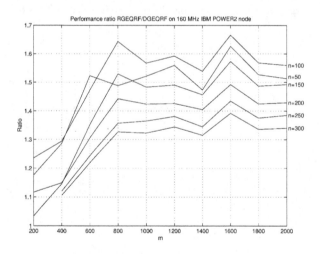

Fig. 7. Performance ratio of RGEQRF to DGEQRF for tall and thin matrices on the 160 MHz POWER2 processor.

In Figures 9 and 10 we present results for tall and extremely thin matrices, i.e., n goes from 4 to 20. Also here, RGEQRF outperforms DGEQRF for all choices of m and n.

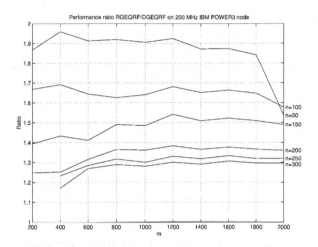

Fig. 8. Performance ratio of RGEQRF to DGEQRF for tall and thin matrices on the 200 MHz POWER3 processor.

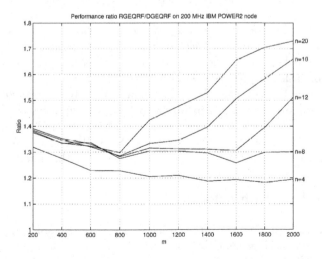

Fig. 9. Performance ratio of RGEQRF to DGEQRF for tall and very thin matrices on the 160 MHz POWER2 processor.

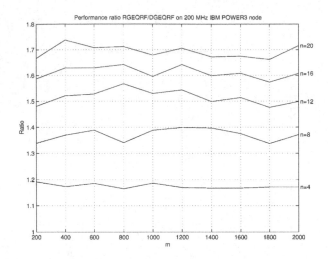

Fig. 10. Performance ratio of RGEQRF to DGEQRF for tall and very thin matrices on the 200 MHz POWER3 processor.

4   The Parallel RGEQRF

The parallel version of RGEQRF for the next release of the IBM ESSL library is a slight modification of the software version described in [5,6]. The mini blocking techniques described for the serial RGEQRF is included also in the parallel version. The $n > m$ case is handled in a straight-forward way. The updates in RGEQRF of the right hand side of the matrix are performed by first partitioning the matrix into a number of block columns. The submatrix computations to update these blocks, and to factorize the next block as soon as it is updated, form a pool of tasks. A processor is assigned a new job from the available tasks as soon as its previous job is completed. This is basically performed as described in [5,6] with some modification in order to allow for updates of columns with column indices $> m$. In [5] parallel speedups of 1.97, 2.99, and 3.97 was obtained for 2, 3, and 4 processors on a 332MHz 4-way SMP PPC604e node. Due to lack of space we do not include any performance figures for the new ESSL version.

5   Concluding Remarks

The new IBM ESSL QR factorization routines for uniprocessor and SMP systems have been presented. This contribution builds on the results on recursive QR factorization algorithms in [5,6] and the main new feature is improved performance especially for small matrices. The improvement comes from pruning the recursion tree, or equivalently, stopping the recursion at $n = mb$, and then performing the QR factorization of the mb wide block column as a direct calculation. We call the concept mini blocking or register blocking. The improved performance provided by the mini blocking leads to a significant performance improvement for small matrices, but as a positive side effect, the overall performance for large matrices is also improved. Notably, the new RGEQRF with mini blocking outperforms the corresponding LAPACK routine for all combinations of m and n tested. The performance gain is at least 10% and at best it is around 100%.

Acknowledgements. The authors especially thank Joan McComb of ESSL Development in Poughkeepsie. In particular she was able to track down a subtle timing problem our algorithm encountered using the IBM XLF SMP compiler. We again thank Bo Kågström and Carl Tengwall for setting up our collaboration.

This research was conducted using the resources of the High Performance Computing Center North (HPC2N) and the Danish Computing Centre for Research and Education (UNI•C). Financial support has been provided by the Swedish Research Council for Engineering Sciences under grant TFR 98-604.

References

1. R.C. Agarwal, F.G. Gustavson, and M. Zubair. Exploiting functional parallelism of POWER2 to design high-performance numerical algorithms. *IBM J. Res. Develop*, 38(5):563–576, September 1994.
2. E. Anderson, Z. Bai, C. Bischof, J. Demmel, J. Dongarra, J. Du Croz, A. Greenbaum, S. Hammarling, A. McKenney, S. Ostrouchov, and D. Sorensen. *LAPACK Users' Guide - Release 2.0*. SIAM, Philadelphia, 1994.
3. C. Bischof and C. Van Loan. The WY representation for products of householder matrices. *SIAM J. Scientific and Statistical Computing*, 8(1):s2–s13, 1987.
4. A. Chalmers and J. Tidmus. *Practical Parallel Processing*. International Thomson Computer Press, UK, 1996.
5. E. Elmroth and F. Gustavson. Applying Recursion to Serial and Parallel QR Factorization Leads to Better Performance. *IBM Journal of Research and Development*, 44, No. 4, 605–624, 2000.
6. E. Elmroth and F. Gustavson. New serial and parallel recursive QR factorization algorithms for SMP systems. In B. Kågström et al., editors, *Applied Parallel Computing, Large Scale Scientific and Industrial Problems*, Lecture Notes in Computer Science, No. 1541, pages 120–128, 1998.
7. F. Gustavson. Recursion Leads to Automatic Variable Blocking for Dense Linear-Algebra Algorithms. *IBM Journal of Research and Development*, Vol. 41, No. 6, 1997.
8. F. Gustavson, A. Henriksson, I. Jonsson, B. Kågström and P. Ling. Superscalar GEMM-based Level 3 BLAS – The On-going Evolution of a Portable and High-Performance Library. In Kågström et al. (eds), *Applied Parallel Computing. Large Scale Scientific and Industrial Problems*, Lecture Notes in Computer Science, Vol. 1541, pp 207-215, Springer-Verlag, 1998.
9. F. Gustavson and I. Jonsson. High Performance Cholesky Factorization via Blocking and Recursion that uses Minimal Storage. *This Proceedings*.
10. R. Schreiber and C. Van Loan. A storage efficient WY representation for products of householder transformations. *SIAM J. Scientific and Statistical Computing*, 10(1):53–57, 1989.
11. S. Toledo. Locality of reference in LU decomposition with partial pivoting. *SIAM J. Matrix. Anal. Appl.*, 18(4), 1997.

Parallel Triangular Sylvester-Type Matrix Equation Solvers for SMP Systems Using Recursive Blocking

Isak Jonsson and Bo Kågström

Department of Computing Science and HPC2N, Umeå University,
SE-901 87 Umeå, Sweden.
{isak,bokg}@cs.umu.se

Abstract. We present recursive blocked algorithms for solving triangular Sylvester-type matrix equations. Recursion leads to automatic blocking that is variable and "squarish". The main part of the computations are performed as level 3 general matrix multiply and add (GEMM) operations. We also present new highly optimized superscalar kernels for solving small-sized matrix equations stored in level 1 cache. Hereby, a larger part of the total execution time will be spent in GEMM operations. In turn, this leads to much better performance, especially for small to medium-sized problems, and improved parallel efficiency on shared memory processor (SMP) systems. Uniprocessor and SMP parallel performance results are presented and compared with results from existing LAPACK routines for solving this type of matrix equations.

Keywords: Sylvester-type matrix equations, recursion, automatic blocking, superscalar, GEMM-based, level 3 BLAS

1 Introduction

Recursion is a key concept for matching an algorithm and its data structure. A recursive algorithm leads to automatic blocking which is variable and "squarish" [4]. This hierarchical blocking allows for good data locality, which makes it possible to approach peak performance on today's memory tiered processors. Using the current standard data layouts and applying recursive blocking have led to faster algorithms for the Cholesky, LU and QR factorizations [4,2,3]. Even better performance can be obtained when recursive dense linear algebra algorithms are expressed using a recursive data structure and recursive kernels [5].

In this contribution, we present new recursive blocked algorithms for solving triangular Sylvester-type matrix equations. Examples include the Sylvester, Lyapunov and Stein matrix equations, and their generalized counterparts, which all appear in different control theory applications. Typically, the second major step in their solution is to solve a triangular matrix equation. For example, the coefficient matrices A and B in the Sylvester equation $AX - XB = C$ are in Schur form (upper triangular or upper quasi-triangular form) [9,13]. For the generalized counterpart, we assume that the matrix pairs (A, D) and (B, E) in

T. Sørevik et al. (Eds.): PARA 2000, LNCS 1947, pp. 64–73, 2001.

$(AX - YB, DX - YE) = (C, F)$ are in generalized Schur form. The triangular matrix equations also appear naturally in estimating the condition numbers of matrix equations and different eigenspace problems [12,9,10,13,11].

In the following, we focus on recursive blocked algorithms for the triangular standard and generalized (coupled) Sylvester equations. Our approach and methods also apply to the earlier mentioned Sylvester-type matrix equations, but this is out of scope for this paper.

We remark that the standard (generalized coupled) Sylvester equation has a unique solution if and only if A and B ($A - \lambda D$ and $B - \lambda E$) have disjoint spectra [14].

2 Recursive Standard Triangular Sylvester Solvers

We consider the standard real Sylvester matrix equation of the form

$$\text{op}(A)X \pm X\text{op}(B) = C,$$

where $\text{op}(A) = A$ or A^T, and A and B are upper triangular or quasi-upper triangular, i.e., in real Schur form. A quasi-upper triangular matrix is block upper triangular with 1×1 and 2×2 diagonal blocks, which correspond to real and complex conjugate pairs of eigenvalues, respectively. A is $M \times M$ and B is $N \times N$; the right hand side C and the solution X are of size $M \times N$.

In the following, we let $\text{op}(A) = A$ and sign $= -$. Typically, the solution overwrites the right hand side ($C \leftarrow X$). Depending on the sizes of M and N, we consider three alternatives for doing a *recursive splitting*.

Case 1 ($1 \le N \le M/2$). We split A by rows and columns, and C by rows only:

$$\begin{bmatrix} A_{11} & A_{12} \\ & A_{22} \end{bmatrix} \begin{bmatrix} X_1 \\ X_2 \end{bmatrix} - \begin{bmatrix} X_1 \\ X_2 \end{bmatrix} B = \begin{bmatrix} C_1 \\ C_2 \end{bmatrix},$$

or equivalently

$$A_{11}X_1 - X_1B = C_1 - A_{12}X_2,$$
$$A_{22}X_2 - X_2B = C_2.$$

The original problem is split in two triangular Sylvester equations. First, we solve for X_2 and after updating the right hand side of the first equation with respect to X_2 ($C_1 = C_1 - A_{12}X_2$), we can solve for X_1.

Case 2 ($1 \le M \le N/2$). We split B by rows and columns, and C by columns only:

$$A\begin{bmatrix} X_1 & X_2 \end{bmatrix} - \begin{bmatrix} X_1 & X_2 \end{bmatrix}\begin{bmatrix} B_{11} & B_{12} \\ & B_{22} \end{bmatrix} = \begin{bmatrix} C_1 & C_2 \end{bmatrix},$$

or equivalently

$$AX_1 - X_1B_{11} = C_1,$$
$$AX_2 - X_2B_{22} = C_2 + X_1B_{12}.$$

Now, we first solve for X_1 and then after updating the right hand side of the second equation $(C_2 = C_2 + X_1 B_{12})$, X_2 can be solved for.

Case 3 $(N/2 \leq M \leq 2N)$. We split A, B and C by rows and columns:

$$\begin{bmatrix} A_{11} & A_{12} \\ & A_{22} \end{bmatrix} \begin{bmatrix} X_{11} & X_{12} \\ X_{21} & X_{22} \end{bmatrix} - \begin{bmatrix} X_{11} & X_{12} \\ X_{21} & X_{22} \end{bmatrix} \begin{bmatrix} B_{11} & B_{12} \\ & B_{22} \end{bmatrix} = \begin{bmatrix} C_{11} & C_{12} \\ C_{21} & C_{22} \end{bmatrix}$$

This recursive splitting results in the following four triangular Sylvester equations:

$$A_{11}X_{11} - X_{11}B_{11} = C_{11} - A_{12}X_{21},$$
$$A_{11}X_{12} - X_{12}B_{22} = C_{12} - A_{12}X_{22} + X_{11}B_{12},$$
$$A_{22}X_{21} - X_{21}B_{11} = C_{21},$$
$$A_{22}X_{22} - X_{22}B_{22} = C_{22} + X_{21}B_{12}.$$

We start by solving for X_{21} in the third equation. After updating the right hand sides of the first and fourth equations with respect to X_{21}, we can solve for X_{11} and X_{22}. Both updates and the triangular Sylvester solves are independent operations and can be executed concurrently. Finally, we update the right hand side of the second matrix equation with respect to X_{11} and X_{22}, and solve for X_{12}.

The recursive splittings described above are now applied to all "half-sized" triangular Sylvester equations and so on. We end the recursion when the new problem sizes M and N are smaller than a certain block size, $blks$, which is chosen such that at least the three submatrices involved in the current Sylvester equation fit in the first level cache memory. For the solution of the small-sized kernel problems, we apply a standard triangular Sylvester algorithm.

We remark that all updates of right hand sides in the three cases above are general matrix multiply and add (GEMM) operations $C \leftarrow \beta C + \alpha AB$, where α and β are real scalars [7,8,6].

In Algorithm 1, we present a Matlab-style function $[X] = \mathbf{rtrsylv}(A, B, C, blks)$ for our blocked recursive solver. The function $[C] = \mathbf{gemm}(A, B, C)$ implements the GEMM-operation $C = C + AB$, and the function $[X] = \mathbf{trsylv}(A, B, C)$ implements an algorithm for solving triangular Sylvester kernel problems.

2.1   Implementation Issues and Performance Results for the Blocked Recursive Standard Sylvester Solver

There are two main alternatives for doing the recursive splits: (i) Always split the largest dimension in two (Cases 1 and 2); (ii) Split both dimensions simultaneously when the dimensions are within a factor two from each other (Case 3).

The first alternative gives less code but a long and narrow recursion tree—just a trunk with no branches. When we split all matrices in both dimensions

---

Algorithm 1: rtrsylv

Input: A $(M \times M)$ and B $(N \times N)$ in upper quasi-triangular (Schur) form. C $(M \times N)$ dense matrix. $blks$, block size that specifies when to switch to an algorithm for solving small-sized triangular Sylvester equations.

Output: X $(M \times N)$, the solution of the triangular Sylvester equation $AX - XB = C$. X is allowed to overwrite C.

function $[X] = \mathbf{rtrsylv}(A, B, C, blks)$
if $1 \leq M, N \leq blks$ **then**
    $X = \mathbf{trsylv}(A, B, C)$
elseif $1 \leq N \leq M/2$
    % Case 1: Split A (by rows and columns), C (by rows only)
    $X_2 = \mathbf{rtrsylv}(A_{22}, B, C_2, blks)$;
    $C_1 = \mathbf{gemm}(-A_{12}, X_2, C_1)$;
    $X_1 = \mathbf{rtrsylv}(A_{11}, B, C_1, blks)$;
    $X = [X_1; X_2]$;
elseif $1 \leq M \leq N/2$
    % Case 2: Split B (by rows and columns), C (by columns only)
    $X_1 = \mathbf{rtrsylv}(A, B_{11}, C_1, blks)$
    $C_2 = \mathbf{gemm}(X_1, B_{12}, C_2)$;
    $X_2 = \mathbf{rtrsylv}(A, B_{22}, C_2, blks)$;
    $X = [X_1, X_2]$;
else
    % Case 3: Split A, B and C (all by rows and columns)
    $X_{21} = \mathbf{rtrsylv}(A_{22}, B_{11}, C_{21}, blks)$;
    $C_{22} = \mathbf{gemm}(X_{21}, B_{12}, C_{22})$; $C_{11} = \mathbf{gemm}(-A_{12}, X_{21}, C_{11})$;
    $X_{22} - \mathbf{rtrsylv}(A_{22}, B_{22}, C_{22}, blks)$; $X_{11} = \mathbf{rtrsylv}(A_{11}, B_{11}, C_{11}, blks)$;
    $C_{12} = \mathbf{gemm}(-A_{12}, X_{22}, C_{12})$;
    $C_{12} - \mathbf{gemm}(X_{11}, B_{12}, C_{12})$;
    $X_{12} = \mathbf{rtrsylv}(A_{11}, B_{22}, C_{12}, blks)$;
    $X = [X_{11}, X_{12}; X_{21}, X_{22}]$;
end

Algorithm 1: Recursive blocked algorithm for solving the triangular standard Sylvester equation.

we get a shorter but wider recursion tree—two new branches for each Sylvester equation—which provides more parallel tasks.

For problems smaller than the block size, we also apply a recursive algorithm with a superscalar kernel. The dimension is split in two until the problem is of size 2×2 or 4×4, when the remaining subsystems (the leaf nodes) are solved using the *Kronecker product representation* of $AX - XB = C$:

$$(I_n \otimes A - B^T \otimes I_m)\mathrm{vec}(X) = \mathrm{vec}(C).$$

This system is permuted in order to make the problem more upper triangular and then solved using LU factorization with row pivoting.

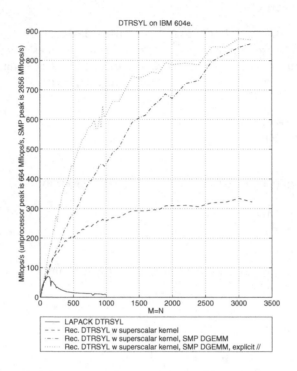

Fig. 1. Results for DTRSYL-variants on IBM 604e, 332 MHz.

The graphs in Figure 1 show the performance of different variants of the routine DTRSYL. The LAPACK-equivalent is mainly a level 2 routine and this explains its poor performance behavior. Our recursive blocked routine shows very good performance with a speedup between 2.2–2.7 on a four processor PowerPC 604E node. The strong dependences in the dataflow execution graph limit the parallel speedup for the triangular Sylvester equation.

3   Recursive Generalized Triangular Sylvester Solvers

We consider the generalized coupled real Sylvester matrix equation of the form

$$AX - YB = C, \quad C \leftarrow X \ (M \times N),$$
$$DX - YE = F, \quad F \leftarrow Y \ (M \times N),$$

where the matrix pairs (A, D) and (B, E) are in generalized Schur form with A, B (quasi-)upper triangular and D, E upper triangular. A, D are $M \times M$ and

B, E are $N \times N$; the right hand sides C, D and the solution matrices X, Y are of size $M \times N$.

Any 2×2 diagonal block in the generalized Schur form corresponds to a conjugate pair of eigenvalues. The 1×1 diagonal blocks correspond to real eigenvalues. For example, if $d_{ii} \neq 0$, then a_{ii}/d_{ii} is a finite eigenvalue of (A, D). Otherwise $(d_{ii} = 0)$, the matrix pair has an infinite eigenvalue.

As for the standard case, we consider three alternatives for doing a *recursive splitting*. Below, we illustrate the two first cases.

Case 1 $(1 \leq N \leq M/2)$. We split (A, D) by rows and columns, and (C, F) by rows only:

$$\begin{bmatrix} A_{11} & A_{12} \\ & A_{22} \end{bmatrix} \begin{bmatrix} X_1 \\ X_2 \end{bmatrix} - \begin{bmatrix} Y_1 \\ Y_2 \end{bmatrix} B = \begin{bmatrix} C_1 \\ C_2 \end{bmatrix},$$

$$\begin{bmatrix} D_{11} & D_{12} \\ & D_{22} \end{bmatrix} \begin{bmatrix} X_1 \\ X_2 \end{bmatrix} - \begin{bmatrix} Y_1 \\ Y_2 \end{bmatrix} E = \begin{bmatrix} F_1 \\ F_2 \end{bmatrix}.$$

The splitting results in the following two generalized Sylvester equations:

$$A_{11}X_1 - Y_1 B = C_1 - A_{12}X_2,$$
$$D_{11}X_1 - Y_1 E = F_1 - D_{12}X_2,$$
$$A_{22}X_2 - Y_2 B = C_2,$$
$$D_{22}X_2 - Y_2 E = F_2.$$

First, we solve for (X_2, Y_2) in the second pair of matrix equations. After updating the right hand sides of the first pair of matrix equations with respect to X_2 (two GEMM operations that can execute in parallel), we solve for (X_1, Y_1).

Case 2 $(1 \leq M \leq N/2)$. We split (B, E) by rows and columns, and (C, F) by columns only:

$$A \begin{bmatrix} X_1 & X_2 \end{bmatrix} - \begin{bmatrix} Y_1 & Y_2 \end{bmatrix} \begin{bmatrix} B_{11} & B_{12} \\ & B_{22} \end{bmatrix} = \begin{bmatrix} C_1 & C_2 \end{bmatrix},$$

$$D \begin{bmatrix} X_1 & X_2 \end{bmatrix} - \begin{bmatrix} Y_1 & Y_2 \end{bmatrix} \begin{bmatrix} E_{11} & E_{12} \\ & E_{22} \end{bmatrix} = \begin{bmatrix} F_1 & F_2 \end{bmatrix},$$

or equivalently

$$AX_1 - Y_1 B_{11} = C_1,$$
$$DX_1 - Y_1 E_{11} = F_1,$$
$$AX_2 - Y_2 B_{22} = C_2 + Y_1 B_{12},$$
$$DX_2 - Y_2 E_{22} = F_2 + Y_1 E_{12}.$$

Now we first solve for (X_1, Y_1), and after updating the right hand sides of the second pair of generalized Sylvester equations with respect to Y_1 (also two independent GEMM operations), we solve for (X_2, Y_2).

In Case 3, we split all the six matrices $(A, B, C, D, E$ and $F)$ by rows and columns, which results in four pairs of coupled Sylvester equations to be solved. Several of the associated operations can be performed concurrently, including GEMM-updates of right hand sides and two of the generalized triangular Sylvester solves.

A Matlab-style function, $[X] = \mathbf{rtrgsylv}(A, B, C, D, E, F, blks)$, which implements our blocked recursive generalized Sylvester algorithm is presented in Algorithm 2. As in the standard case, the recursion is only applied down to a certain block size, when the function $[X] = \mathbf{trgsylv}(A, B, C, D, E, F)$ is applied for solving triangular generalized Sylvester kernel problems.

3.1 Implementation Issues and Performance Results for the Blocked Recursive Generalized Sylvester Solver

Following the standard case, we use a recursive algorithm with a superscalar kernel for solving the small-sized generalized Sylvester equations. The recursion is done down to 2×2 or 4×4 blocks, when the leaf nodes are solved using the *Kronecker product representation* of the *coupled Sylvester equation* [12]:

$$\begin{bmatrix} I_n \otimes A & -B^T \otimes I_m \\ I_n \otimes D & -E^T \otimes I_m \end{bmatrix} \begin{bmatrix} \text{vec}(X) \\ \text{vec}(Y) \end{bmatrix} = \begin{bmatrix} \text{vec}(C) \\ \text{vec}(F) \end{bmatrix}.$$

In Figure 2, results for our blocked recursive algorithm and the LAPACK routine DTGSYL with two different kernels are presented. DTGSYL implements a blocked algorithm [10,1], which mainly performs level 3 operations. Uniprocessor and multiprocessor performance graphs are presented in the left and right parts, respectively. **CP-S** is the same as the LAPACK kernel DTGSY2 [10,1], which makes use of both complete pivoting and scaling to prevent over/underflow [10]. Our superscalar partial pivoting kernel **SS-PP** solves generalized Sylvester equations using LU factorization with row pivoting on the Kronecker product representation [12].

Using the **SS-PP** kernel we gain between 40–100% in performance for the LAPACK DTGSYL, and we get 25–70% improved performance for the recursive blocked solver. This is mainly due to a much faster but possibly less robust and reliable kernel solver. Moreover, the recursive blocked algorithm outperforms the LAPACK DTGSYL with 25–35% using the **CP-S** kernel and with 10–20% using the **SS-PP** kernel for medium to large-sized problems.

From the multiprocessor results, we conclude that the best performance (around 3 times speedup on a four processor PowerPC 604E node) is obtained by our recursive blocked algorithm with the **SS-PP** kernel, using an SMP DGEMM routine and exploiting the explicit parallelism in the recursive task tree.

4   Conclusions and Future Work

Our recursive approach is very efficient for solving triangular Sylvester-type matrix equations on today's hierarchical memory computer systems. The recursion

Algorithm 2: rtrgsylv

Input: (A, D) of size $M \times M$, and (B, E) of size $N \times N$ in upper quasi-triangular (generalized real Schur) form. (C, F) of size $M \times N$ dense matrix pair. *blks*, block size that specifies when to switch to a standard algorithm for solving small-sized triangular coupled Sylvester equations. *blks* is chosen such that at least the six submatrices involved in the small-sized matrix equation fit in the first level cache memory.

Output: (X, Y) of size $M \times N$, the solution of the generalized Sylvester equation $(AX - YB, DX - YE) = (C, F)$. X, Y are allowed to overwrite C, F.

function $[X, Y] = \mathbf{rtrgsylv}(A, B, C, D, E, F, blks)$
if $1 \le M, N \le blks$ **then**
　　$[X, Y] = \mathbf{trgsylv}(A, B, C, D, E, F)$
elseif $1 \le N \le M/2$
　　% Case 1: Split (A, D) (by rows and columns), (C, F) (by rows only)
　　$[X_2, Y_2] = \mathbf{rtrgsylv}(A_{22}, B, C_2, D_{22}, E, F_2, blks)$;
　　$C_{11} = \mathbf{gemm}(-A_{12}, X_2, C_1)$;　$F_{11} = \mathbf{gemm}(-D_{12}, X_2, F_1)$;
　　$[X_1, Y_1] = \mathbf{rtrgsylv}(A_{11}, B, C_1, D_{11}, E, F_1, blks)$;
　　$X = [X_1; X_2]$;　$Y = [Y_1; Y_2]$;
elseif $1 \le M \le N/2$
　　% Case 2: Split (B, E) (by rows and columns), (C, F) (by columns only)
　　$[X_1, Y_1] = \mathbf{rtrgsylv}(A, B_{11}, C_1, D, E_{11}, F_1, blks)$
　　$C_2 = \mathbf{gemm}(Y_1, B_{12}, C_2)$;　$F_2 = \mathbf{gemm}(Y_1, E_{12}, F_2)$;
　　$[X_2, Y_2] = \mathbf{rtrgsylv}(A, B_{22}, C_2, D, E_{22}, F_2, blks)$;
　　$X = [X_1, X_2]$;　$Y = [Y_1, Y_2]$
else
　　% Case 3: Split (A, D), (B, E) and (C, F) (all by rows and columns)
　　$[X_{21}, Y_{21}] = \mathbf{rtrgsylv}(A_{22}, B_{11}, C_{21}, D_{22}, E_{11}, F_{21}, blks)$;
　　$C_{22} = \mathbf{gemm}(Y_{21}, B_{12}, C_{22})$;　$C_{11} = \mathbf{gemm}(-A_{12}, X_{21}, C_{11})$;
　　$F_{22} = \mathbf{gemm}(Y_{21}, E_{12}, F_{22})$;　$F_{11} = \mathbf{gemm}(-D_{12}, X_{21}, F_{11})$;
　　$[X_{22}, Y_{22}] = \mathbf{rtrgsylv}(A_{22}, B_{22}, C_{22}, D_{22}, E_{22}, F_{22}, blks)$;
　　$[X_{11}, Y_{11}] = \mathbf{rtrgsylv}(A_{11}, B_{11}, C_{11}, D_{11}, E_{11}, F_{11}, blks)$;
　　$C_{12} = \mathbf{gemm}(-A_{12}, X_{22}, C_{12})$;　$F_{12} = \mathbf{gemm}(-D_{12}, X_{22}, F_{12})$;
　　$C_{12} = \mathbf{gemm}(Y_{11}, B_{12}, C_{12})$;　$F_{12} = \mathbf{gemm}(Y_{11}, E_{12}, F_{12})$
　　$[X_{12}, Y_{12}] = \mathbf{rtrgsylv}(A_{11}, B_{22}, C_{12}, D_{11}, E_{22}, F_{12}, blks)$;
　　$X = [X_{11}, X_{12}; X_{21}, X_{22}]$;　$Y = [Y_{11}, Y_{12}; Y_{21}, Y_{22}]$;
end

Algorithm 2: Recursive blocked algorithm for solving the triangular generalized coupled Sylvester equation.

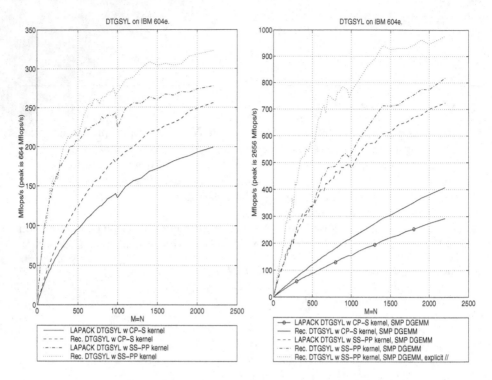

Fig. 2. Uniprocessor and multiprocessor performance results for generalized Sylvester–IBM 604E ($M = N$).

is ended when the remaining subproblems to be solved are smaller than a given block size, which is determined by the size of the level 1 cache memory. This is the only architecture-dependent parameter in our algorithms. To complete recursion until element level would cause too much overhead and a drop in performance. Our solution is to develop new high-performance kernels for the small triangular Sylvester-type matrix equations, which efficiently utilize the superscalar processors of today [6,5]. This implies that a larger part of the total execution time will be spent in GEMM operations [7,8], which leads to much better performance, especially for small to medium-sized problems, and improved parallel efficiency on SMP systems.

Acknowledgements. This research was conducted using the resources of the High Performance Computing Center North (HPC2N) and the Danish Computing Centre for Research and Education (UNI•C), Lyngby, Denmark. We also thank Fred Gustavson and our recursive pals in the Umeå group for stimulating and fruitful discussions.

Financial support has been provided by the Swedish Research Council for Engineering Sciences under grant TFR 98-604.

References

1. E. Anderson, Z. Bai, J. Demmel, J. Dongarra, J. DuCroz, A. Greenbaum, S. Hammarling, A. McKenny, S. Ostrouchov, and D. Sorensen *LAPACK Users Guide, Third Edition*. SIAM Publications, 1999.
2. E. Elmroth and F. Gustavson. New Serial and Parallel Recursive QR Factorization Algorithms for SMP Systems, In Kågström et al. (eds), *Applied Parallel Computing, PARA'98*, Lecture Notes in Computer Science, Vol. 1541, pp 120–128, Springer-Verlag, 1998.
3. E. Elmroth and F. Gustavson. Applying Recursion to Serial and Parallel QR Factorization Leads to Better Performance. *IBM Journal of Research and Development*, 44, No. 4, 605–624, 2000.
4. F. Gustavson. Recursion leads to automatic variable blocking for dense linear algebra. *IBM J. Res. Develop*, 41(6):737–755, November 1997.
5. F. Gustavson, A. Henriksson, I. Jonsson, B. Kågström and P. Ling. Recursive Blocked Data Formats and BLAS's for Dense Linear Algebra Algorithms. In Kågström et al. (eds), *Applied Parallel Computing, PARA'98*, Lecture Notes in Computer Science, Vol. 1541, pp 195–206, Springer-Verlag, 1998.
6. F. Gustavson, A. Henriksson, I. Jonsson, B. Kågström and P. Ling. Superscalar GEMM-based Level 3 BLAS – The On-going Evolution of a Portable and High-Performance Library. In Kågström et al. (eds), *Applied Parallel Computing, PARA'98*, Lecture Notes in Computer Science, Vol. 1541, pp 207-215, Springer-Verlag, 1998.
7. B. Kågström, P. Ling, and C. Van Loan. GEMM-based level 3 BLAS: High-performance model implementations and performance evaluation benchmark. *ACM Trans. Math. Software*, 24(3):268–302, 1998.
8. B. Kågström, P. Ling, and C. Van Loan. GEMM-based level 3 BLAS: Portability and optimization issues. *ACM Trans. Math. Software*, 24(3):303–316, 1998.
9. B. Kågström and P. Poromaa. Distributed and Shared Memory Block Algorithms for the Triangular Sylvester Equation with sep^{-1} Estimator. *SIAM Journal on Matrix Analysis and Application*, 13(1):90–101, January 1992.
10. B. Kågström and P. Poromaa. LAPACK–Style Algorithms and Software for Solving the Generalized Sylvester Equation and Estimating the Separation between Regular Matrix Pairs. *ACM Trans. Math. Software*, 22(1):78–103, March 1996.
11. B. Kågström and P. Poromaa. Computing Eigenspaces with Specified Eigenvalues of a Regular Matrix Pair (A, B) and Condition Estimation: Theory, Algorithms and Software. *Numerical Algorithms*, 12:369–407, 1996.
12. B. Kågström and L. Westin. Generalized Schur methods with condition estimators for solving the generalized Sylvester equation. *IEEE Trans. Autom. Contr.*, 34(4):745–751, 1989.
13. P. Poromaa. Parallel Algorithms for Triangular Sylvester Equations: Design, Scheduling and Scalability Issues. In Kågström et al. (eds), *Applied Parallel Computing, PARA'98*, Lecture Notes in Computer Science, Vol. 1541, pp 438–446, Springer-Verlag, 1998.
14. G.W. Stewart and J-G. Sun. *Matrix Perturbation Theory*. Academic Press, 1990.

On the Efficiency of Scheduling Algorithms for Parallel Gaussian Elimination with Communication Delays

Raimondas Čiegis[1], Vadimas Starikovičius[2], and Jerzy Waśniewski[3]

[1] Institute of Mathematics and Informatics, Vilnius Gediminas Technical University, Akademijos 4, LT-2600 Vilnius, Lithuania,
rc@fm.vtu.lt
[2] Vilnius University, Naugarduko 24, LT-2600 Vilnius, Lithuania
vs@sc.vtu.lt
[3] The Danish Computing Center for Research and Education, UNI-C, Bldg. 305, DK-2800 Lyngby, Denmark
jw@uni-c.dk

Abstract. We consider the Gaussian elimination method on parallel distributed memory computers. A theoretical model for the performance prediction is developed. It takes into account the workload distribution and the communication overhead. We investigate the efficiency of the parallel Gaussian algorithm with matrices distributed in 1D and 2D block and cyclic layouts. The results are generalized for block-block and block-cyclic distributions. We find the condition when communications are overlapped by the computations. Using this analysis we propose a simple heuristic for scheduling Gaussian elimination tasks. We compare the efficiency of this scheduling algorithm with the efficiency of the block and cycling data distributions.

1 Introduction

Scheduling of the tasks of weighted directed acyclic graphs is one of the most important problems in parallel computing. Finding a scheduling algorithm that minimizes the parallel execution time is an NP-complete problem. The main research efforts in this area are focused on heuristic methods for obtaining near-optimal solutions in a reasonable time (see, e.g., [1], [3]).

This paper discusses the problem of solving a system of linear equations

$$Ax = f.$$

Here A is a dense $n \times n$ matrix, f is an $n \times 1$ vector, and x is the solution vector. The Gaussian elimination algorithm has two stages. First, through a series of algebraic manipulations , the system of equations is reduced to an upper triangular system

$$Ux = F,$$

T. Sørevik et al. (Eds.): PARA 2000, LNCS 1947, pp. 74–81, 2001.

where U is an upper-triangular matrix in which all principal diagonal entries are equal to one. In the second stage of solving a system of linear equations, the upper-triangular system is solved for the variables in reverse order.

We want to develop the performance prediction tool for parallel Gaussian elimination algorithm. Such model must characterize accurately the workload and communication overheads. General requirements for performance evaluation are formulated in [4].

Performance models of the parallel Gaussian elimination algorithm will permit us to answer the following questions:

1. What performance can be achieved ?
2. What data distribution should be used ?
3. How many processors should be used for a given size of the matrix ?

The implementation of the parallel Gaussian elimination has been intensively studied for various types of parallel computers. The performance analysis of the ScaLAPACK library for 2D block-cyclic distributions is made in [2], the performance prediction of 1D block distribution algorithms is given in [1].

Theoretical model. Most performance prediction tools are based on the estimation of computation costs and communication overheads. Our performance tool simulates the performance of each processor. This enables us to estimate accurately the overlapping of communications by useful computations.

The computation costs of each subproblem are estimated by counting the number of floating point operations that needed to be executed. The speed of one operation is estimated by running test subproblems on a specific machine. The execution time per floating point operation of each instruction is denoted by γ.

The target computer is made up of an arbitrary number p of homogeneous processors. Each processor contains specialized hardware which enables it to perform both computation and communication in parallel, and to use all its communication channels simultaneously (see [3]). The cost of broadcasting n items of data along a row or column of the mesh of processors is estimated by

$$T_b = K(r, p)(\alpha + \beta n),\qquad(1)$$

where α denotes the latency, β denotes the inverse of the bandwidth, r is the number of processors in a row/column, and $K(r, p)$ depends on the topology of interconnection networks.

We investigated the completely connected network with the cost function $K(r, p) = 1$, the hypercube network with the cost function $K(r, p) = \log r$, and the LAN type network with costs $K(r, p) = r - 1$.

The remainder of the paper is organized as follows. In section 2 we present results on the efficiency of parallel Gaussian elimination algorithm, when only computational costs are taken into account. In section 3 we study communication costs of the algorithm for block and cyclic matrix mapping onto processors. In section 4 we propose an heuristic for scheduling the matrix. Finally, in section 5 we present simulation results for the comparison of the scheduling heuristic and block and cyclic matrix decompositions.

2   Efficiency of Parallel Gaussian Elimination Algorithm

In this section we briefly summarize results on the efficiency of parallel Gaussian elimination algorithm, when communication delays are neglected.

First we assume that the coefficient matrix is distributed among processors using 1D and 2D block decomposition schemes.

Let denote by $m = n/p$ the number of contiguous columns assigned to each processor in the case of 1D matrix decomposition, and let $M = n/\sqrt{p}$ be the block size of the matrix assigned to each processor in the case of 2D block decomposition.

Lemma 1. *In the case of 1D block matrix decomposition the speedup S_p and the efficiency E_p of the parallel Gaussian algorithm are given by:*

$$S_{p,1D} = \frac{n(2n^2 + 3n + 1)}{m(3n^2 - m^2 + 3n + 1)}, \quad E_{p,1D} = \frac{2n^2 + 3n + 1}{3n^2 - m^2 + 3n + 1}, \quad (2)$$

and in the case of 2D block matrix decomposition we have:

$$S_{p,2D} = \frac{n(2n^2 + 3n + 1)}{M(6nM - 4M^2 + 3M + 1)}, \quad E_{p,2D} = \frac{2nM + 3M + M/n}{6nM - 4M^2 + 3M + 1}. \quad (3)$$

Proof. We consider the 2D block decomposition, since the 1D case is very similar. The sequential run time of the Gaussian elimination algorithm is

$$T_0 = \frac{n(n+1)(2n+1)}{6}\gamma.$$

Now consider the case in which the matrix is mapped onto $\sqrt{p} \times \sqrt{p}$ mesh of processors. The parallel run time of the Gaussian elimination algorithm is determined by the total computation time of the last processor, hence we get

$$T_p = \left(M^2(n - M) + \frac{M(M+1)(2M+1)}{6} \right)\gamma$$

$$= \frac{M(6nM - 4M^2 + 3M + 1)}{6}\gamma.$$

The estimates for S_p and E_p follow trivially from the equalities given above. The lemma is proved.

For sufficiently large number of processors p, i.e., when $m \ll n$ or $M \ll n$, we get the following estimates of the efficiency coefficients:

$$E_{p,1D} \approx 0.67, \quad E_{p,2D} \approx 0.33.$$

The inefficiency of Gaussian elimination with block data distribution is due to processor idling resulting from an uneven load distribution during computations, despite even initial load distribution.

The efficiency of the parallel Gaussian algorithm is increased if the matrix is partitioned among processors using the *cyclic* data mapping.

Lemma 2. *In the case of 1D cyclic matrix decomposition the speedup S_p and the efficiency E_p of the parallel Gaussian algorithm are given by:*

$$S_{p,1D} = \frac{(n+1)(2n+1)}{(m+1)(2n+1.5-0.5p)}, \quad E_{p,1D} = \frac{(n+1)(2n+1)}{(n+p)(2n+1.5-0.5p)}, \quad (4)$$

and in the case of 2D cyclic matrix decomposition we have:

$$S_{p,2D} = \frac{p(n+1)(2n+1)}{(n+\sqrt{p})(2n+\sqrt{p})}, \quad E_{p,2D} = \frac{(n+1)(2n+1)}{(n+\sqrt{p})(2n+\sqrt{p})}. \quad (5)$$

Proof. As in the case of a block mapping, we restrict to the analysis of 2D matrix decomposition. The parallel run time of the Gaussian elimination algorithm is

$$T_p = \gamma\sqrt{p}\left(M^2 + (M-1)^2 \cdots + 1\right)$$
$$= \frac{n((n+\sqrt{p})(2n+\sqrt{p})}{6p}\gamma.$$

Then equalities (5) follow after simple computations. The lemma is proved.

For a fixed number of processors p and for sufficiently large n we get from (4), (5) the following estimates of the efficiency coefficients:

$$E_{p,1D} \approx 1, \quad E_{p,2D} \approx 1.$$

3   Communication Complexity Analysis

In this section we prove that the block and cyclic distribution schemes have quite different communication costs.

We assume that the processors work asynchronously, that is no processor waits for the others to finish an iteration before starting the next one. Each processor firstly sends any data destined for other processors and only then performs the remaining computations using the data it has. Hence the processor is idle only if it waits to receive data to be used.

Let assume that we use the completely connected communication network, hence $K(r,p) = 1$ in the equality (1). We denote by γ the time for computation of one *saxpy* operation.

First we prove a general theorem, which gives conditions when data communication is overlapped with useful computations.

Theorem 1. *For 1D decomposition the broadcast communication is overlapped with computations if the following condition*

$$m \geq 1 + \frac{\beta}{\gamma} + \frac{\alpha}{(n-k)\gamma} \quad (6)$$

is satisfied , where m is the number of columns distributed to pivot processor, k is the step of Gaussian elimination algorithm.

Proof. Let consider the kth step of the parallel Gaussian elimination algorithm. A processor, which has the pivot column, updates it and broadcasts this column to the other processors. The total computation and communication time during both steps is

$$t_1 = (n - k)\,\gamma + \alpha + (n - k)\,\beta.$$

The remaining processors are performing the elimination step with their part of the matrix, the run time of this step is

$$t_2 = m(n - k)\,\gamma.$$

The communication time is overlapped with useful computations if $t_2 \geq t_1$ and the condition (6) follows after simple computations. The theorem is proved.

It follows from (6) that the latency coefficient α can be neglected for large n. In the following example we use the computer parameters given in [1]:

$$\alpha = 136\,\mu s, \quad \beta = 3.2\,\mu s, \quad \gamma = 0.36\,\mu s. \tag{7}$$

Then for $n = 1024$ we get from (6) that the communication is overlapped with computations if $m \geq 10$.

Theorem 2. *In the case of 2D square grid of processors the overlapping of the broadcast communication with computations takes place if the following condition*

$$M \geq \frac{\beta}{\gamma} + 1.5 + \sqrt{\left(\frac{\beta}{\gamma} + 1.5\right)^2 + \frac{2\alpha}{\gamma}} \tag{8}$$

is satisfied.

Proof. In this case the Gaussian iteration involves a rowwise and a columnwise communication of $M = n/\sqrt{p}$ values.

Let consider the kth step of the algorithm. A processor containing a part of the pivot column communicates it along its row, this step takes the time

$$t_1 = \alpha + \frac{n}{\sqrt{p}}\,\beta\,.$$

Then a processor containing a part of the pivot row updates it and communicates this row along its column. This step takes the time

$$t_2 = \frac{n}{\sqrt{p}}\,\gamma + \alpha + \frac{n}{\sqrt{p}}\,\beta\,.$$

The other processors perform elimination step with their data, the computation time is

$$t_3 = \frac{n^2}{p}\,\gamma\,.$$

The communication time is overlapped with useful computations if $t_3 \geq t_1 + t_2$. We get the quadratical inequality

$$M^2\,\gamma \geq (3\gamma + 2\beta)M + 2\alpha$$

and the condition (8) follows after simple computations. The theorem is proved.

The important consequence from Theorem 2 is that for 2D data partitioning the latency coefficient α can not be neglected even for large n.

In the following example we use the computer parameters given in (7). Then we get from (8) that the communication is overlapped with computations if $M \geq 41$ and the latency coefficient α contributes the main part to this condition.

Next we study total communication costs during the implementation of all steps of the Gaussian elimination algorithm.

Theorem 3. *If conditions of Theorem 1 are not satisfied, then the non over-lapped communication overhead occurs during each step of Gaussian algorithm with the cyclic matrix decomposition scheme, and it occurs only once per each block, i.e., $p - 1$ times, when the block data decomposition is used.*

Proof. We consider the block data distribution case. Let denote by t_j the time moments when processors get pivot columns broadcasted by the pivot processor. It is sufficient to consider the time moments t_1 and t_2. Simple computations give

$$t_1 = (n - k)\gamma + \alpha + (n - k)\beta,$$
$$t_2 = (n - k - 1)\gamma + m(n - k)\gamma + \alpha + (n - k)\beta.$$

Processors finish the kth elimination step of the Gaussian algorithm at the time moment

$$t_3 = t_1 + m(n - k)\gamma.$$

It is easy to see, that $t_3 < t_2$, hence processors are idle only when the pivot processor is changed. The theorem is proved.

It follows from Theorem 3 that the block partitioning of the matrix is more efficient with respect to communication costs.

4   Heuristic for Scheduling Tasks

In this section we propose a simple heuristic for scheduling the tasks of Gaussian elimination algorithm onto processors. We restrict to the case of 1D partitioning of the matrix. This heuristic is based on the performance prediction tool for the parallel Gaussian elimination algorithm. Our tool characterize accurately the workload and communication overheads of each processor.

Using results of Theorem 1 and Theorem 3 we propose to distribute the last J columns of the matrix according the block partitioning scheme. The remaining $n - J$ columns are distributed according the cyclic partitioning scheme. Using the proposed performance tool we find the optimal value of J.

We note that this heuristic algorithm coincides with the block partitioning if $J = n$, and it coincides with the cyclic partitioning if $J = 1$.

5   Simulation Results

We have carried out a number of simulations by scheduling an 1024×1024 matrix. The following computer parameters were used in our analysis [1]

$$\alpha = 136\mu s, \quad \beta = 3.2\mu s, \quad \gamma = 0.18\mu s.$$

In figure 1 we compare the speed-ups of the block, cyclic and heuristic distributions for the completely connected network.

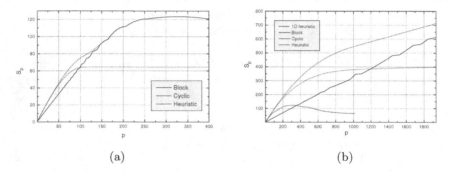

(a)   (b)

Fig. 1. Speed-up of the parallel Gaussian elimination algorithm for the completely connected network as a function of number of processors: 1D (a) and 2D (b) cases.

We see that the completion time of the heuristic scheduling is optimal for all values of n.

The speed-ups of the parallel Gaussian elimination algorithm for the hypercube network and the same three data distribution algorithms are given in figure 2. Finally, the simulation results for the LAN are plotted in figure 3.

6   Conclusions

In this paper we have presented an efficient heuristic for the appropriate task scheduling of the parallel Gaussian elimination algorithm. It based on the performance tool, which estimates accurately the workload and communication overheads of each processor. The distribution strategy is based on the analysis of communication and computation costs of different data mappings.

References

1. Amoura, A., Bampis, E., König, J.: Scheduling algorithms for parallel Gaussian elimination with communication costs, IEEE Transactions on Parallel and Distributed systems, **9** (1998) 679–686.

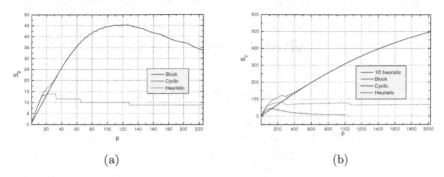

(a)   (b)

Fig. 2. Speed-up of the parallel Gaussian elimination algorithm as a function of number of processors for the hypercube network 1D (a) and 2D (b).

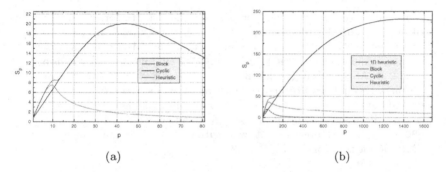

(a)   (b)

Fig. 3. Speed-up of the parallel Gaussian elimination algorithm as a function of number of processors for the LAN 1D (a) and 2D (b).

2. Choi, J., Demmel, J., Dhillon, I., Dongarra, J. et al.: ScaLAPACK: a portable linear algebra library for distribution memory computers - design issues and performance, LAPACK Working Note 95, University of Tennessee, 1995.
3. Djordjević, G., Tošić, M.: A heuristic for scheduling task graphs with communication delays onto multiprocessors, Parallel Computing, **22** (1996) 1197–1214.
4. Xu, Z., Hwang, K.: Early prediction of MPP performance: the SP2, T3D, and Paragon experiences, Parallel Computing, **22** (1996) 917–942.

High Performance Cholesky Factorization via Blocking and Recursion That Uses Minimal Storage

Fred Gustavson[1] and Isak Jonsson[2]

[1] IBM T.J. Watson Research Center,
P.O. Box 218, Yorktown Heights, NY 10598, U.S.A.
gustav@watson.ibm.com
[2] Department of Computing Science, Umeå University,
SE-901 87 Umeå, Sweden.
isak@cs.umu.se

Abstract. We present a high performance Cholesky factorization algorithm, called BPC for Blocked Packed Cholesky, which performs better or
equivalent to the LAPACK DPOTRF subroutine, but with about the same
memory requirements as the LAPACK DPPTRF subroutine, which runs at
level 2 BLAS speed. Algorithm BPC only calls DGEMM and level 3 kernel
routines. It combines a recursive algorithm with blocking and a recursive
packed data format. A full analysis of overcoming the non-linear addressing overhead imposed by recursion is given and discussed. Finally, since
BPC uses GEMM to a great extent, we easily get a considerable amount of
SMP parallelism from an SMP GEMM.

Keywords: Cholesky factorization, recursive algorithm, recursive data
structure, packed format, level 3 BLAS parallelism

1 Introduction

In [1], a new recursive packed format for Cholesky factorization was described. It
is a variant of the triangular format described in [5], and suggested in [4]. For a
reference to recursion in linear algebra, see [4,5]. The advantage of using packed
formats instead of full format is the memory savings possible when working with
large matrices. The disadvantage is that the use of high-performance standard
library routines such as the matrix multiply and add operation, GEMM, is inhibi-
ted. We combine blocking with recursion to produce a practical implementation
of a new algorithm called BPC, for blocked packed Cholesky [3, pages 142-147].
BPC first transforms standard packed lower format to packed recursive lower row
format. Then, during execution, algorithm BPC only calls standard DGEMM and
level 3 kernel routines. This method has the benefit of being transparent to the
user. No extra assumptions about the input matrix needs be made. This is important since an algorithm like Cholesky using a new data format would not
work on previous codes.

T. Sørevik et al. (Eds.): PARA 2000, LNCS 1947, pp. 82–91, 2001.

1.1 Preliminary Remarks and Rationale

Existing codes for Cholesky and LDL^T factorization use either full storage or packed storage data format. For machines with a uniform memory hierarchy, packed format is usually the method of choice as it conserves memory and still performs about as well as a full storage implementation. However, with today's machines' deep memory hierarchy, this is not the case.

ESSL, [8], and LAPACK, [2], support both data formats so a user can choose either for his application. Also, ESSL packed storage implementation has always been level 3, some times at great programming cost. On the other hand, LAPACK and some other libraries do not produce level 3 implementations for packed data formats.

For performance reasons, users today generally use the full data format to represent their symmetric matrix. Nonetheless, saving half the storage is an important consideration especially for those applications which will run with packed storage and fail to run with full storage.

Another important user consideration is migration. Many existing codes use one or the other or both of these formats. Producing an algorithm like Cholesky using a new data format has little appeal since existing massive programs cannot use the new algorithm. The approach described here is to redo its packed Cholesky factorization code and instead use the new recursive packed data structure.

The idea behind recursive factorization of are symmetric matrix stored in packed recursive format is simple: Given AP holding symmetric A in lower packed storage mode, overwrite AP with A in the recursive packed row format. To do so requires a temporary array of size $\frac{1}{8}n^2$, which we allocate and then deallocate. Next, we execute the new recursive level 3 Cholesky algorithm. Even when one includes the cost of converting the data from conventional packed to recursive packed format, the performance, as we will see, is better than LAPACK's level 3 DPOTRF.

In summary, the users can now get full level 3 performance using packed format and thereby save about half of their storage.

2 Algorithmic and Implementation Issues

2.1 The Packed Recursive Format

Here we describe the new recursive packed formats. It was first described by Andersen, Gustavson, and Waśniewski in [1]. The reader is encouraged to read [1] for an algorithm which transforms from conventional packed format to recursive packed format using a small temporary buffer. We now show that there are four instances of this new data structure. The four ways are distinguished by the names column lower, column upper, row lower, and row upper, see Figure 1. These packed recursive data formats are hybrid triangular formats consisting of $(n-1)$ full format rectangles of varying sizes and n triangles of size 1 by 1 on the diagonal. They use the same amount of data storage as the ordinary packed triangular format, i.e. $n(n+1)/2$. Because the rectangles (square submatrices)

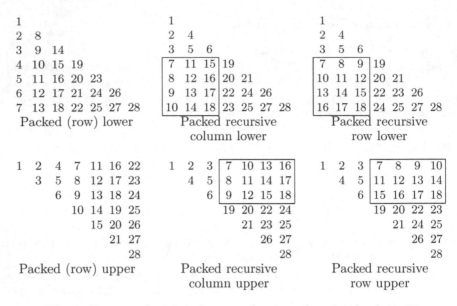

Fig. 1. Six ways of storing elements of an isosceles triangle of size 7.

are in full format it is possible to use high performance level 3 BLAS on the square submatrices. The difference between the formats is shown in Figure 1 for the special case $n = 7$.

Notice that the triangles are split into two triangles of sizes $n_1 = n/2$ and $n_2 = n - n_1$ and that a rectangle of size n_2 by n_1 for lower format and n_1 by n_2 for upper format. The elements in the upper left triangle are stored first, the elements in the rectangle follows and the elements in the lower right triangle are stored last. The order of the elements in each triangle is again determined by the recursive scheme of dividing the sides n_1 and n_2 by 2 and ordering these sets of points in the order triangle, square, triangle. The elements in the rectangle are stored in full format, either by row or by column.

Notice that we can store the elements of a rectangle in two different ways. The first is by column (standard Fortran order) and the second is by row (standard C order). Assume that A is in lower recursive packed format. Then the rectangle is size n_2 by n_1, $n_1 \leq n_2$. The storage layout for the matrix in L format is identical to the storage layout of the transpose matrix in U format. Our implementation uses this property to produce algorithms which accesses the elements in the rectangles with stride 1.

2.2   The Block Version of Recursive Packed Format Cholesky

In [1], Andersen, Gustavson and Waśniewski show an implementation of recursive Cholesky. The recursive formulation comes from the block factorization of positive definite A,

$$A = \begin{pmatrix} A_{11} & A_{21}^T \\ A_{21} & A_{22} \end{pmatrix} = LL^T = \begin{pmatrix} L_{11} & 0 \\ L_{21} & L_{22} \end{pmatrix} \begin{pmatrix} L_{11}^T & L_{21}^T \\ 0 & L_{22} \end{pmatrix} \tag{1}$$

which consists of two Cholesky factorizations (2,5), one triangular equation system (3), and one symmetrical rank k update (4),

$$A_{11} = L_{11}L_{11}^T \tag{2}$$
$$L_{21}L_{11}^T = A_{21} \tag{3}$$
$$\tilde{A}_{22} = A_{22} - L_{21}L_{21}^T \tag{4}$$
$$\tilde{A}_{22} = L_{22}L_{22}^T. \tag{5}$$

Now, the triangular equation system can in turn be written as

$$(X_1\ X_2) \begin{pmatrix} A_{11}^T & A_{21}^T \\ 0 & A_{22}^T \end{pmatrix} = (B_1\ B_2),$$

which is the same as

$$X_1 A_{11}^T = B_1 \tag{6}$$
$$\tilde{B}_2 = B_2 - X_1 A_{21}^T \tag{7}$$
$$X_2 A_{22}^T = \tilde{B}_2, \tag{8}$$

i.e., two triangular equation systems and one matrix-matrix multiply. In the same fashion, the symmetrical rank k update is written

$$\begin{pmatrix} \tilde{C}_{11} & 0 \\ \tilde{C}_{21} & \tilde{C}_{22} \end{pmatrix} = \begin{pmatrix} C_{11} & 0 \\ C_{21} & C_{22} \end{pmatrix} + \begin{pmatrix} A_1 \\ A_2 \end{pmatrix} \cdot (A_1^T\ A_2^T)\ ,\ \text{i.e.,}$$

$$\tilde{C}_{11} = C_{11} + A_1 A_1^T \tag{9}$$
$$\tilde{C}_{21} = C_{21} + A_2 A_1^T \tag{10}$$
$$\tilde{C}_{22} = C_{22} + A_2 A_2^T \tag{11}$$

The difference between the implementation in [1] and the one described here is that we combine recursion with blocking. They let the recursion go down to a single element. Because of the overhead of the recursive calls, the implementation has a significant performance loss for small problems. However, for large problems most of the computation takes place in the DGEMM, so that performance loss, while significant for smaller problems, is minor.

In our implementation, however, we combine blocking and recursion to produce a blocked version of their recursive algorithm. We only apply recursion above a fixed block size $2nb$. To solve problems, for sizes less than $2nb$, we use a variety of algorithmic techniques, some new and some old ([6]), to produce optimized unrolled kernel routines. These techniques are used both for the Cholesky

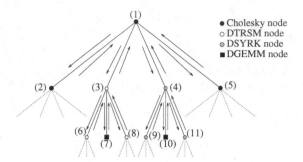

Fig. 2. The recursive tree for recursive Cholesky factorization

factorization routine and the recursive TRSM and recursive SYRK routines. The main technique we use to produce these fast kernels is register and level 1 cache blocking. The specific kernels are discussed in Section 2.5.

Next we give details about our block version and demonstrate that blocking the recursion tree via pruning completely eliminates performance problems for small matrix sizes.

2.3   Algorithms for Block Cholesky, Block TRSM, and Block SYRK

In this short section we give blocked algorithms of the recursive formulations sketched in Section 2.2. In each case, we have a blocking factor nb which determines what problem sizes will call a non-recursive kernel. This blocking factor determines when to continue recursion and when to solve the problem in one single kernel call. The algorithms, together with the blocking logic, are given in Figures 3, 4, 5.

The need of the recursive TRSM and SYRK stems from the recursive packed format. The factorization algorithm calls TRSM and SYRK with triangular matrix operands stored in recursive packed format, and with rectangular matrix operands stored in full format. Dividing the recursive packed matrices in TRSM and SYRK gives rise to two recursive packed triangular matrices and a rectangular matrix stored in full format, which becomes an argument to DGEMM.

Blocking reduces a large recursion tree to a recursion tree of small size. The Cholesky algorithm of size N has a binary tree with N leaves and $N-1$ interior nodes. The overhead at any node increases with n, where n is the size of the Cholesky subproblem, as each interior node performs the same Cholesky computation. To see this, note that each call is recursive so a call at size $2n$ include two calls at size n plus calls to recursive routines RTRSM and RSYRK. What is important for high performance is the ratio r of the cost of the recursive call overhead to the computation cost (number of flops). It turns out that for n large, this ratio is tiny and can be safely neglected. This is shown in Figure 6 where the ratio between the number of calls to subroutines and arithmetic work is plotted. Notice that the recursive code handles problems where $n \geq nb$ as we provide special kernels for small problems where $1 \leq n < nb$.

$A(1:n, 1:n) = \text{BC}(A(1:n, 1:n))$
if $(n < 2nb)$ **then**
   $A(1:n, 1:n) = \text{BCK}(A(1:n, 1:n))$! Block Cholesky factor A
else ! $n \geq 2nb$! with a level 3 kernel routine
   $n_1 = n/2$ and $j_1 = n_1 + 1$
   $A(1:n_1, 1:n_1) = \text{BC}(A(1:n_1, 1:n_1))$! (2) in Figure 2
   $A(j_1:n, 1:n_1) = \text{BT}(A(1:n_1, 1:n_1), A(j_1:n, 1:n_1))$! (3) in Figure 2
   $A(j_1:n, j_1:n) = \text{BS}(A(j_1:n, 1:n_1), A(j_1:n, j_1:n))$! (4) in Figure 2
   $A(j_1:n, j_1:n) = \text{BC}(A(j_1:n, j_1:n))$! (5) in Figure 2
endif

Matrix A is stored in packed recursive lower format, PRLF.

Fig. 3. Algorithm for Block Recursive Cholesky Factorization

$B(1:m, 1:n) = \text{BT}(A(1:n, 1:n), B(1:m, 1:n))$
if $(n < 2nb)$ **then**
   $B(1:m, 1:n) = \text{BTK}(A(1:n, 1:n), B(1:m, 1:n))$! Solve $XA^T = B$ using
else ! $n \geq 2nb$! level 3 RUNN kernel
   $n_1 = n/2$ and $j_1 = n_1 + 1$
   $B(1:m, 1:n_1) = \text{BT}(A(1:n_1, 1:n_1), B(1:m, 1:n_1))$! (6) in Figure 2
   Update $B(1:m, j_1:n) = B(1:m, j_1:n) - B(1:m, 1:n_1) \cdot A(j_1:n, 1:n_1)^T$! (7)
   $B(1:m, j_1:n) = \text{BT}(A(j_1:n, j_1:n), B(1:m, j_1:n))$! (8) in Figure 2
endif

A is in PRLF and B is in row major order with leading dimension n.

Fig. 4. Algorithm for Block Recursive Triangular Solve

2.4 The Cholesky Algorithm Applied to Recursive Lower Row Storage Produces Stride 1 Storage Access throughout

Firstly, we note that the recursive algorithm works for all four data formats shown in Figure 1.

In order to demonstrate the validity of the title of this Section it is necessary to establish the contents of Section 2.2. So, we now turn to demonstrating why transposing recursive lower column storage leads to stride 1 performance throughout. This is true for Cholesky factorization and many other symmetric algorithms. Assume we use full recursion. Then the factor part of the code becomes n square root calculations. The rest of the code consists of calls to RTRSM and RSYRK. But both RTRSM and RSYRK are recursive and when used with full recursion they always consist of calls to DGEMM and level 1 calls to DSCAL and DDOT. Now take a DGEMM call from RTRSM. Note that $C = C - A \cdot B^T$ is computed, where C is m by n, A is m by k and B is n by k. Similarly a DGEMM call from RSYRK has the same form $C = C - A \cdot B^T$. Now transpose this generic DGEMM computation to obtain $C^T = C^T - B \cdot A^T$ and assume that L is stored in re-

$C(1:m,1:m) = \text{BS}(A(1:m,1:n), C(1:m,1:m))$
if $(m < 2nb)$ then
    $C(1:m,1:m) = \text{BSK}(A(1:m,1:n), C(1:m,1:m))$! Perform rank n update
else ! $m \geq 2nb$! using level 3 LT kernel
    $m_1 = m/2$ and $j_1 = m_1 + 1$
    $C(1:m_1,1:m_1) = \text{BS}(A(1:m_1,1:n), C(1:m_1,1:m_1))$! (9) in Figure 2
    Update $C(j_1:m_1,1:m) = C(j_1:m,1:m) - A(j_1:m,1:n) \cdot A(j_1:m,1:n)^T$! (10)
    $C(j_1:m,j_1:m) = \text{BS}(A(j_1:m,1:n), C(j_1:m,j_1:m))$! (11) in Figure 2
endif

A is in row major order with leading dimension n and C is in PRLF.

Fig. 5. Algorithm for Block Recursive Triangular Rank n Update

cursive lower row-wise storage. Since storing a full matrix rowwise is identical to storing its transpose column-wise we see that $C^T = C^T - B \cdot A^T$ becomes $D = D - E^T \cdot F$ where $D = C^T$, $E = B^T$, and $A = F^T$. Note that each computation problem for C and D consists of doing mn dot products each of size k. The form $C = C - A \cdot B^T$ computes dot products stride lda, ldb while the form $D = D - E^T \cdot F$ computes dot products stride 1.

2.5 Kernel Issues

The operands of the kernel routines BCK, BTK, and BSK, are stored in recursive packed format. BCK operates solely on recursive format, while BTK and BSK do DTRSM and DSYRK on mixed format, see Section 2.3, page 86.

In our implementation, two types of kernel routines have been considered. What distinguishes these kernel is how they deal with the non-linear addressing of the packed recursive triangles. The first one is called a mapping kernel, and the technique is used for the Cholesky factorization kernel. Recall that this kernel operates solely on a triangular matrix, which is stored in packed recursive lower format. The ratio of the number of triangular matrix accesses to the number of operations is small, so the performance loss of copying the triangle to a full matrix in order to simplify addressing is not feasible. Also, the triangle is updated, so two copy operations would have been necessary, one before the factorization and one after, to store the triangle back in packed recursive format. Instead, an address map of the elements' structure in memory is preconstructed. Now, for every problem size N, there will only be two kernel problem sizes, so only two maps need to be generated, M_1 for problem sizes $n_i + 1$ and M_2 for problem sizes n_i. These maps are initialized before the recursion starts.

This map design leads to one extra memory lookup and two extra index operations per reference. However, the performance impact should be negligible. The extra operations are performed by the Integer Unit[1], a unit which has many spare cycles in these floating point intensive subroutines. (This statement,

---

[1] Sometimes referred to as the Fixed Point Unit, FXU.

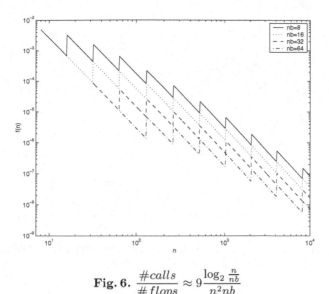

Fig. 6. $\dfrac{\#calls}{\#flops} \approx 9\dfrac{\log_2 \frac{n}{nb}}{n^2 nb}$

however, is not true for processors which have a combined Integer Unit and Load/Store Unit.) Therefore, these operations can be done in parallel with the floating point operations. Secondly, many integer registers are available for use by the kernels, so several pieces of mapping data can be stored in registers for later reuse.

For the TRSM and SYRK kernels, more floating point operations are performed, which reduces the ratio of the number of accesses to the number of operations. This suggests that it would be beneficial to copy the triangle to a buffer, and store the triangle in full array format. This is especially so for the BT kernel BTK as there the triangle is only read and so copying is done only once. For the BS kernel, BSK, the triangle is both read and written and so data copying must be done twice. In fact, since the number of operations depends on the size of the rectangular operand as well, we use this size as a threshold. If the rectangle is large enough, the triangle is copied to a buffer.

3   Performance

In this Section, we give performance results for our algorithm for an IBM RS/6000 SP POWER3 SMP Thin Node, with two IBM POWER3 CPUs running at 200 MHz. Each processor has a peak rate of 800 Mflops/s. The main motivations for these graphs is a) to show the good overall performance of the recursive algorithm and b) to show the good performance for small problems due to blocking. Other results for larger problems for various machines are found in [7] and [1]. As can be seen in Figure 7, the performance of the recursive algorithm is about 5 % better than *full storage* LAPACK routine DPOTRF for large problems, and much better than DPOTRF for small problem. The recursive

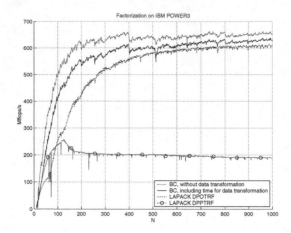

Fig. 7. The performance for Cholesky factorization on the IBM POWER3. Matrix order is 10 to 1000 in steps of 1. Peak rate is 800 Mflops/s.

algorithm is much better than the LAPACK routine `DPPTRF`, which operates on packed storage.

For SMP machines, a fair speedup is obtained by simply linking to a SMP version of `DGEMM`, see Figure 8. This is due to the geometric nature of the sizes of the arguments to `DGEMM`. In the recursive algorithm, most of the operations tends to be performed in large, "squarish" `DGEMM` operations which are easily parallelized.

4   Conclusions

Our novel algorithm for Cholesky factorization has three attractive features. Firstly, it uses minimal storage. Secondly, it attains level three performance due to mostly calling matrix-matrix multiplication routines during execution. The new algorithm outperforms the standard LAPACK routines `DPPTRF` and `DPOTRF`. Thirdly, the new code is portable and migratable, so existing codes can use the new algorithm.

The use of blocking ensures that the good performance is maintained also for smaller problems. We described two ways to efficiently address the non-linearly stored elements of a block.

A variant of the algorithm presented here is now part of the IBM ESSL.

References

1. B. Andersen, F. Gustavson, and J. Waśniewski. A recursive formulation of Cholesky factorization of a matrix in packed storage. Technical Report CS-00-441, University of Tennessee, Knoxville, TN, Computer Science Dept., May 2000. Also LAPACK Working Note number 146 (`lawn146.ps`), and submitted to the *ACM Transaction of Mathematical Software*.

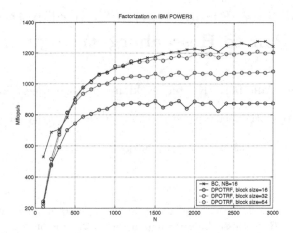

Fig. 8. The performance for Cholesky factorization on the IBM POWER3. Matrix order is 100 to 3000 in steps of 100. Peak rate is 2 × 800 Mflops/s.

2. E. Anderson, Z. Bai, C. Bischof, S. Blackford, J. Demmel, J. Dongarra, J. Du Croz, A. Greenbaum, S. Hammarling, A. McKenney, and D. Sorensen. *LAPACK User's Guide*. SIAM, Philadelphia, third edition, 1999.

3. Gene H. Golub and Charles Van Loan. *Matrix Computations*. Johns Hopkins, third edition, 1996.

4. F.G. Gustavson. Recursion leads to automatic variable blocking for dense linear-algebra algorithms. *IBM Journal of Research and Development*, 41(6), November 1997.

5. F.G. Gustavson, A. Henriksson, I. Jonsson, B. Kågström, and P. Ling. Recursive Blocked Data Formats and BLAS's for Dense Linear Algebra Algorithms. In B. Kågström, J. Dongarra, E. Elmroth, and J. Waśniewski, editors, *Applied Parallel Computing, PARA'98*, volume 1541 of *Lecture Notes in Computer Science*, pages 195–206. Springer-Verlag, 1998.

6. F.G. Gustavson, A. Henriksson, I. Jonsson, B. Kågström, and P. Ling. Superscalar GEMM-based Level 3 BLAS – The On-going Evolution of a Portable and High-Performance Library. In B. Kågström, J. Dongarra, E. Elmroth, and J. Waśniewski, editors, *Applied Parallel Computing, PARA'98*, volume 1541 of *Lecture Notes in Computer Science*, pages 207–215. Springer-Verlag, 1998.

7. Fred Gustavson and Isak Jonsson. Minimal Storage High Performance Cholesky Factorization via Blocking and Recursion. Submitted to *IBM Journal of Research and Development* in June 2000.

8. IBM Corporation. *Engineering and Scientific Subroutine Library for AIX, Guide and Reference*, third edition, October 1999.

Parallel Two-Stage Reduction of a Regular Matrix Pair to Hessenberg-Triangular Form

Björn Adlerborn, Krister Dackland, and Bo Kågström

Department of Computing Science and HPC2N, Umeå University, SE–901 87 Umeå, Sweden.
{c95ban, dacke, bokg}@cs.umu.se

Abstract. A parallel two-stage algorithm for reduction of a regular matrix pair (A, B) to Hessenberg-triangular form (H, T) is presented. Stage one reduces the matrix pair to a block upper Hessenberg-triangular form (H_r, T), where H_r is upper r-Hessenberg with $r > 1$ subdiagonals and T is upper triangular. In stage two, the desired upper Hessenberg-triangular form is computed using two-sided Givens rotations. Performance results for the ScaLAPACK-style implementations show that the parallel algorithms can be used to solve large scale problems effectively.

1 Introduction

The Hessenberg-triangular reduction, named DGGHRD in the LAPACK library [1], is one of the most time-consuming steps when solving the generalized eigenvalue problem $Ax = \lambda Bx$, where A and B are square real $n \times n$ matrices. After an inital reduction of (A, B) to Hessenberg-triangular form (H, T), the generalized Schur form of the regular matrix pair is computed by applying the QZ algorithm to the condensed matrix pair [10,7]. The Hessenberg-triangular reduction is also useful when solving matrix pencil systems $(A - sB)x = y$ for different values of the parameter s and possibly several right hand sides y [8].

This contribution builds on previous work by the Umeå group including, for example, the design of a blocked (LAPACK-style) two-stage variant of DGGHRD and a ScaLAPACK-style implementation of the first stage of the reduction [7,6, 3,5,4]. The focus in this contribution is on the second stage of the algorithm, the reduction from blocked Hessenberg-triangular to Hessenberg-triangular form. The paper is organized as follows: In Section 2, we present our parallel two-stage algorithm for the Hessenberg-triangular reduction. Section 2.1 reviews the ScaLAPACK programming style and conventions [2]. In Section 2.2, we give an overview of the parallel implementation of the first stage. Section 2.3 describes the parallel algorithm for the second stage. Finally, in Section 3, we present and discuss measured performance results of our parallel implementation of DGGHRD from test runs on an IBM SP system.

T. Sørevik et al. (Eds.): PARA 2000, LNCS 1947, pp. 92–102, 2001.
© Springer-Verlag Berlin Heidelberg 2001

2 Parallel Two-Stage Reduction to Hessenberg-Triangular Form

The parallel ScaLAPACK-style algorithm presented here is based on the blocked *two-stage* algorithm described in [7].

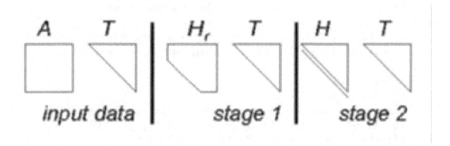

Fig. 1. The two stages in the parallel Hessenberg-triangular reduction algorithm.

Stage one reduces the matrix pair to a block upper Hessenberg-triangular form (H_r, T) form using two-sided Householder reflections and the compact WY representation [11] of the Householder matrices. The matrix H_r is upper r-Hessenberg with $r > 1$ subdiagonals and T is upper triangular. In the second stage of the reduction algorithm, all but one of the subdiagonals of the block Hessenberg A-part are set to zero while keeping T upper triangular. The annihilation of elements along the subdiagonals of H_r and fill-in elements in T are performed using two-sided Givens rotations.

2.1 Data Distribution

Our parallel implementation of the two-stage reduction follows the ScaLAPACK software conventions [2]. The P processors (or virtual processes) are viewed as a rectangular processor grid $P_r \times P_c$, with $P_r \geq 1$ processor rows and $P_c \geq 1$ processor columns such that $P = P_r \cdot P_c$. The data layout of dense matrices on a rectangular grid is assumed to be done by the two-dimensional block-cyclic distribution scheme. The block size used in the parallel algorithms is $NB = r$, where r is the number of subdiagonals in the block Hessenberg-triangular reduction (Stage 1). This enables good load balancing by splitting the work reasonably evenly among the processors throughout the algorithms. Moreover, the block-cyclic data layout enables the use of blocked operations and the level 3 BLAS for high-performance computations on single processors.

2.2 Stage 1: Parallel Reduction to (H_r, T) Form

The parallel implementation of the first stage is presented and analyzed in [6,3]. This includes a scalability analysis based on an hierarchical performance model

[5] and real experiments. Here, we only review the parallel implementation of the blocked (H_r, T) reduction algorithm outlined in Figure 2, which is based on existing parallel operations in the ScaLAPACK library [2].

```
function [A, B, Q, Z] = HrT (A, B, r, p)
  k = n/r;                      # blocks in the first block column.
  for j = 1:r:n-r
    k = max(k - 1, 2);      # blocks to reduce in current block column j.
    l = ceil((k-1)/(p-1));# steps required for the reduction.
    i = n;
    for step = 1:l
      nb = min(p*r, i-j-r+1);
      Phase 1: Annihilation of p r × r blocks in block column j of A.
      [q, A(i-nb+1:i,j:j+r-1)] = qr(A(i-nb+1:i,j:j+r-1));
      A(i-nb+1:i,j+r:n) = q'*A(i-nb+1:i,j+r:n);
      B(i-nb+1:i,i-nb+1:n) = q'*B(i-nb+1:i,i-nb+1:n);
      Q(:,i-nb+1:i) = Q(:,i-nb+1:i)*q;  Q = Iₙ initially.
      Phase 2: Restore B - annihilation of fill-in.
      [z, B(i-nb+1:i,i-nb+1:i)] = rq(B(i-nb+1:i,i-nb+1:i));
      A(1:n,i-nb+1:i) = A(1:n,i-nb+1:i)*z;
      B(1:i-nb,i-nb+1:i) = B(1:i-nb,i-nb+1:i)*z;
      Z(:,i-nb+1:i) = Z(:,i-nb+1:i)*z;  Z = Iₙ initially.
      i = i - nb + r;      Pointer for next block annihilation.
    end
  end
```

Fig. 2. Matlab-style algorithm for the blocked (H_r, T) reduction of (A, B) (Stage 1).

On entry to the HrT algorithm, $B \in R^{n \times n}$ is assumed to be in upper triangular form. If this is not the case, the ScaLAPACK routines PDGEQRF and PDLARFB are used to perform a QR factorization of B and to update the matrix A accordingly. On exit, A is upper r−Hessenberg, B is upper triangular, and Q, Z are the accumulated orthogonal transformation matrices such that $A = Q^H A Z$ and $B = Q^H B Z$.

The A-matrix is reduced by QR factorizations of rectangular $pr \times r$ blocks ($p \geq 2$) and B is restored by RQ factorizations of square $pr \times pr$ blocks, using the ScaLAPACK routines PDGEQR2 and PDGERQ2, respectively. All updates are performed using a combination of the ScaLAPACK routines PDLARFT for extraction of the triangular matrix T and PDLARFB for application of the Householder transformations represented in compact WY form.

Since the fill-in overlap for consecutive iterations, we apply the RQ factorizations to blocks of size $(p - 1)r \times pr$ in all iterations except the last one in each block column.

2.3   Stage 2: Parallel Reduction to Hessenberg-Triangular Form

The second stage is to annihilate the remaining $r - 1$ subdiagonals of H_r to get an upper Hessenberg matrix H, while keeping the B-part upper triangular. We solve this problem by implementing a parallel variant of the blocked algorithm described in [7].

The operations to be performed for the (H_r, T) to (H, T) reduction are summarized in Figure 3. On entry to HrT2HT, $A, B \in \mathcal{R}^{n \times n}$, A is upper r-Hessenberg, and B is upper triangular. On exit, A is upper Hessenberg and B is upper triangular.

```
function [A,B] = HrT2HT(A, B, r)
  [m,n]=size(A)
  for k = 1:n-2
    for l = min(k+r-1, n-1):-1:k+1
      [c,s]=givens(A(l:l+1,k))
      A(l:l+1, k:n) = row_rot(A(l:l+1, k:n),c,s)
      for i = l:r:n-1
        B(i:i+1,i:n) = row_rot(B(i:i+1,i:n),c,s)
        [c,s]=givens(B(i+1,i:i+1))
        B(1:i+1, i:i+1) = col_rot(B(1:i+1, i:i+1),c,s)
        m = min(i+r+1,n)
        A(1:m,i:i+1) = col_rot(A(1:m,i:i+1),c,s)
        if (i+r+1 <= n)
          [c,s]=givens(A(i+r:i+r+1,i))
          A(i+r:i+r+1, i:n) = row_rot(A(i+r:i+r+1, i:n),c,s)
        end
      end
    end
end
```

Fig. 3. Unblocked reduction to upper Hessenberg-triangular form (Stage 2).

Algorithm HrT2HT annihilates $A_{l+1,k}$ using a Givens rotation and applies the rotation to rows $l, l+1$ of A. To preserve the eigenvalues of (A, B) the rotation is applied to rows $l, l+1$ of B as well. This application introduces a non-zero element $B_{l+1,l}$. We zero this fill-in by a column rotation (applied from right), which in turn, when applied to A, introduces a new non-zero element $A_{l+r+1,l}$. The i-loop chases the unwanted non-zero elements down the $(r + 1)$-th subdiagonal of A and the subdiagonal of B. To complete the reduction of column k this procedure is repeated $r - 1$ times. Similar operations are applied to the remaining columns $(k + 1, \ldots, n - 2)$ to produce the desired (H, T) form.

In the blocked variant of the (H_r, T) to (H, T) reduction, the data locality in each sweep (one iteration of the k loop in HrT2HT) is improved as follows [7]:

1. All $r - 1$ subdiagonal elements in column k are reduced before the chasing of unwanted non-zero fill-in elements starts.
2. A super-sweep that reduces m columns of A per iteration in the outer k loop is introduced.
3. The updates of A and B are restricted to r consecutive columns at a time. We store all rotations (in vectors) to enable delayed updates with respect to Givens rotations generated for previous updated columns in the current super-sweep.

These three items are important for reducing the data traffic in a single node memory hierarchy. However, when moving to a (distributed) parallel environment the delayed updates (see item three above) cause most of the processors to be idle and only a few working at the same time. So, instead of delaying updates all rotation vectors are broadcasted rowwise (or columnwise) in the processor mesh after a complete reduction of a column in A (or row in B) has been made. Thereby, we enable other processors to participiate in the computations as soon as possible.

In the rest of this section, we describe the parallel blocked variant of HrT2HT more in detail.

When reducing the k-th column of A, the submatrix pair $(A_{:,k+1:n}, B_{:,k+1:n})$ is partitioned in $s = \lceil (n-k)/r \rceil$ column blocks of size $n \times r$ (the last one of size $n \times \mod((n-k), r)$ when r is not a factor of $n - k$). Each block column pair i is further divided into square $r \times r$ upper triangular blocks denoted $A_i^{(t)}$, $B_i^{(t)}$, and rectangular blocks denoted $A_i^{(r)}$ and $B_i^{(r)}$:

$$\begin{bmatrix} A_i^{(r)} \\ A_i^{(t)} \\ 0 \end{bmatrix}, \qquad \begin{bmatrix} B_i^{(r)} \\ B_i^{(t)} \\ 0 \end{bmatrix}, \tag{1}$$

where $A_i^{(r)}$ is of size $(i \cdot r + k) \times r$, and $B_i^{(r)}$ is $((i-1) \cdot r + k) \times r$. Notice that the zero-block is not present in block column $s - 1$ of $A_{:,k+1:n}$ and in block column s of $B_{:,k+1:n}$. Moreover, the last block column of $A_{:,k+1:n}$ has neither a zero nor a triangular block, i.e., it consists of $A_s^{(r)}$ only. We also remark that $A_{s-1}^{(t)}$ is upper trapezoidal when it has fewer than r rows.

This block partitioning is illustrated in Figure 4 when the first column ($k = 1$) of a matrix pair of size 12×12 is reduced, where A has $r = 4$ subdiagonals. The blocks labeled 5 and 9 are $A_1^{(t)}$ and $A_2^{(t)}$, and the blocks labeled 4, 8, and 12 correspond to $A_1^{(r)}$, $A_2^{(r)}$ and $A_3^{(r)}$, respectively. Similarly, the diagonal blocks 2, 6 and 10 are $B_1^{(t)}$, $B_2^{(t)}$, and $B_3^{(t)}$, and finally, the blocks 3, 7, and 11 are $B_1^{(r)}$, $B_2^{(r)}$, and $B_3^{(r)}$, respectively.

We define row_i as the set of all row eliminations (rotations) required to annihilate $r - 1$ elements of A, and similarly col_i is the set of all column rotations needed to annihilate $r - 1$ elements of B. The row_1-set reduces $r - 1$ subdiagonal elements of the k-th column of A, while row_i for $i \geq 2$ annihilates fill-in

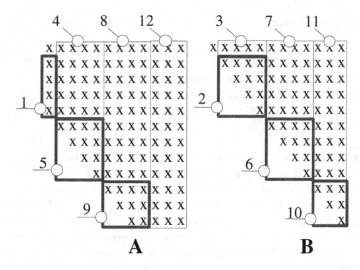

Fig. 4. Block partitioning and reference pattern in blocked (H_r, T) to (H, T) reduction (Stage 2).

introduced in the subdiagonal of $A_{i-1}^{(t)}$. The sets col_i zero fill-in introduced in the subdiagonals of $B_i^{(t)}$. By $row_{1:i}$ and $col_{1:i}$ we denote the row and column sets 1 to i, respectively. Notice that when $A_i^{(t)}$ and $B_i^{(t)}$ have $r' < r$ rows, we annihilate $r' - 1$ elements only and the associated rotation sets contain $r' - 1$ rotations.

The labeling of the blocks in Figure 4 follows the blocked spiral reference pattern of a matrix pair (A, B) in the blocked implementation [7]. The access pattern of the blocks are different in our parallel implementation, which is obvious from the algorithm description below. The label sets within brackets show which blocks in Figure 4 that are referenced in the operations for each block iteration i.

A *sweep* reducing column k of A proceeds as follows:
Reduce:
The set row_1 is generated ([1]) and broadcasted along the current processor row. The set is then applied to $B_1^{(t)}$ ([2]), $B_{2:s}^{(r)}$ ([7, 11]), $A_{1:s}^{(r)}$ ([4, 8, 12]). By generating col_1 the resulting fill-in is annihilated in $B_1^{(t)}$ ([2]). col_1 is broadcasted along current processor column and $B_1^{(r)}$ is updated with respect to col_1 ([3]).

Chase:

> **for** $i = 1 : s$
>   Apply col_i to $A_i^{(r)}$.   ([4], [8], [12])
>   **if** $i < s$
>   Apply col_i to. $A_i^{(t)}$.   ([5], [9])
>   Zero fill-in of $A_i^{(t)}$, i.e., generate row_{i+1}.   ([5], [9])

Broadcast row_{i+1} along the current processor row.

Apply row_{i+1} to $A^{(r)}_{i+1:s}$.   ([8, 12], [12])

Apply row_{i+1} to $B^{(t)}_{i+1}$, $B^{(r)}_{i+2:s}$.   ([6, 11], [10])

Zero fill-in of $B^{(t)}_{i+1}$, i.e., generate col_{i+1}.   ([6], [10])

Broadcast col_{i+1} along the current processor column.

Apply col_{i+1} to $B^{(r)}_{i+1}$.   ([7], [11])

end

In case of empty row or column rotation sets, no action is taken in the updates.

As in the blocked implementation [7], the above described procedure is extended to allow m columns to be reduced and chased in each sweep (called *super-sweep*). To distinguish *row*-sets belonging to different reduced columns of A, we use a superscript $j = 1, \ldots, m$. For example, $row^1_1, row^2_1, \ldots, row^m_1$ denote the first *row*-set of each of the m columns reduced in a super-sweep. Column sets belonging to a super-sweep are denoted analogously.

In the reduce-part of a super-sweep, the sets $row^i_{1:m-i+1}$ and $col^i_{1:m-i}$ ($i = 1, \ldots, m$) are generated. The chase-part of the super-sweep iteratively advances the sweeps one block column ahead in a pipelined fashion, starting with the leading block.

To find the optimal value of m several parameters must be considered, including the number of non-zero subdiagonals r ($= NB$), the matrix size N, processor grid configuraton and the memory hierarchy of the processors.

Typically, the annihilation of elements and the resulting fill-in require cooperation between four processors since the current window (a virtual block of size $NB \times NB$) most of the times is spread among different processors. No more than four processors can share a virtual block since the cyclic distribution is done with a blocking factor of NB. The sharing of blocks means that boarder elements have to be exchanged for the application of *row*- and *col*-sets and the calculation of new *row*- and *col*-sets for reduction of the fill-in since the update/reduction always operates on two consecutive rows or columns.

In Figure 5, we illustrate how a row rotation in B and reduction of the resulting fill-in is done on a $NB \times NB$ block ($NB = 6$), which is shared by four processors. The procedure for a column rotation is similar. A rowwise broadcast is made before application of the row rotatons begins, that is before step 1, and during steps 5–7 a columnwise broadcast is made. This is done to ensure that all processors that need rotation values have them before the associated updates start.

3   Performance Results and Discussion

We present measured performance results of our ScaLAPACK-style algorithms in Table 1, using up to 64 processors on the IBM Scalable POWERparallel (SP) system at High Performance Computing Center North (HPC2N). All results reported are excluding the time and the flopcounts for the accumulation of the transformation matrices Q and Z .

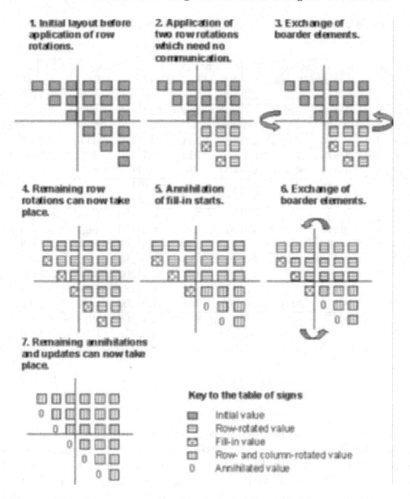

Fig. 5. Application of row rotation and annihilation of fill-in across processor boarders (Stage 2).

The first three columns define the problem size (N) and the processor grid configuration ($P = P_r \times P_c$). We start with $N = 1024$ on one processor and scale-up the problem size along with P so that we keep a fixed local matrix size ($\approx 1024 \times 1024$ entries of A and B per processor). We vary P between 2 and 64 in multiples of 2 and for a fixed P we investigate different grid configurations. For each processor grid we vary the block size NB to find a near to optimal value, which provides the best performance of the parallel algorithm. These best NB values together with the performance measured in Million floating point operations per second (*Mflops/s*) and the scaled (constant memory usage) speedup (S_P) are listed in columns 3–6 for the parallel Stage 1 reduction. S_P is computed as the ratio between the performance (measured in Mflops/s) obtained on P processors and one processor, respectively. The S_P values shown are rounded to one decimal's accuracy.

Table 1. Performance results on 1, 2, 4, 8, 16, 32, and 64 IBM SP Thin Nodes (120 MHz).

Configuration			Stage 1			Stage 2			Stage 1+2			Ratios	
N	P_r	P_c	NB	$Mflops/s$	S_P	NB	$Mflops/s$	S_P	NB	$Mflops/s$	S_P	F	T
1024	1	1	160	276	1.0	64	89	1.0	64	153	1.0	0.7	1.9
1448	2	1	150	479	1.7	150	184	2.1	150	292	1.9	0.7	1.9
1448	1	2	150	446	1.6	200	163	1.8	170	253	1.7	0.8	2.1
2048	2	2	160	734	2.7	180	244	2.7	180	414	2.7	0.7	2.1
2048	4	1	170	734	2.7	170	236	2.7	170	407	2.7	0.7	2.2
2048	1	4	160	473	1.7	180	40	0.4	150	90	0.6	0.7	8.1
2816	2	4	180	910	3.3	180	321	3.6	180	550	3.6	0.7	1.9
2816	4	2	140	1189	4.3	180	380	4.3	180	673	4.4	0.7	2.1
4096	4	4	180	1791	6.5	200	573	6.4	200	1034	6.8	0.6	2.0
4096	8	2	170	1588	5.8	200	546	6.1	170	968	6.3	0.6	1.8
4096	2	8	200	1076	3.9	200	394	4.4	200	679	4.4	0.6	1.7
5792	16	2	180	1775	6.4	200	613	6.9	180	1092	7.1	0.6	1.6
5792	2	16	200	1186	4.3	200	453	5.1	200	781	5.1	0.6	1.6
5792	8	4	180	2573	9.3	180	752	8.4	180	1430	9.3	0.6	2.1
5792	4	8	180	2041	7.4	200	707	7.9	200	1260	8.2	0.6	1.8
8192	8	8	170	3581	13.0	200	989	11.1	200	1919	12.5	0.6	2.3
8192	16	4	150	3146	11.4	200	922	10.4	170	1736	11.3	0.6	2.1
8192	4	16	180	2095	7.6	200	716	8.0	180	1269	8.3	0.6	1.8

Similar results for the parallel Stage 2 reduction are listed in columns 7–9. The results for the complete parallel two-stage Hessenberg-triangular reduction are displayed in columns 10–14. In Stage 1, the number of blocks annihilated in each blocked QR factorization is chosen as $p = \max(2, P_r)$. In Stage 2, the number of columns in a supersweep is kept fixed ($m = 2$). We have performed testing with larger values of m, but without observing any substantial performance improvements. Some results are shown in Table 2.

Table 2. Performance results using different values of m in Stage 2 on 8 IBM SP Thin Nodes (120 MHz).

Configuration				Stage 1+2		
N	P_r	P_c	m	NB	$Mflops/s$	S_P
2816	4	2	1	180	656	4.3
2816	4	2	2	180	673	4.4
2816	4	2	3	180	679	4.4
2816	4	2	4	180	661	4.4
2816	4	2	5	180	671	4.4

The last two columns in Table 1 show the ratios between the number of floating point instructions (*flops*) in Stage 2 and Stage 1 (F) and their corresponding execution time ratios (T). The number of flops in the two stages are determined by N, NB and p (only in Stage 1).

Although Stage 2 involves much less flops ($0.6 \leq F \leq 0.8$), the best execution time for Stage 2 is roughly twice as long as for the corresponding Stage 1 reduction. This can be explained by the implicit nature of the Stage 2 parallel reduction. It has a more fine-grained and costly communication compared to Stage 1, which is ruled by the data dependences of Stage 2. Moreover, we mainly perform level 3 operations in Stage 1, while there are lower level operations in Stage 2 (mostly level 1–2 and some level 2.5).

The processor grid configuration affects the data distribution and thereby the communication overhead and the execution rate of the parallel algorithms. The results in Table 1 show that choosing $P_r = P_c$ (when possible) gives the best performance. Otherwise, the best results are obtained for $P_r > P_c$, with $P_c > 1$ as large as possible. For a given configuration (N, $P_r \times P_c$), the block size NB giving the best performance of the combined Stage 1+2 algorithm is, as we expected, in between the block sizes for the parallel Stage 1 and Stage 2 algorithms. Typically, Stage 2 and Stage 1+2 have the same "best" NB values, which also show the impact of Stage 2 to the overall performance of the two-stage algorithm. Future work includes the design of a performance model of the parallel two-stage algorithm that can be used for automatic selection of algorithm-architecture parameters, e.g., block sizes and grid sizes.

References

1. E. Anderson, Z. Bai, J. Demmel, J. Dongarra, J. Du Croz, A. Greenbaum, S. Hammarling, A. McKenney, O. Ostrouchov and D. Sorensen. *LAPACK Users' Guide*, Third Edition. SIAM Publications, Philadelphia, 1999.
2. S. Blackford, J. Choi, A. Clearly, E. D'Azevedo, J. Demmel, I. Dhillon, J. Dongarra, S. Hammarling, G. Henry, A. Petit, K. Stanley, D. Walker, and R.C. Whaley. *ScaLAPACK Users' Guide*. SIAM Publications, Philadelphia, 1997.
3. K. Dackland. Parallel Reduction of a Regular Matrix Pair to Block Hessenberg-Triangular Form - Algorithm Design and Performance Modeling. Report UMINF-98.09, Department of Computing Science, Umeå University, S-901 87 Umeå, 1998.
4. K. Dackland and B. Kågström. Reduction of a Regular Matrix Pair (A, B) to Block Hessenberg-Triangular Form. In Dongarra et.al., editors, *Applied Parallel Computing: Computations in Physics, Chemistry and Engineering Science, PARA95*, Lecture Notes in Computer Science, Springer, Vol. 1041, pages 125–133, 1995.
5. K. Dackland and B. Kågström. An Hierarchical Approach for Performance Analysis of ScaLAPACK-based Routines Using the Distributed Linear Algebra Machine. In Wasniewski et.al., editors, *Applied Parallel Computing in Industrial Computation and Optimization, PARA96*, Lecture Notes in Computer Science, Springer, Vol. 1184, pages 187–195, 1996.

6. K. Dackland and B. Kågström. A ScaLAPACK-Style Algorithm for Reducing a Regular Matrix Pair to Block Hessenberg-Triangular Form. In Kågström et.al., editors, *Applied Parallel Computing: Large Scale Scientific and Industrial Problems*, *PARA 98*, Lecture Notes in Computer Science, Springer, Vol. 1541, pages 95–103, 1998.
7. K. Dackland and B. Kågström. Blocked Algorithms and Software for Reduction of a Regular Matrix Pair to Generalized Schur Form. *ACM Trans. Math. Software*, Vol. 25, No. 4, 425–454, 1999.
8. W. Enright and S. Serbin. A Note on the Efficient Solution of Matrix Pencil Systems. *BIT* 18, 276–281, 1978.
9. G. H. Golub and C. F. Van Loan. *Matrix Computations*, Second Edition. The John Hopkins University Press, Baltimore, Maryland, 1989.
10. C. B. Moler and G. W. Stewart. An Algorithm for Generalized Matrix Eigenvalue Problems. *SIAM J. Num. Anal.*, 10:241–256, 1973.
11. R. Schreiber and C. Van Loan. A Storage Efficient WY Representation for Products of Householder Transformations. *SIAM J. Sci. and Stat. Comp.*, 10:53-57, 1989.

A Fast Minimal Storage Symmetric Indefinite Solver [*]

Fred Gustavson[1], Alexander Karaivanov[2], Minka Marinova[2],
Jerzy Waśniewski[2], and Plamen Yalamov[2]

[1] IBM T.J. Watson Research Center, P.P. Box 218, Yorktown Heights, NY 10598.
USA, email: gustav@watson.ibm.com
[2] The Danish Computing Centre for Research and Education (UNI•C), Technical
University of Denmark, Building 304, DK-2800 Lyngby, Denmark,
emails: Alexander.Karaivanov@uni-c.dk, Minka.Marinova@uni-c.dk,
Jerzy.Wasniewski@uni-c.dk or yalamov@ami.ru.acad.bg

Abstract. In LAPACK there are two types of subroutines for solving
problems with symmetric matrices: routines for full and packed storage.
The performance of full format is much better as it allows the usage
of Level 2 and 3 BLAS whereas the memory requirement of the packed
format is about 50% of full. We propose a new storage layout which combines the advantages of both algorithms: its factorization performance
is better than that of full storage layout, and its memory requirement is
percentage-wise slightly larger than packed storage.
Our new algorithms, called DBSSV, DBSTRF, and DBSTRS are now
part of ESSL[9]. On three recent IBM RS/6000 platforms, Power3, Po-
wer2 and PowerPC 604e DBSTRF outperforms LAPACK's DSYTRF
by about 20%, and DBSTRS, with 100 RHS, outperforms LAPACK's
DSYTRS by more than 100%. These performance results are decidedly
unfair to our new algorithms: we compare against Level 3 algorithms as
opposed to Level 2 packed algorithms.

1 Introduction

Nowadays performance of numerical algorithms depends significantly on the engineering of the underlying computer architecture. Modern processors have a
hierarchical memory which, if utilized appropriately, can achieve several times
better performance.

One of the ways to effectively use the different levels of memory in the algorithms of numerical linear algebra is to introduce blocking into their algorithms.
In this way effectively designed BLAS (Basic Linear Algebra Subroutines) [2]
can be used. This significantly improves performance. This approach is used by
LAPACK (Linear Algebra PACKage) [2]. In most algorithms of LAPACK Level

---

[*] This research is supported by the UNI•C collaboration with the IBM T.J. Watson
Research Center at Yorktown Heights. The research of the last author was supported
by Grants MM-707 and I-702 of the Bulgarian Ministry of Education and Science.

T. Sørevik et al. (Eds.): PARA 2000, LNCS 1947, pp. 103–112, 2001.

3 (matrix-matrix operations) and Level 2 (matrix-vector operations) BLAS are used.

A different approach has been taken in [1,3,5,6,7], where recursive algorithms have been developed for a number of important numerical linear algebra problems. There it is shown that recursion leads to automatic blocking, and to Level 3 BLAS usage in most cases. Thus recursive algorithms turn out to be more effective for some problems because they work on large blocks and only use Level 3 calls.

In this paper we consider the solution of a linear system $Ax = b, A \in \mathcal{R}^{n \times n}$, where A is symmetric indefinite. The most popular algorithm for the solution of this problem uses the Bunch-Kaufman pivoting [4, §4.4], [8, §10.4.2] for the LDL^T decomposition of matrix A. There are two type of subroutines in LAPACK implementing this method. In the first the matrix is stored in a two-dimensional array, and this is called full storage.

Practical problems can be very large, and in this case conserving memory is an important issue. Clearly in full storage we use about twice the necessary memory. Therefore, a second type of storage called packed storage has been designed. With this type of storage we keep in a one-dimensional array only the essential part of the matrix needed for the computations.

The disadvantage of full storage is that it uses more memory but its advantage is that it allows the usage of Level 3 and Level 2 BLAS calls which essentially speeds up the computation. With packed storage we can only use Level 1 and packed Level 2 BLAS and consequently performance suffers drastically on RISC systems for large problems.

In the present paper we propose a new type of packed storage. The columns of the matrix are divided into blocks. The blocks are kept in packed storage; i. e. the blocks are stored successively in the memory. Several successive columns of the matrix are kept inside each block as if they were in full storage. The result of this storage is that it allows the usage of Level 3 BLAS. Of course, we need slightly more memory than for packed format but this memory is about 5% more for problems of practical interest. Thus the new storage scheme combines the two advantages of the storage formats in LAPACK, the smaller memory size of LAPACK's DSPSV, and the better performance of LAPACK's DSYSV [2].

This new storage layout is a generalization of both packed and full format. In fact, we modify this new format slightly to produce lower packed blocked overlapped (LPBO) format in order to handle efficiently the 1×1 or 2×2 blocks that occur during symmetric indefinite factorization.

Our new algorithms introduce a new form of back pivoting which we call intermediate or block back pivoting. Block back pivoting allows the usage of Level 3 BLAS in the triangular solves with multiple right hand sides. LAPACK's full format multiple solve algorithm, DSYTRS operates instead on a LINPACK style factorization. The choice of factorization format impacts performance greatly. In fact, our new multiple solve algorithm DBSTRS outperforms LAPACK's full DSYTRS algorithm by more than a factor of two.

Section 2 reviews traditional storage formats, introduces our new storage format and describes symmetric indefinite factorization. Section 3 describes new LAPACK algorithms in ESSL [9] called DBSSV, DBSTRF, and DBSTRS where BS stands for blocked symmetric. Section 4 discusses innovations leading to programming efficiencies. Section 5 discusses right and left looking algorithms and how they relate in the context of symmetric indefinite factorization. Section 6 describes performance results and Section 7 gives a brief conclusion.

2 Symmetric Matrix Format and Sketch of Our Block Algorithm

2.1 The Traditional Symmetric Formats

Because of space limitations we only give the lower symmetric format. Upper formats are detailed in [2]. A small example will suffice to give the general picture. Let N=10 and LDA=12. In the full format Fortran representation of a symmetric matrix A each column is LDA apart from the preceding column (see Figure 1a). In lower packed format the i-th column of the matrix is stored as a vector of length $n + 1 - i$, $1 \le i \le n$ (see Figure 1b). In Figure 1 the integers represent where in the arrays A and AP the (i, j) element of the matrix is stored. In Figure 1a, an * represents a storage location of array A that must be allocated but also cannot be used by the algorithm.

```
1  *  *  *  *  *  *  *     *    *      1
2 14  *  *  *  *  *  *     *    *      2 11
3 15 27  *  *  *  *  *     *    *      3 12 20
4 16 28 40  *  *  *  *     *    *      4 13 21 28
5 17 29 41 53  *  *  *     *    *      5 14 22 29 35
6 18 30 42 54 66  *  *     *    *      6 15 23 30 36 41
7 19 31 43 55 67 79  *     *    *      7 16 24 31 37 42 46
8 20 32 44 56 68 80 92     *    *      8 17 25 32 38 43 47 50
9 21 33 45 57 69 81 93 105      *      9 18 26 33 39 44 48 51 53
10 22 34 46 58 70 82 94 106 118       10 19 27 34 40 45 49 52 54 55
*  *  *  *  *  *  *  *     *    *
*  *  *  *  *  *  *  *     *    *
```

1a)   1b)

Fig. 1. Full and Lower Packed Format

2.2 A Generalization of the Traditional Symmetric Formats

Let $1 \leq NB \leq N$ where NB is the block size. Let $n_1 = \lceil N/NB \rceil$ and $n_2 = N + NB - n_1 NB$

The new format, called lower packed block format (LPB) partitions the N columns of A into $(n_1 - 1)$ block columns consisting of NB successive columns of A and a last block consisting of n_2 columns. The LDA of block i is $LDA - (i-1)NB$. An example of LPB format with NB=4 is given in Figure 2a. Note that when NB=N one gets full format whereas when NB=1 and LDA=N one gets packed format. Hence, the LPB format generalizes both traditional formats.

2.3 Very Brief Description of Symmetric Indefinite Factorization

First note that a block factorization stage produces

$$PAP^T = LDL^T = WL^T = LW^T, \text{ with } W = LD,$$

where P is a permutation matrix, L is unit lower triangular, and D is block diagonal consisting of 1×1, or 2×2 blocks. More detailed descriptions can be found in [4, §4.4], [8, §10.4.2]. Our algorithm follows the LAPACK's _SYTRF algorithm in that it produces identical results.

2.4 A Modification of LPB Format Called Lower Packed Block Overlapped Format (LPBO)

Symmetric Indefinite Factorization produces 1×1 or 2×2 pivot blocks in an arbitrary order depending on the numerical values of A. The LPB data format does not allow efficient handling of the either 1×1 or 2×2 pivot block that occurs at the boundary between blocks of columns. To overcome this inefficiency, we introduce the LPBO format. To obtain the new LPBO format (see Figure 2b) we pad, at the end of the first $n_1 - 1$ block columns an extra NB storage locations.

We now illustrate how this padding allows us to handle efficiently a block factorization and its level 3 update. The problem arises when we factor column j and j is multiple of NB. If column j is a 1×1 pivot then there will be NB or NB-1 columns in the factor block. If column j is a 2×2 pivot then there will be NB+1 or NB columns in the factor block. This means that the first update block will have NB or NB-1 columns. To illustrate let $j = NB = 4$. Column j is either a 1×1 or a 2×2 pivot block. Let us first assume the pivot block is 2×2. We shall update columns 6:9 of blocks 2 and 3 with columns 1:5 consisting of block 1 and column 1 of block 2. The updates are

$A(6:9,6:9) = A(6:9,6:9) - W(6:9,1:5) * L^T(6:9,1:5)$

and

$A(10:10,10:10) = A(10:10,10:10) - W(10:10,1:5) * L^T(6:9,1:5)$

In case 1, we need to update block 2 with block 1. This is the easier case. The updates are:

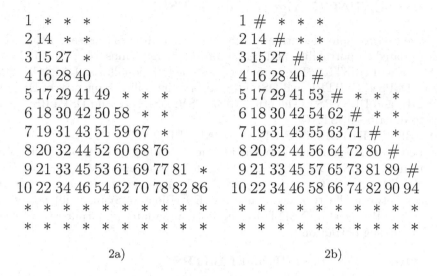

Fig. 2. Lower Packed Block and Lower Packed Block Overlapped Format

$A(5:8,5:8) = A(5:8,5:8) - W(5:8,1:4) * L^T(5:8,1:4)$

and

$A(9:10,5:8) = A(9:10,5:8) - W(9:10,1:4) * L^T(6:8,1:4)$

In either case it should be clear from studying the LPBO storage layout that the above block computations are laid out in full storage format: therefore, level 3 BLAS can be applied. Also, in both cases above we also must update the last block, either A(10:10,10:10) or A(9:10,9:10). Again, it should be clear how this is done using LPBO format.

The total memory we need for LPBO storage is exactly

$$S_{LPBO} = A_{LPBO} + WORK$$

where

$$A_{LPBO} = A_{N1} + (LDA - N + N2) * N2$$

$$A_{N1} = (N1 - 1) * (LDA * NB + NB) - ((N1 - 1) * (N1 - 2)/2) * NB^2$$

$$N1 = (N + NB - 1)/NB$$

$$N2 = N + NB - N1 * NB$$

$$WORK = (LDA + 1) * NB$$

For large N the size of A_{LPBO} is percentage-wise only slightly more than the size of packed storage.

3 New LAPACK Algorithms in ESSL

We have packaged our algorithms in ESSL. We now discuss these algorithms and issues related to portability and migration. Our algorithms are called DBSSV, DBSTRF and DBSTRS. They provide the same functionality as LAPACK algorithms DSPSV, DSPTRF and DSPTRS. In fact the calling sequences of DBSSV, DBSTRF and DBSTRS are identical to DSPSV, DSPTRF and DSPTRS. It suffices to describe DBSTRF.

SUBROUTINE DBSTRF(UPLO, N, AP, IPIV, NSINFO)

The first four arguments have the identical meaning as does DSPTRF's first four arguments. DSPTRF fifth and last argument is output only argument INFO. For DBSTRF, NSINFO is both an input and output argument. Hence the differences between the two algorithms are BS (Blocked Symmetric) verses SP (Symmetric Packed) and NSINFO being both an input and output variable verses INFO being output only.

3.1 Overview of DBSTRF and DBSTRS

Let NT=N(N+1)/2 be the size of array AP. DBSTRF accepts lower or upper packed format AP (exactly like DSPTRF) depending on value of UPLO. After argument checking DBSTRF compute NSO the size of the LPBO format array plus the size of the W buffer (used by DBSTRF and also by DSYTRF). If NSINFO<NSO then NSINFO is set to -NSO and no computation is done. Otherwise DBSTRF moves in place packed AP(1:NT) to ABPO format. Then DBSTRF performs a Level 3 Bunch Kaufman pivoting on APBO = AP(1:NSO).

DBSTRS and DSPTRS have identical calling sequences:

SUBROUTINE DBSTRS(UPLO, N, NRHS, AP, IPIV, B, LDB, INFO)

The only difference between DBSTRS and DSPTRS is the name (BS instead SP) and the data format of AP. Depending on N, AP is either in packed format or in LPBO format. Note that DBSTRS is only called if DBSTRF successfully completes factorization of AP. The crossover value of using packed format or LPBO format is decided by DBSTRF.

3.2 Migration and Portability Consideration

We designed our subroutines DBSTRF and DBSTRS to minimize migration to existing codes that already call DSPTRF and DSPTRS. Changes are indeed minimal. For each call to DSPTRF and DSPTRS change SP to BS. Also, DBSTRF requires that the size of array AP be inputted in argument NSINFO. If necessary, one can use the return of -NSO in NSINFO to allocate additional storage.

4 Innovations Leading to Program Efficiencies

4.1 Intermediate Back Pivoting

A difference between LINPACK and LAPACK is that LAPACK usually uses full back pivoting whereas LINPACK does no back pivoting. For DBSTRF we have introduced intermediate or block back pivoting. Intermediate pivoting saves pivoting operations and has almost the same effect on efficiency (speed) as does full back pivoting. In the context of using LPBO this is especially true. Block back pivoting is necessary for a Level 3 factorization and full back pivoting is necessary for Level 3 multiple solve.

4.2   The LPBO Format Allows for a Level 3 Multiple Solve

We remark that LAPACK DSYTRS is not a level 3 implementation.
A 2×2 pivot apparently causes a storage problem with full format. However, the LPBO format does not have this problem. Consider 2×2 pivot block D and its associated block of L :

$$D = \begin{pmatrix} a & b \\ b & c \end{pmatrix}, L = \begin{pmatrix} 1 & 0 \\ 0 & 1 \end{pmatrix}$$

In LAPACK A(2,1) cannot both hold b and zero. However with LPBO format we are free to use A(2,1) and A(1,2) (see # of LPBO format in Figure 2b) and we represent D (UPLO='L') as:

$$D = \begin{pmatrix} a & b \\ 0 & c \end{pmatrix}$$

In any case LAPACK's full DSYTRS uses LINPACK type pivoting and hence is not a level 3 implementation.

4.3 WL^T Is More Efficient Than LW^T

The level 3 part of DSYTRF and DBSTRF is the update computation $A \leftarrow A - LDL^T$. A buffer or work array W is used to compute $W = LD$. Now $WL^T = LW^T$ as $D = D^T$. LAPACK computes $A \leftarrow A - LW^T$ as opposed to $A \leftarrow A - WL^T$. The reason WL^T is to be preferred over LW^T (at least on IBM platforms) is that (during DGEMM) the left operand matrix (W in this case) fills the L1 cache as opposed to right operand matrix (L^T in this case) . Now W is at the control of the algorithm implementer whereas L is an input array. Thus WL^T is to be preferred over LW^T.

4.4   The Workspace Size of W Can Be NB^2 and Not $N * NB$

During the factor part of DBSTRF we only need to compute the diagonal block of L and not the entire panel of L. However, we must store W on top of A since the buffer size is only NB^2. This means the benefit of section 4.3 might be lost. When factorization is complete we overwrite the diagonal part of W with L from the buffer. Also, during update $A \leftarrow A - WL^T$ we can update a block panel at a time. Again, we only need to compute a block of L to do this. After a block panel is updated the L created in the buffer is then copied over the corresponding part of W.

4.5   The UPLO='U' Case

Using stride one gives the best performance. The UPLO='L' case uses stride one for the most part. Stride LDA can be very costly especially on IBM processors. This very costly cases occurs when UPLO='U' and one factors A in the usual "top down, left right" manner. However, one can just as well factor A "bottom up, right left". When this is done the UPLO='U' case "converts to" the UPLO='L' case. Now both LINPACK and LAPACK almost follow this strategy. Our reason to follow them is based on performance; we have not seen this reasoning for doing so presented in the literature.

5   Left Looking Verses Right Looking Algorithms

Many LAPACK algorithms are right looking (RL). However, left looking (LL) algorithm have better performance on todays processors with deep memory hierarchies. This is because LL algorithms only read from memory whereas RL algorithms both read from and write to memory. To see this note that LL algorithms updates a single panel by both reading from and writing to this panel by reading data from all panels to its left. A RL algorithm reads from a single panel and uses that data to update (read and write) all panels to its right. Nonetheless, for symmetric indefinite factorization a RL algorithm is the method of choice because a LL algorithm, depending on A's numerical values, require more flops than a RL algorithm. It is, however, necessary to use a LL panel factor if one wants to obtain a level 3 update. Now larger NB means better performance; i.e. , DGEMM runs at a faster MFLOP rate. It also means less overall memory traffic as the matrix is only updated $\lceil N/NB \rceil$ times. However, by using a larger NB in the LL panel factor one should potentially obtain a higher FLOP count. Hence the current algorithm of choice DSYTRF and our algorithm DBSTRF has to balance these opposing factors.

6   Performance

We tested our algorithms on three recent IBM platforms: Power3, Power2, and PowerPC 604e. In each figure we compare our new BS routines verses LAPACK's SY routines. All plots measure MFLOPS verses matrix order which runs from 10 to 1000 in steps of 10. In Figure 3a we consider factorization. Initially both programs have identical performance because the algorithms are identical. There is crossover point where DBSTRF switches from packed to LPBO format. The first big drop (caused by program loading) in Figure 3a is where this crossover occurs. It can be seen that DBSTRF outperforms DSYTRF by up to say 40% and this includes the cost of converting from packed to LPBO format. Also, note that for N>300 there are three dips in DSYTRF performance (see section 4.3 for an explanation). In Figure 3b, we compare DBSTRS verses DSYTRS for multiple and single right hand sides. Here we see a dramatic improvement of using intermediate block factorization format over using LINPACK factorization

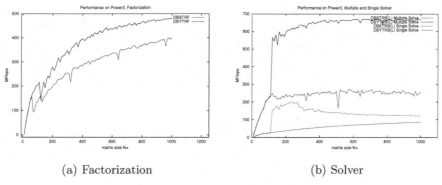

(a) Factorization   (b) Solver

Fig. 3. Performance on IBM Power3

format. Figure 4a shows factorization performance on an IBM 160 MHz Power2 machine. The new feature are the large performance dips that occur for the DSYTRF algorithm. These dips occur when the LDA of A (chosen here to equal N) does not produce a good data mapping into the four way set associative L1 cache. Now DBSTRF does not have these dips (see section 4.3). In Figure 4b we did not choose LDB−N because they produced similar dips. Figure 5a shows factorization performance on a 332 MHz PowerPC 604e machine. Relative to Power3 and Power2 this machine lacks floating point performance. Figure 5b shows how good level 3 performance can be relative to the level 2.5 performance of DSYTRS (about a factor of 5).

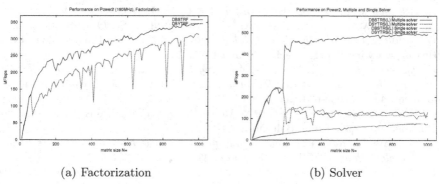

(a) Factorization   (b) Solver

Fig. 4. Performance on IBM Power2

7   Conclusions

In our view this paper convincingly demonstrates that our new LPBO data format can be used to produce a high performance storage efficient implementation of symmetric indefinite factorization. To us, it appears to combine the best features of LAPACK's SP and SY routines.

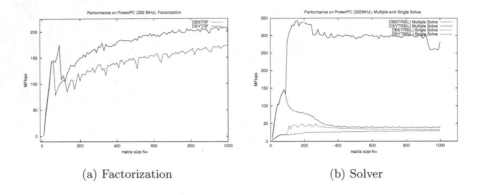

(a) Factorization   (b) Solver

Fig. 5. Performance on IBM PowerPC 604e

References

1. B. ANDERSEN, F. GUSTAVSON, AND J. WAŚNIEWSKI, *A recursive Formulation of the Cholesky Factorization Operating on a Matrix in Packed Storage Form*, University of Tennessee, Knoxville, TN, Computer Science Dept. Technical Report CS-00-441, May 2000, also LAPACK Working Note number 146 (http://www.netlib.org/lapack/lawns/lawn146.ps), and submitted to the Transaction of Mathematical Software (TOMS) of the ACM.
2. E. ANDERSON, Z. BAI, C. BISCHOF, J. DEMMEL, J. DONGARRA, J. DU CROZ, A. GREENBAUM, S. HAMMARLING, A. MCKENNEY, S. OSTROUCHOV, AND D. SORENSEN, *LAPACK Users' Guide Release 3.0*, SIAM, Philadelphia, 1999, (http://www.netlib.org/lapack/lug/lapack_lug.html).
3. E. ELMROTH AND F. GUSTAVSON, *Applying Recursion to Serial and Parallel QR Factorization Leads to Better Performance*, IBM Journal of Research and Development, 44, #4, 605-624, 2000
4. G. H. GOLUB AND C. F. VAN LOAN, *Matrix Computations*, 3rd edition, The John Hopkins University Press, Baltimore, 1996.
5. F. GUSTAVSON, *Recursion leads to automatic variable blocking for dense linear-algebra algorithms*, IBM J. Res. Develop., 41 (1997), pp. 737–755.
6. F. GUSTAVSON, A. KARAIVANOV, J. WAŚNIEWSKI, AND P. YALAMOV, *Recursive formulation of algorithms for symmetric indefinite linear systems* (manuscript).
7. F. GUSTAVSON, A. HENRIKSSON, I. JONSSON, B. KÅGSTRÖM, AND P. LING, *Recursive Blocked Data Formats and BLAS's for Dense Linear Algebra Algorithms*, in Applied Parallel Computing, B. Kågström et. al., eds., Lecture Notes in Computers Science, v. 1541, Springer, 1998, pp. 120–128.
8. N. J. HIGHAM, *Accuracy and Stability of Numerical Algorithms*, SIAM, Philadelphia, 1996.
9. *IBM Engineering Scientific Subroutine Library for AIX, (ESSL) V3, rel2 Pgm #5765-C42, July 2000*

A Scalable Parallel Assembly for Irregular Meshes Based on a Block Distribution for a Parallel Block Direct Solver [*]

David Goudin[1] and Jean Roman[1]

LaBRI, Université Bordeaux I - ENSERB, CNRS UMR 5800,
351, cours de la Libération, 33405 Talence Cedex, France
{goudin, roman}@labri.u-bordeaux.fr

Abstract. This paper describes a distribution of elements for irregular finite element meshes as well as the associated parallel assembly algorithm, in the context of parallel solving of the resulting sparse linear system using a direct block solver. These algorithms are integrated in the software processing chain EMILIO being developped at LaBRI for structural mechanics applications. Some illustrative numerical experiments on IBM SP2 validate this study.

1 Introduction

The OSSAU code of the CEA/CESTA is an industrial vectorized structural mechanics code, non linear in time and in two or three dimensions. It is a computational finite element code which solves plasticity (or thermo-plasticity, possibly coupled with large displacements) problems; it therefore computes the response of a structure to various physical constraints (external forces, pressure, ...).

The response of the structure is characterized by the node displacements of the finite element discretization. The total load applied to the structure is divided into a certain number of increments that are successively applied; each increment solves the non linear problem by a Newton-Raphson technique that leads to the iterative process $K(U_i) \cdot \Delta U_i = R(U_i)$ where ΔU_i represents the unknown incremental displacement vector, $K(U_i)$ the consistent stiffness matrix and $R(U_i)$ the right-hand-side vector. The positive definite symmetric matrix K and the vector R are computed by assembling elementary matrices $Ke(U_i)$ and elementary residual vectors $Re(U_i)$ associated with the finite elements e of the mesh respectively:

$$K(U_i) = \sum_{e \in \Omega} Ke(U_i), \; R(U_i) = \sum_{e \in \Omega} Re(U_i) \; . \tag{1}$$

---

[*] This work is supported by the French *Commissariat à l'Énergie Atomique* CEA/CESTA under contract No. 7V1555AC, and by the GDR ARP (iHPerf group) of the CNRS.

T. Sørevik et al. (Eds.): PARA 2000, LNCS 1947, pp. 113–120, 2001.

When dealing with complex 3D objects whose meshes imply a very great number of finite elements, several difficult problems arise [5]. Since the matrix of these systems does not have good properties (mainly due to the use of the Serendipity family of finite elements), classical iterative methods do not behave well. Therefore, as we want to obtain a robust versatile industrial software, we must use direct linear out-of-core solvers that take into account the sparsity of the system. When the number of Degrees Of Freedom becomes too large (specially above 150,000 DOFs), the out-of-core capabilities are not yet sufficient and parallel direct sparse solvers must be used, for reasons of memory capabilities and acceptable solving time.

As described before, the assembly of matrix K consists in computing the elementary matrix of each element of the mesh and then adding this elementary matrix to the stiffness matrix K. An elementary matrix describes the physical characteristics of the element, and the complexity of its calculation essentially depends on the constraints applied to the structure. In certain test cases that we must deal with (3D Cologne Challenge for example), the assembly cost in time of the stiffness matrix K is equivalent to the one of the matrix factorization. Therefore, the parallelization of the assembly step is a crucial problem.

The parallel assembly is a step in the software processing chain EMILIO (figure 1) developped at LaBRI and which gives efficient solutions to the problem of the parallel resolution of large sparse linear systems by direct methods [2, and included references]. This chain is organized into two major parts: a sequential pre-processing with a *global strategy* (ordering of the finite element graph [7], block symbolic factorization [1], repartitionning and distribution of the blocks of the factored matrix for the parallel block solver, distribution of the elements for the parallel assembly) and the parallel phase (parallel assembly and parallel resolution of the system). To avoid costly redistributions, we propose in section 1 an original distribution scheme of the elements so that the induced parallel assembly (described in section 2) will use efficiently the distribution of the matrix blocks for the EMILIO LDL^T block solver [3,4]. This distribution of the matrix blocks can be performed according to a 1D or a 2D scheme. Moreover, the informations regarding the geometrical description of the mesh are duplicated on the processors; indeed, this amount of data is negligeable with respect to the storage needed by K.

Section 3 will provide some numerical experiments on an IBM SP2 for a large class of sparse matrices coding regular and irregular meshes, including performance results and analysis. We will conclude with remarks on the benefits of this study and our future works.

2   A Parallel Assembly Induced by a Nested Dissection Strategy

2.1   Distribution of the Mesh Elements

The ordering step of our pre-processing computes a symmetric permutation of the matrix K such that its factorization will exhibit as much concurrency as

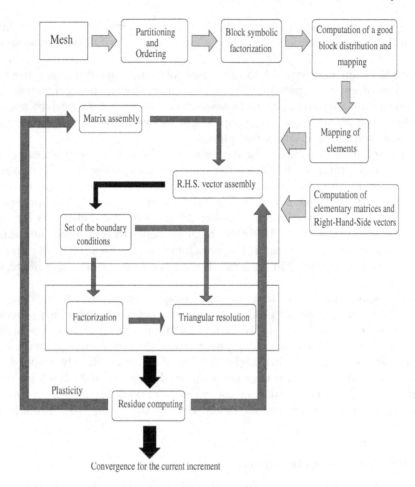

Fig. 1. Software processing chain EMILIO.

possible while incurring low fill-in. We use a tight coupling of the Nested Dissection and Approximate Minimum Degree algorithms. The *partition of the finite element graph* associated with K into supernodes is achieved by merging the partition of separators computed by the Nested Dissection algorithm and the supernodes amalgamated for each subgraph ordered by Halo Approximate Minimum Degree (see [7,6] for more details). This topological separation of the finite element graph induces a natural topological decomposition of the set of elements of the mesh into independent subdomains. The unknowns belonging to independent subgraphs that will be eliminated in parallel in the factorization process belong to elements located in independent subdomains of the mesh; the assembly of these elements can be performed in parallel too. Therefore, our distribution of elements will make use of both the mesh after the reordering, as well as the distribution of column blocks computed in the partitionning and mapping phase for the factored matrix. In this paper, we restrict our attention to a parallel assembly induced by a 1D column block distribution for the solver.

Our algorithm balances the total amount of computation in the sense that each processor must compute an equivalent number of elementary matrices. The communications corresponding to frontier elements are not taken into account for the balance, because they are supposed to be overlapped by the computations. The algorithm proceeds in two phases.

In a first phase, one builds for each element the list of the processors onto which the nodes (and so the associated columns) composing this element have been mapped by the block column distribution. If this list consists in only one processor P, the element is *totally local* to P, and *non totally local* otherwise. So, we can compute the minimum number M of totally local elements for all of the processors, and we map M totally local elements to each processor. Then, we store the remaining elements (the remaining totally local elements and the non totally local elements) in an array T with the number of owner processors for each of them.

In the second phase, we first sort this array T by decreasing order according to the number of owner processors; then, we determine on which processor each element must be mapped, favouring first the processor owning the highest number of nodes belonging to the element (locality criterion of external contributions), and second according to the number of elements already mapped. For the local elements of T which were not mapped during the first phase, we try to optimize in addition the locality of the communications between the processor that computes the elementary matrix and the one that owns the column blocks to update.

2.2   Parallel Assembly Algorithm

The parallel assembly uses the distribution of the matrix column blocks and the distribution of the elements of the mesh. The latter not only defines the list of the elements whose elementary matrices will be computed on each processor but also some additional information for communication (the sizes of the messages to send, the number of the messages to receive...).

For each processor, we begin by processing the non totally local elements which have been mapped on it. Before computing the elementary matrix of a non local element, we see if one or several messages are not waiting for reception. In that case, we make a blocking receive (we agree to receive at most a number N of messages to schedule the computations and the reception activities of the processor), and we carry out, according to the contents of the messages, the addition of contributions in the blocks of the matrix; otherwise, we compute the elementary matrix of a non totally local element, we prepare the messages to send to the processors involved, and we process the addition of local contributions, if there are any.

When the processing of the "non totally local elements" is complete, we then process the totally local elements, that is to say we compute their elementary matrices, and then add the contributions in the local blocks. This step is fully parallel, without any communication. However, before calculating one such elementary matrix, we test if a message containing local update is still waiting, and if it is the case, it is processed first.

This strategy leads to a complete *overlap of the communications* induced by the treatment of non totally local elements by the computations due to the totally local elements. We use a similar strategy for the right-hand-side vector assembly.

3   Numerical Experiments and Analysis

The results we present in this section were obtained on an 192 node IBM SP2 at CINES (Centre Informatique National de l'Enseignement Supérieur, Montpellier, France), whose nodes are 120 MHz Power2SC thin nodes (480 MFlops peak performance) having 256 MBytes of physical memory each. Switch interrupts are enabled with default delay to perform non-blocking communication efficiently.

These experiments are performed on a collection of regular and irregular meshes : a 3D 39 × 39 × 39 cube (1 DOF per node) and some irregular meshes from CEA/CESTA problems (3 DOFs per node).

Table 1 gives the characteristics of these test meshes, the dimension of the associated matrices K, and the time used by the algorithm to distribute the elements on a number of processors varying from 4 to 64. This sequential time on one processor grows linearly with the number of elements and very slowly (logarithmically) with the number of processors; therefore, our distribution algorithm is very little time consuming.

Table 1. Characteristics of the test meshes.

Name	Elements	Nodes	Columns	Processors				
				4	8	16	32	64
COL30	1260	6791	20373	0.014	0.022	0.025	0.032	0.046
COL75	3150	16736	50208	0.037	0.040	0.052	0.074	0.112
CUBE39	54872	59319	59319	0.281	0.332	0.491	0.914	1.553
BILLES	5184	5539	16617	0.035	0.044	0.066	0.103	0.152
CEL3DM	14400	10360	31080	0.107	0.142	0.216	0.249	0.401

Table 2. Assembly performance results (time in seconds) on the IBM SP2.

Name	Processors					Elt time
	4	8	16	32	64	
COL30	3.50	1.93	0.98	0.56	0.32	0.01s
COL75	8.79	4.64	2.38	1.36	0.80	-
CUBE39	139.52	69.71	35.15	17.89	9.02	-
BILLES	13.40	6.72	3.45	1.86	1.09	-
CEL3DM	37.32	18.72	9.55	4.92	2.72	-
COL30	16.56	8.55	4.58	2.31	1.24	0.05
COL75	41.13	20.72	10.67	5.88	3.01	-
CUBE39	690.42	344.44	172.55	86.60	43.47	-
BILLES	65.30	32.72	16.45	8.40	4.46	-
CEL3DM	181.89	91.29	45.64	22.93	11.49	-
COL30	34.42	16.77	8.43	4.28	2.23	0.1s
COL75	80.90	40.60	20.67	10.68	5.48	-
CUBE39	1374.69	687.19	344.06	172.10	87.41	-
BILLES	130.34	65.22	32.91	16.56	8.32	-
CEL3DM	362.83	181.81	90.91	45.80	22.86	-

Table 2 presents the execution times of parallel assembly with 4 to 64 processors for three elementary matrix computation times : 0.01 s, 0.05 s and 0.1 s. These times come from profiled real applications (3D Cologne Challenge for example). On both regular (CUBE39) and irregular problems (test cases from CEA), the results achieved by our parallel algorithm are very scalable; moreover, varying elementary matrix computation time has no impact on this scalability. This confirms the quality of our element distribution and the fact that our parallel assembly implementation fully overlaps communications due to non totally local elements by the computation of the elementary matrices of totally local elements. Indeed, we can see on figure 2 that communications (leftmost part of the charts) are all better overlapped by local computations (rightmost part of the charts) and even more that the elementary matrix computation time is important.

4   Conclusion and Prospects

This paper presents promising results on the parallelization of the assembly of a matrix in a structural computational code. We can now assemble matrices with a great number of DOFs in a acceptable CPU time. This work regarding the parallel assembly is still in progress; we are currently developing a more general strategy corresponding to a 2D block distribution of the matrix and studying the problem of non-linear plasticity which requires us to redistribute some elements during the increments according to their computation cost.

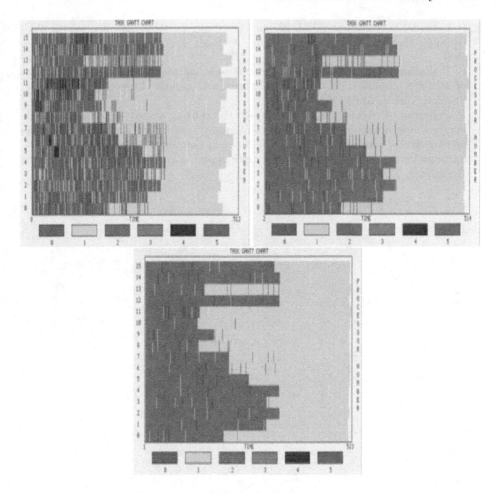

Fig. 2. Gantt charts for COLOGB75; times = 0.01 s, 0.05 s and 0.1 s.

References

1. P. Charrier and J. Roman. Algorithmique et calculs de complexité pour un solveur de type dissections emboîtées. *Numerische Mathematik*, 55:463–476, 1989.
2. D. Goudin, P. Hénon, F. Pellegrini, P. Ramet, J. Roman, and J.-J. Pesqué. Parallel sparse linear algebra and application to structural mechanics. *accepted in Numerical Algorithms, Baltzer Science Publisher*, 2000, to appear.
3. P. Hénon, P. Ramet, and J. Roman. A Mapping and Scheduling Algorithm for Parallel Sparse Fan-In Numerical Factorization. In *Proceedings of EuroPAR'99*, number 1685 in Lecture Notes in Computer Science, pages 1059–1067. Springer Verlag, 1999.
4. P. Henon, P. Ramet, and J. Roman. PaStiX: A Parallel Sparse Direct Solver Based on a Static Scheduling for Mixed 1D/2D Block Distributions. In *Proccedings of Irregular'2000*, number 1800 in Lecture Notes in Computer Science, pages 519–525. Springer Verlag, May 2000.

5. P. Laborde, B. Toson, and J.-J. Pesqué. On the consistent tangent operator algorithm for thermo-pastic problems. *Comp. Methods Appl. Mech. Eng.*, 146:215–230, 1997.
6. F. Pellegrini and J. Roman. Sparse matrix ordering with SCOTCH. In *Proceedings of HPCN'97*, number 1225 in Lecture Notes in Computer Science, pages 370–378. Springer Verlag, April 1997.
7. F. Pellegrini, J. Roman, and P. Amestoy. Hybridizing nested dissection and halo approximate minimum degree for efficient sparse matrix ordering. In *Proceedings of Irregular'99*, number 1586 in Lecture Notes in Computer Science, pages 986–995. Springer Verlag, April 1999. *Extended paper appeared in Concurrency: Practice and Experience, 12:69-84, 2000.*

MUMPS: A General Purpose Distributed Memory Sparse Solver[*]

Patrick R. Amestoy[1], Iain S. Duff[2], Jean-Yves L'Excellent[3], and Jacko Koster[4]

[1] ENSEEIHT-IRIT, 2 rue Camichel, 31071 Toulouse cedex, France, and NERSC,
Lawrence Berkeley National Laboratory, 1 Cyclotron Road, Berkeley CA 94720
amestoy@enseeiht.fr
[2] Rutherford Appleton Laboratory, Chilton, Didcot, Oxon, OX11 0QX England,
and CERFACS, Toulouse, France
I.Duff@rl.ac.uk
[3] NAG LTD, Wilkinson House, Jordan Hill Road, Oxford, OX2 8DR England
jeanyves@nag.co.uk
[4] Parallab, University of Bergen, 5020 Bergen, Norway
jak@ii.uib.no

Abstract. MUMPS is a public domain software package for the multifrontal solution of large sparse linear systems on distributed memory computers. The matrices can be symmetric positive definite, general symmetric, or unsymmetric, and possibly rank deficient. MUMPS exploits parallelism coming from the sparsity in the matrix and parallelism available for dense matrices. Additionally, large computational tasks are divided into smaller subtasks to enhance parallelism. MUMPS uses a distributed dynamic scheduling technique that allows numerical pivoting and the migration of computational tasks to lightly loaded processors. Asynchronous communication is used to overlap communication with computation. In this paper, we report on recently integrated features and illustrate the present performance of the solver on an SGI Origin 2000 and a CRAY T3E.

1 Introduction

MUMPS is a MUltifrontal Massively Parallel Solver [2,3] that is capable of solving systems of the form $\mathbf{Ax} = \mathbf{b}$, where $\mathbf{A}$ is an $n \times n$ symmetric positive definite, general symmetric, or unsymmetric sparse matrix that is possibly rank deficient, $\mathbf{b}$ is the right-hand side vector, and $\mathbf{x}$ is the solution vector to be computed.

The development of the MUMPS software package started as part of the European project PARASOL. Although this project finished in June 1999, the functionality of MUMPS is still being extended and its performance improved. It has been fully integrated in the parallel domain decomposition solver DD that is being developed by Parallab (Bergen, Norway) and that is presented elsewhere in this volume [6].

Several aspects of the algorithms used in MUMPS combine to give us a package which is unique amongst sparse direct solvers. These include:

---

[*] This work has been partially supported by the PARASOL Project (EU ESPRIT IV LTR Project 20160).

T. Sørevik et al. (Eds.): PARA 2000, LNCS 1947, pp. 121–130, 2001.

- partial threshold pivoting (to control the growth in the factors) during the numerical factorization of unsymmetric and general symmetric matrices,
- the ability to automatically adapt to processor load variations during the numerical phase,
- high performance, by exploiting the independence of computations due to sparsity and that available for dense matrices, and
- the capability of solving a wide range of problems, using either an **LU** or **LDL**^{T} factorization.

To address all these factors, MUMPS is a fully asynchronous algorithm with dynamic data structures and a distributed dynamic scheduling of tasks. Other available codes are usually based on a static mapping of the tasks and data and, for example, do not allow task migration during numerical factorization.

Besides these features, the current version of the MUMPS package provides a large range of options, including the possibility of inputting the matrix in assembled format either on a single processor or distributed over the processors. Additionally, the matrix can be input in elemental format (currently only on one processor). MUMPS can also determine the rank and a null-space basis for rank-deficient matrices, and can return a Schur complement matrix. It contains classical pre- and postprocessing facilities; for example, matrix scaling, iterative refinement, and error analysis.

In Section 2, we introduce the main characteristics of the algorithms used within MUMPS and discuss the sources of parallelism. In Section 3, we discuss the performance on an SGI Origin 2000 and a CRAY T3E. We analyse the effect of new algorithmic aspects to improve the scalability and present preliminary comparisons with some other sparse direct solvers.

2   Distributed Memory Multifrontal Factorization

We refer the reader to the literature (e.g., [7,10]) for an introduction to multifrontal methods. Briefly, the multifrontal factorization of a sparse matrix can be described by an **assembly tree**, where each node corresponds to the computation of a Schur complement matrix, and each edge represents the transfer of this matrix (the contribution block) from the child node to the parent node (or father) in the tree. This parent node assembles (or sums) the contribution blocks from all its child nodes with entries from the original matrix.

Because of numerical pivoting, it is possible that some variables cannot be eliminated from a frontal matrix. The rows and columns that correspond to such variables are added to the contribution block that is sent to the parent node resulting in a larger than predicted frontal matrix. In general this introduces additional (numerical) fill-in in the factors and requires the use of dynamic data structures.

The parallel code solves the system $\mathbf{Ax} = \mathbf{b}$ in three steps:

1. **Analysis.** One of the processors (the host) computes an approximate minimum degree ordering based on the symmetrized matrix pattern $\mathbf{A} + \mathbf{A}^T$ and carries out the symbolic factorization. The ordering can also be provided by the user. The host computes a mapping of the nodes of the assembly tree to the processors and sends symbolic information to the other processes, each of which estimates the work space required for its part of the factorization and solution.

2. **Factorization.** The original matrix is first preprocessed and distributed (or redistributed) to the processes. Each process allocates space for contribution blocks and factors. The numerical factorization on each frontal matrix is performed by a process determined by the analysis phase and potentially one or more other processes that are determined dynamically (see below).

3. **Solution.** The right-hand side vector $\mathbf{b}$ is broadcast from the host to the other processes. They compute the solution vector $\mathbf{x}$ using the distributed factors computed during the factorization phase. The solution vector is then assembled on the host.

Operations at a pair of nodes in the assembly tree where neither is an ancestor of the other are independent. This makes it possible to obtain parallelism from the tree (so-called **tree parallelism**). To reduce communication between processes during the factorization and solution phases, the analysis phase computes a mapping that assigns a complete subtree of the assembly tree to a single process.

Tree parallelism is poor near the root of the tree. It is thus necessary to obtain further parallelism within the large nodes near the root. Therefore, a second level of parallelism (**node parallelism**) is introduced by using a one-dimensional row block partitioning of the frontal matrices for (type 2) nodes near the root that have a large contribution block (see Figure 1). The process (master) that was assigned such a node during the analysis phase, determines a set of (slave) processors dynamically, based on work load estimates that the processes broadcast regularly. The (fully summed) rows of the frontal matrix that contain the potential pivots are assigned to the master process, while the rest of the matrix is partitioned over the slave processes. The master and slave processes then use parallel blocked versions of higher Level BLAS routines for the factorization of the frontal matrix. In case the number of fully summed rows is large relative to the size of the contribution block, MUMPS will split (during the analysis phase) the corresponding tree node into several smaller nodes to avoid the master process from having too much work in the factorization of this node.

Finally, if the frontal matrix of the (type 3) root node is large enough, this matrix is also factorized in parallel. MUMPS uses a two-dimensional block cyclic distribution of the root matrix and subroutines from ScaLAPACK or the vendor equivalent for the actual factorization.

MUMPS allows the input matrix to be held initially on one processor, but also allows the input matrix to be already distributed over the processors. This gives more freedom to the user, but also reduces the time for redistribution of data, because MUMPS can use asynchronous all-to-all (instead of one-to-all)

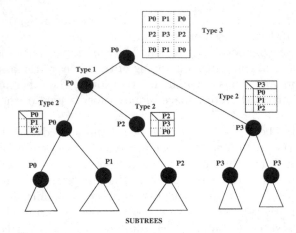

Fig. 1. Example of a (condensed) assembly tree where the leaves represent subtrees with a possible distribution of the computations over the four processes **P0**, **P1**, **P2**, and **P3**.

communications. Additionally, MUMPS can format the input data and sort tasks in parallel. Furthermore, we can expect to solve larger problems since storing the complete matrix on one processor limits the size of the problem that can be solved on a distributed memory computer.

3   Performance Analysis

MUMPS is a portable code that has been used successfully on SGI Cray Origin 2000, IBM SP2, SGI Power Challenge and Cray T3E machines. The software is written in Fortran 90, uses MPI for message passing and the BLAS, LAPACK, BLACS, and ScaLAPACK libraries for (parallel) dense matrix operations.

3.1   Environment

In this paper we report on experiments on the SGI Origin 2000 at Parallab (2×64 R10000 processors, each running at 195 MHertz and with a peak of 400 MFlops per second) and on the Cray T3E from NERSC at Lawrence Berkeley National Laboratory (512 DEV EV-5 processors, each with a peak of 900 MFlops per second).

MUMPS (latest release 4.1) has been tested extensively on problems provided by industrial partners in the PARASOL project. The corresponding matrices are available at http://www.parallab.uib.no/parasol/. Other test problems that we have used come from the forthcoming Rutherford-Boeing collection [8] and from the EECS Department of UC Berkeley. Table 1 summarizes some characteristics of the test problems that are used in this paper. The test problems include unsymmetric, symmetric, and symmetric unassembled matrices. The structural symmetry (column "StrSym") is the ratio of the number of nonzeros that are matched by nonzeros in symmetric locations, divided by the total number of entries in the matrix. For symmetric problems, the column "Entries" represents the number of nonzeros in only the lower triangular part of the matrix.

Table 1. Test matrices.

Real Unsymmetric Assembled				
Matrix name	Order	Entries	StrSym	Origin
ECL32	51993	380415	0.93	EECS Dept. of UC Berkeley
BBMAT	38744	1771722	0.54	Rutherford-Boeing, CFD
INVEXTR1	30412	1793881	0.97	Polyflow (PARASOL)
Real Symmetric Assembled				
Matrix name	Order	Entries	StrSym	Origin
CRANKSG2	63838	7106348	1.00	MSC.Software (PARASOL)
BMWCRA_1	148770	5396386	1.00	MSC.Software (PARASOL)
HOOD	220542	5494489	1.00	INPRO (PARASOL)
BMW3_2	227362	5757996	1.00	MSC.Software (PARASOL)
INLINE_1	503712	18660027	1.00	MSC.Software (PARASOL)
LDOOR	952203	23737339	1.00	INPRO (PARASOL)
Real Symmetric Unassembled				
Matrix name	Order	Entries	elements	Origin
SHIP_003.RSE	121728	9729631	45464	Det Norske Veritas (PARASOL)
SHIPSEC5.RSE	179860	11118602	52272	Det Norske Veritas (PARASOL)

The ordering currently available inside MUMPS is the Approximate Minimum Degree ordering [1]. It has been observed (see for example [3]) that hybrid ND orderings (i.e., orderings based on nested dissection combined with local heuristics for smaller subgraphs) are usually more efficient than purely local heuristics for large problems. Figure 2 illustrates this. The figure shows that the number of floating-point operations for the factorization is significantly smaller with hybrid ND orderings. Furthermore, such orderings can lead to a better load balance in a parallel environment. Since MUMPS has the facility to use orderings that are computed externally (prior to calling MUMPS), the experiments reported in this paper use hybrid ND orderings. (Either METIS [12], a combination of SCOTCH and Halo-AMD [14], or an in-house Det Norske Veritas ordering is used).

Finally, the times reported are measured in seconds and represent elapsed times for the numerical factorization phase, normally the most costly part of the solver.

3.2 Performance on an SGI Origin 2000

Table 2 presents times for the factorization of some large symmetric PARASOL test matrices (assembled and unassembled) on up to 32 processors of the SGI

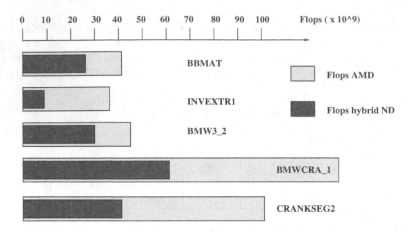

Fig. 2. Number of floating-point operations during MUMPS factorization using hybrid ND and Approximate Minimum Degree orderings.

Origin 2000. Speedups are up to 13.8. We consider the absolute GFlops rate obtained on this machine good for a sparse direct solver.

Table 2. Time (in seconds) for factorization using MUMPS Version 4 on some PARASOL test problems on an SGI Origin 2000.

| Matrix | Flops $(\times 10^9)$ | \multicolumn{6}{c}{Number of processors} |
		1	2	4	8	16	32
LDOOR	74.5	416	228	121	68	39	31 (2.4 GFlops)
BMWCRA_1	61.0	307	178	82	58	36	27 (2.3 GFlops)
BMW3_2	28.6	151	96	53	33	18	15 (1.9 GFlops)
INLINE_1	143.2	757	406	225	127	76	55 (2.6 GFlops)
SHIP_003.RSE	73.0	392	237	124	108	51	43 (1.7 GFlops)
SHIPSEC5.RSE	51.7	281	181	103	62	37	29 (1.8 GFlops)

We note that the largest problem that we have solved to date on this platform is a symmetric problem provided by the PARASOL partner MSC.Software of order 943695 with 39.3 million entries in its lower triangular part. The number of entries in the factors is 1.4×10^9 (11.1 Gbytes) and the number of floating-point operations required by the numerical factorization is 5.9×10^{12}; MUMPS required 6.2 hours on 2 processors and 17 Gbytes of memory to factorize the matrix. Note that because of the increase in the total memory requirement, we could not solve the problem on more processors.

3.3   Performance Tuning on a CRAY T3E

While MUMPS was initially developed on machines with a limited number of processors, access to the 512 processor CRAY T3E from NERSC gave us new targets
in terms of scalability.

MUMPS has a set of platform-dependent parameters (e.g., block sizes) controlling the uniprocessor and multiprocessor efficiency. First, those parameters were
experimentally tuned to meet the characteristics of the architecture. Because
of the relatively higher ratio between communication bandwidth and processor
performance, the granularity of the tasks arising from type 2 node parallelism
(see Section 2) could be reduced, thus allowing also the use of more processors.
In this context, task management became even more crucial and the dynamic
scheduling strategy was modified to allow the master of a type 2 node to choose
dynamically *not only* the set of processors to which generated tasks should be
assigned, but also the number of tasks and the size of each task. In this dynamic
situation, one rule of thumb is that a process should avoid giving work to a more
loaded process.

Table 3.   Factorization time (in seconds) for various buffer sizes of the matrix
CRANKSG2 on 8 processors of the T3E. (* denotes the default size on the T3E)

type of receive	MPI buffer size (bytes)						
	0	512	1K	4K*	64K	512K	2Mb
standard (MPI_RECV)	37.0	37.4	38.3	37.6	32.8	28.3	26.4
immediate (MPI_IRECV)	27.3	26.5	26.6	26.4	26.2	26.2	26.4

Another important modification was the introduction of immediate communication primitives (MPI_IRECV) on the receiver side, allowing transfer of data to
start earlier, as well as more overlap between computation and communication.
On the T3E, depending on the size of the MPI internal reception buffer, it can
happen that a message reaches a process before the corresponding MPI_RECV
is posted. On the other hand, if a message is too large to fit in the MPI internal
reception buffer, communication cannot start. Posting an MPI_IRECV earlier
allows communication to start as soon as possible, independently of the size of
the MPI buffer. This effect is shown in Table 3, where one can see that a significant gain is obtained for the default size of the MPI buffer (26.4 vs. 37.6
seconds).

In Table 4, MUMPS 4.0 denotes the version of the code where only uniprocessor parameters are tuned for the T3E, and MUMPS 4.1 denotes the version in
which all the modifications described above are integrated. We see in Table 4
that the new version performs significantly better, both for a symmetric and an
unsymmetric test problem.

Table 4. Times (in seconds) for factorization by MUMPS 4.0 and MUMPS 4.1 on two PARASOL test problems. ('—' denotes not enough memory)

Matrix	MUMPS	Number of processors						
		1	2	4	8	16	32	64
HOOD	4.0	—	—	25.1	12.1	6.3	3.7	3.5
$(8.2 \times 10^9$ flops)	4.1	—	—	15.0	8.2	4.4	4.0	2.4
INVEXTR1	4.0	—	23.1	18.0	15.6	14.9	14.4	14.5
$(8.1 \times 10^9$ flops)	4.1	—	23.1	11.8	8.5	7.5	7.0	6.8

3.4   Preliminary Comparisons with Other Codes

Table 5 shows a preliminary comparison between the performances of MUMPS and PSPASES (version 1.0.2) [11]. PSPASES solves symmetric positive definite systems and performs a Cholesky ($\mathbf{LL}^T$) factorization based on a static mapping of the computational tasks to the processors. (MUMPS uses an $\mathbf{LDL}^T$ factorization.) The table shows that MUMPS behaves well in absolute performance and scalability even though PSPASES is especially designed for symmetric positive definite problems. Note that on a larger number of processors (128 or more), one should expect PSPASES to more significantly outperform MUMPS for which the 1D partitioning of the frontal matrices (due to numerical pivoting) will limit the scalability.

Table 5. Time (in seconds) for factorization by MUMPS and PSPASES of two symmetric test cases (BMWCRA_1 and INLINE_1) on the SGI Origin 2000. The ordering used for both solvers is based on METIS. PSPASES requires 2^k, $k > 0$, processors.

BMWCRA_1	Flops $(\times 10^9)$	Number of processors						
		1	2	4	8	16	32	
PSPASES 1.0.2	between 62 and 71	—	157	95	66	39	21	
MUMPS 4.1	61	307	178	82	58	36	27	
INLINE_1	Flops $(\times 10^9)$	1	2	4	8	16	32	64
PSPASES 1.0.2	between 147 and 157	—	574	399	225	128	75	41
MUMPS 4.1	143	757	406	225	127	76	55	42

Table 6 shows a preliminary comparison of MUMPS and SuperLU [13] on the CRAY T3E for two unsymmetric test cases (BBMAT and ECL32). SuperLU uses a supernodal right-looking formulation with a static approach. Numerical stability issues are handled with "static pivoting" in the sense that, if a pivot is smaller than $\sqrt{\epsilon} \parallel \mathbf{A} \parallel$, the pivot is set to $\sqrt{\epsilon} \parallel \mathbf{A} \parallel$ and computations can continue exactly as forecast by the analysis. This means that SuperLU actually computes the factorization of a modified matrix. However, an accurate solution to the initial system is obtained by using (i) iterative refinement, and (ii) the MC64 code [9], used as a preprocessing step to permute large entries onto the diagonal of the matrix.

Although the results presented here are only preliminary, an extensive comparison of MUMPS and SuperLU on a CRAY T3E has now been performed in the context of a France-Berkeley funded project and results of this study are available in [4].

Table 6. Time (in seconds) for factorization by MUMPS and SuperLU for two unsymmetric matrices (BBMAT and ECL32) on the T3E. The same ND ordering is used for both solvers.

		Number of processors						
BBMAT	Flops ($\times 10^9$)	4	16	32	64	128	256	512
SuperLU	23.5	137.8	31.2	25.2	17.3	12.4	14.3	14.7
MUMPS 4.1	25.7	39.4	13.2	11.9	9.9	9.2	9.4	11.6
ECL32	Flops ($\times 10^9$)	4	16	32	64	128	256	512
SuperLU	20.7	49.0	16.7	12.0	9.9	8.8	9.9	9.5
MUMPS 4.1	20.9	24.7	9.7	7.7	6.9	7.0	7.0	8.9

In fact, the only code we are aware of which is comparable to MUMPS in terms of functionality is SPOOLES [5]. For example, it also allows numerical pivoting during the factorization. It uses an object-oriented approach and has the advantage of being very modular and flexible. However, partial experiments performed by a third party show that the performance of SPOOLES is far behind the performance of the solvers mentioned in this section.

4   Concluding Remarks

MUMPS is a general purpose distributed multifrontal solver with a wide range of functionalities. Experimental results show that the solver has good overall performance and scales reasonably well on various parallel architectures.

On-going and future work include algorithmic developments that will improve the basic performance of the code, the tuning of the code on other architectures (e.g., clusters), further comparisons with other codes, and the development of new features required by MUMPS users.

The MUMPS package is freely available (for non-commercial use) and can be obtained from http://www.enseeiht.fr/apo/MUMPS/.

References

1. P. R. Amestoy, T. A. Davis, and I. S. Duff. An approximate minimum degree ordering algorithm. *SIAM Journal on Matrix Analysis and Applications*, 17:886–905, 1996.
2. P. R. Amestoy, I. S. Duff, and J.-Y. L'Excellent. Multifrontal parallel distributed symmetric and unsymmetric solvers. *Comput. Methods in Appl. Mech. Engrg.*, 184:501–520, 2000. Special issue on domain decomposition and parallel computing.

3. P. R. Amestoy, I. S. Duff, J.-Y. L'Excellent, and J. Koster. A fully asynchronous multifrontal solver using distributed dynamic scheduling. Technical Report RAL-TR-1999-059, Rutherford Appleton Laboratory, 1999. Submitted to *SIAM Journal on Matrix Analysis and Applications*.

4. P. R. Amestoy, I. S. Duff, J.-Y. L'Excellent, and X. S. Li. Analysis, tuning and comparison of two general sparse solvers for distributed memory computers. Technical Report LBNL-45992, NERSC, Lawrence Berkeley National Laboratory, June 2000. FBF report.

5. C. Ashcraft and R. G. Grimes. SPOOLES: An object oriented sparse matrix library. In *Proceedings of the Ninth SIAM Conference on Parallel Processing*, 1999.

6. P. E. Bjørstad, J. Koster, and P. Krzyżanowski. Domain decomposition solvers for large scale industrial finite element problems. In *Proceedings of the PARA2000 Workshop on Applied Parallel Computing, June 18-21, Bergen*. Springer-Verlag, 2000.

7. I. S. Duff, A. M. Erisman, and J. K. Reid. *Direct Methods for Sparse Matrices*. Oxford University Press, London, 1986.

8. I. S. Duff, R. G. Grimes, and J. G. Lewis. The Rutherford-Boeing Sparse Matrix Collection. Technical Report RAL-TR-97-031, Rutherford Appleton Laboratory, 1997. Also Technical Report ISSTECH-97-017 from Boeing Information & Support Services and Report TR/PA/97/36 from CERFACS, Toulouse.

9. I. S. Duff and J. Koster. On algorithms for permuting large entries to the diagonal of a sparse matrix. Technical Report RAL-TR-1999-030, Rutherford Appleton Laboratory, 1999. Submitted to *SIAM Journal on Matrix Analysis and Applications*.

10. I. S. Duff and J. K. Reid. The multifrontal solution of indefinite sparse symmetric linear systems. *ACM Transactions on Mathematical Software*, 9:302–325, 1983.

11. A. Gupta, G. Karypis, and V. Kumar. Highly scalable parallel algorithms for sparse matrix factorization. *IEEE Trans. on Parallel and Distributed Systems*, 8(5):502–520, 1997.

12. G. Karypis and V. Kumar. MEΠS – *A Software Package for Partitioning Unstructured Graphs, Partitioning Meshes, and Computing Fill-Reducing Orderings of Sparse Matrices – Version 4.0*. University of Minnesota, September 1998.

13. X. S. Li and J. W. Demmel. A scalable sparse direct solver using static pivoting. In *Proceedings of the Ninth SIAM Conference on Parallel Processing for Scientific Computing*, San Antonio, Texas, March 22–24 1999.

14. F. Pellegrini, J. Roman, and P. R. Amestoy. Hybridizing nested dissection and halo approximate minimum degree for efficient sparse matrix ordering. In *Proceedings of Irregular'99, San Juan*, Lecture Notes in Computer Science 1586, pages 986–995, 1999.

Runtime Adaptation of an Iterative Linear System Solution to Distributed Environments

Masha Sosonkina

Department of Computer Science, University of Minnesota, Duluth,
320 Heller Hall, 10 University Drive, Duluth, Minnesota 55812-2496
masha@d.umn.edu

Abstract. Distributed cluster environments are becoming popular platforms for high performance computing in lieu of single-vendor supercomputers. However, the reliability and sustainable performance of a cluster are difficult to ensure since the amount of available distributed resources may vary during the application execution. To increase robustness, an application needs to have self-adaptive features that are invoked at the runtime. For a class of computationally-intensive distributed scientific applications, iterative linear system solutions, we show a benefit of the adaptations that change the amount of local computations based on the runtime performance information. A few strategies for efficient exchange of such information are discussed and tested on two cluster architectures.

1 Introduction

Distributed environments are now widely used for computationally-intensive scientific applications. However, efficiency and robustness are still difficult to attain in such environments due to the varying distributed resource availability at any given time. For example, the "best effort" interconnection networks have no mechanism to satisfy application communication requirements all the time. Thus an application needs to have its own adaptive mechanisms. Applications that adjust their quality of service (i.e., the computation and communication demands) to the state of interconnecting network and computing nodes are quite common in multimedia already. In scientific computing, however, the concept of self-adaptation is rather new. For scientific applications with adaptation features, distributed environments may win a performance battle over single-vendor supercomputers, since adaptation makes scientific computing more robust and fault-tolerant.

Linear system solution is usually the most computationally-expensive part of many high-performance computing applications. Thus focusing on its adaptive features will affect significantly the overall performance of applications. Large-scale sparse linear systems are often solved using *iterative solution* techniques, which find an approximate solution given a desired accuracy. These techniques have high degree of parallelism and are easy to implement. Here, we propose a few adaptation strategies for an iterative linear system solution in distributed

T. Sørevik et al. (Eds.): PARA 2000, LNCS 1947, pp. 131–139, 2001.
© Springer-Verlag Berlin Heidelberg 2001

environments. The paper is organized as follows. Section 2 outlines a general framework of a parallel iterative linear system solution method. Next, we describe a few strategies for obtaining run-time information about the system performance as measured by elapsed time. In section 3, numerical experiments are provided for different cluster environments. We summarize the work in Section 4.

2   Distributed Iterative Linear System Solution with Adaptation Features

An iterative solution method can be easily implemented in parallel, yielding a high degree of parallelism. Consider, for example, a parallel implementation of FGMRES [4], a variation of a popular solution method, restarted Generalized Minimum RESidual algorithm (GMRES) [3]. If the classical Gram-Schmidt procedure is used in its orthogonalization phase, an iteration of the parallel algorithm has only two synchronization points, in which all-to-all processor communications are incurred. A drawback of iterative methods is that it is not easy to predict how fast a linear system can be solved to a certain accuracy and whether it can be solved at all by certain types of iterative solvers. This depends on the algebraic properties of the matrix. To accelerate the convergence of an iterative method, a linear system can be transformed into one that has the same solution but for which the iteration converges faster. This transformation process is called preconditioning. With a good preconditioner, the total number of steps required for convergence can be reduced dramatically, at the cost of a slight increase in the number of operations per step, resulting in much more efficient algorithms. In distributed environments, an additional benefit of preconditioning is that it reduces the parallel overhead, and thus decreases the total parallel execution time.

2.1   Distributed Matrix Representation and Block-Jacobi Preconditioning

One way to partition the linear system $Ax = b$ is to assign certain equations and corresponding unknowns to each processor. For a graph representation of sparse matrix, graph partitioner may be used to select particular subsets of equation-unknown pairs (subproblems) to minimize the amount of communication and to produce subproblems of almost equal size. It is common to distinguish three types of unknowns: (1) Interior unknowns that are coupled only with local equations; (2) Local interface unknowns that are coupled with both non-local (external) and local equations; and (3) External interface unknowns that belong to other subproblems and are coupled with local equations. Thus each local vector of unknowns x_i is reordered such that its subvector u_i of internal components is followed by the subvector y_i of local interface components. The right-hand side b_i is conformly split into the subvectors f_i and g_i , i.e.,

$$x_i = \begin{pmatrix} u_i \\ y_i \end{pmatrix} \; ; \quad b_i = \begin{pmatrix} f_i \\ g_i \end{pmatrix} .$$

When block-partitioned according to this splitting, the local matrix A_i residing in processor i has the form

$$A_i = \left(\begin{array}{c|c} B_i & F_i \\ \hline E_i & C_i \end{array} \right), \qquad (1)$$

so the local equations can be written as follows:

$$\begin{pmatrix} B_i & F_i \\ E_i & C_i \end{pmatrix} \begin{pmatrix} u_i \\ y_i \end{pmatrix} + \begin{pmatrix} 0 \\ \sum_{j \in N_i} E_{ij} y_j \end{pmatrix} = \begin{pmatrix} f_i \\ g_i \end{pmatrix}. \qquad (2)$$

Here, N_i is the set of indices for subproblems that are neighbors to the subproblem i. The term $E_{ij}y_j$ reflects the contribution to the local equation from the neighboring subproblem j. The result of the multiplication with external interface components affects only the local interface unknowns, which is indicated by zero in the top part of the second term of the left-hand side of (2).

The simplest of distributed preconditioners is the block-Jacobi procedure (see, e.g., [3] and the references therein). This form of a block-Jacobi iteration, in which blocks refer to the subproblems, is sketched next.

ALGORITHM 1 *Block Jacobi Iteration:*
1. *Obtain external data $y_{i,ext}$*
2. *Compute (update) local residual $r_i = (b - Ax)_i = b_i - A_i x_i - \sum_{j \in N_i} E_{ij} y_j$*
3. *Solve $A_i \delta_i = r_i$*
4. *Update solution $x_i = x_i + \delta_i$.*

The required communication, as well as the overall structure of the routine, is identical to that of a matrix-vector multiplication. In distributed environments, the block-Jacobi preconditioner is quite attractive since it incurs no extra communication and becomes less expensive with increase in processor numbers. However, it is well-known that, in this case, the effect of block-Jacobi on convergence deteriorates requiring more iterations to converge.

2.2 Adaptive Strategies

For the local preconditioning strategies, such as block-Jacobi, the amount of work each processor accomplishes in the preconditioning application is different and depends on the properties of the local submatrices. Since the properties of the local submatrices may vary greatly, the times of the preconditioning phase may also differ substantially leading to a load imbalance among processors. Load imbalance may also occur under the *unequal resource* conditions in one or more processors. We assume that the unequal resource condition arises when nodes differ in computational power either due to to their hardware configuration or due to competing loads that consume a part of node resources. For unbalanced local loads, when the processor synchronizations take place (in the orthogonalization phase of FGMRES and during the matrix-vector product), the processors with small preconditioning workload must wait for the other processors. One way to avoid this idling is to force all the processors to spend the same time in the preconditioning application. The time may be fixed, for example, based on the time required by the processor with the largest workload to apply a preconditioning step. The rationale is that it is better to spend the time, which would be wasted otherwise, to perform more iterations in the "faster" processors. A better accuracy may be achieved in "faster" processors which would eventually propagate to others, resulting in a reduction of the number of iterations to converge.

There are several approaches to control the "fixed-time" condition for a block-Jacobi preconditioning step (Algorithm 1) when an iterative process (e.g., GMRES) is used to solve the linear system in line 3. One of these approaches (tested in [5]) is to change locally the number of inner GMRES iterations at a certain (outer) iteration of FGMRES based on some criterion. The following iteration adjustment parameters have been determined experimentally and applied after each preconditioning step in processor i, $(i = 1, \ldots, p)$:

$$\text{if } (\varDelta_j^i > n_{j-1}^i/3) \quad n_j^i = n_{j-1}^i + \varDelta_j^i,$$

where n_j^i is the number of the inner iterations in the (next) jth iteration of FGMRES; $\varDelta_j^i$ is the number of iterations that processor i can fit into the time to be wasted in idling otherwise at the jth outer iteration of FGMRES. Specifically,

$$\varDelta_j^i = \frac{(T_{\max} - T^i)n^i}{T^i},$$

where $T_{\max}$ is the maximum time among all the processors and T^i is the time for processor i to perform preconditioning operations during $j - 1$ previous outer iterations; n^i is the total number of preconditioning operations performed by processor i so far. The number of inner iterations n_j^i can be updated provided that the limit $n_{\lim}$ on the number of inner iterations is not reached.

The maximum time $T_{\max}$ has been obtained by an all-to-all communication required by the global maximum computation. However, for the distributed environments in which the interconnecting network has large communication latency, obtaining $T_{\max}$ may incur significant parallel overhead. It is, therefore, desirable to use already existing communication points to exchange the timing information, so that no separate message is issued for sharing the performance information.

First, we explain how the communication during a distributed matrix-vector multiplication may be exploited. The value of T^i is appended to the message containing the local interface unknowns sent to a neighbor. Upon receipt of all $T^k(k \in N_i)$, processor i determines locally the maximum $T'_{\max}$ over all T^k and uses it in adaptation decision for the jth iteration. A disadvantage of this strategy is that the information is exchanged only among neighbors, so there is no knowledge of the global $T_{\max}$. To alleviate this drawback, the computation of $\varDelta_j^i$ has been modified to use the information only from the previous $(j - 1)$st iteration.

Second, a strategy that finds the global maximum time can be designed such that no extra communication takes place. An all-to-all communication of the FGMRES orthogonalization phase may be used to compute and disseminate global T_{max}. Since in the distributed orthogonalization, a global sum is computed, we need to introduce a "mixed" reduction operation that determines a vector (sum, max). Message Passing Interface [1], which we use as a communication library, permits such a mixed user-defined reduction operation.

3   Numerical Experiments

The experiments have been performed on the two types of distributed architectures: a cluster of 8 nodes with two Power3 (200MHz) processors connected via Gigabit Ethernet and a cluster of 64 nodes, in a Single PentiumPro (200MHz) processor mode, connected via Fast Ethernet. Each node of Power3 and Pentium-Pro clusters has 1 GB and 256 MB of main memory, respectively. Both clusters are built in Ames Laboratory (Ames, IA). The test problem AF23560 comes from the Davis collection [2]. This problem is unstructured, has 23,560 unknowns and 484,256 nonzeros in the matrix. In the preconditioning phase, block-Jacobi is used such that the linear system solve is handled by ILUT-preconditioned GM-RES(20) with the following parameters: 30 fill in elements per row in the ILUT factorization [3], $n_0^i = 5$, $n_{lim} = 20$, and the relative accuracy of 10^{-2}. This accuracy is increased to 10^{-8} whenever n_j^i is adapted.

Figure 1 compares the time (top) and iteration (bottom), for the following implementations: standard block-Jacobi (Jac) and two versions of adaptive block-Jacobi (Jac_global_1 and Jac_local). Jac_global_1 uses a separate global communication to gather performance information and Jac_local exchanges the timing information among the neighbors only. It can be seen that both adaptive strategies are more robust than the standard algorithm since they converge when standard block-Jacobi fails (for 8 and 10 processors). With increase in processor numbers, the communication cost starts to dominate the execution time. Thus the adaptive version without extra communication point becomes beneficial. However, the neighbor-only balancing strategy is inferior to the global one so the performance of Jac_local is hindered.

Next, we monitor the performance of the proposed adaptive versions under the unequal resource conditions. In certain cluster nodes, competing loads are introduced such that they consume a significant (constant) amount of main memory. We choose to focus on memory resources since the performance of scientific applications is often memory-bound for large-scale problems.

The test problem and solution parameters are taken as in the previous set of experiments. The tests have been performed on four processors. A simple program consuming a constant amount (512 MB) of memory (denoted as load L1) was run on *node0* of the Power3 cluster together with the distributed linear system solution the mapping of which included *node0*. Similarly, a constant amount of memory equal to 150 MB (denoted as load L2) was consumed on *node0* of the PentiumPro cluster while a parallel linear system solution was running

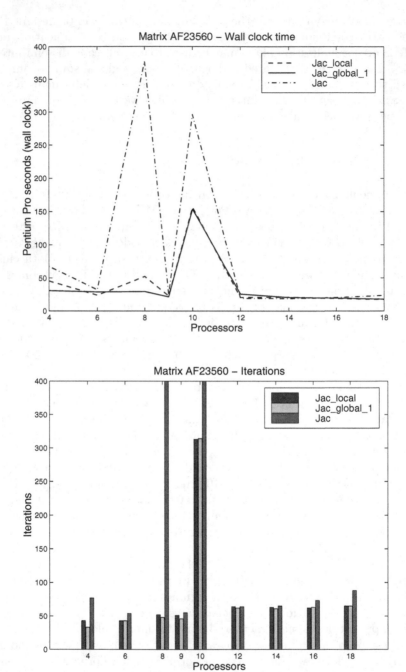

Fig. 1. Time (top) and iteration (bottom) comparisons of three variations of block-Jacobi for subproblems of varying difficulty on increasing processor numbers

on a set of four nodes including *node0*. For the experiments on both clusters, Table 1 provides the number of the outer iterations `outer_it` until convergence and the total solution time `total_time` for the following variations of the solution process: standard block-Jacobi `Jac` with the number of inner iterations $n^i{}_0 = 5$, $n^i{}_0 = n_{\lim} = 20$, and with unequal subproblem sizes `ne_part` (for $n^i{}_0 = 5$); `Jac_local`; `Jac_global_1`; and (for Power3 only) `Jac_global_2`, an adaptive version of block-Jacobi that exchanges performance information via a global communication in the FGMRES orthogonalization phase. To partition a problem into unequal subproblems (the `ne_part` variation of standard block-Jacobi), a graph partitioner was modified such that it takes as an input a vector of relative sizes for each of the subproblems and attempts to produce partitions of corresponding sizes.

We observe that, in general, adapting to the unequal memory condition, either by an *a priori* matching of partition size to the extra loads or by runtime iteration adjustment, improves performance. A drawback of an *a priori* size selection, however, is that it is difficult to choose the relative partition sizes to reflect the unequal resource availability, which may change dynamically. Standard block-Jacobi with $n^i{}_0 = n_{\lim}$ appears to be quite competitive on Power3 processors, since the cost per preconditioner (inner) iteration does not increase much going from 5 to 20 iterations. All adaptive variations show comparable performance. `Jac_local` outperforms other variations on the Power3 cluster. This can be explained by the partition connectivity (Figure 2) of the problem and the dual-processor architecture of the Power3 cluster nodes. Similar situation can be seen in Table 2, which, in the column '2 `nodes` / 2 L2', gives the timing and iteration results on the PentiumPro cluster when both *node0* and *node1* have load L2. The column '2 `nodes` / 3 L2' of Table 2 shows the case when *node0* has load L2 and *node1* has 2×L2. A global strategy `Jac_global_1` appears to react better than the local strategy to varying imbalance conditions.

4   Conclusions

We have showed that parallel iterative linear system solution may need to adapt itself when the distributed resources, such as memory, are not equal among nodes in a distributed environment. Block-Jacobi preconditioning has been taken as an example to dynamically balance unequal resources with the amount of local work per processor. The dynamic adjustment of local iterations reduces the imbalances due to different resource availability and subproblem difficulty, which makes the linear system more robust and decreases the number of iterations to convergence. We observed that an *a priori* adjustment of subdomain sizes is not a viable alternative since both the environment and algorithm performances may vary dynamically. It has been shown that the runtime performance information exchange requiring no separate communications is advantageous in distributed memory environments that have high-latency interconnects.

Table 1. The performance of block-Jacobi variations when one cluster node has amount of memory depleted

Method		Power3		PentiumPro	
		outer_it	time_it	outer_it	time_it
Jac	$n^i_0 = 5$	77	87.47	77	51.75
	$n^i_0 = 20$	30	48.98	30	62.56
	ne_part	74	57.28	83	54.55
Jacobi_local		33	42.33	32	28.82
Jacobi_global_1		33	55.39	38	26.29
Jacobi_global_2		32	54.07		

Table 2. The performance of block-Jacobi variations with increasing memory imbalance in the nodes of the PentiumPro cluster

Method	2 nodes / 2 L2		2 nodes / 3 L2	
	outer_it	time_it	outer_it	time_it
Jac $n^i_0 = 5$	77	66.67	77	99.56
Jacobi_local	33	35.54	32	49.21
Jacobi_global_1	35	38.11	33	48.72

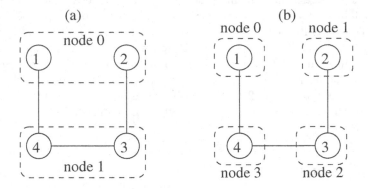

Fig. 2. Partition connectivity and mapping for the test problem solved in four processors of the Power3 (a) and PentiumPro (b) clusters

References

1. MPI Forum. MPI: A message-passing standard. *Intl. J. Supercomut. Applic.*, 8, 1994.
2. T. Davis. University of florida sparse matrix collection. *NA Digest*, 1997. http://www.cise.ufl.edu/~davis/sparse.
3. Y. Saad. *Iterative Methods for Sparse Linear Systems*. PWS publishing, New York, 1996.

4. Y. Saad and A. Malevsky. PSPARSLIB: A portable library of distributed memory sparse iterative solvers. In V. E. Malyshkin et al., editor, *Proceedings of Parallel Computing Technologies (PaCT-95), 3-rd international conference, St. Petersburg, Russia, Sept. 1995*, 1995.
5. Y. Saad and M. Sosonkina. Non-standard parallel solution strategies for distributed sparse linear systems. In A. Uhl P. Zinterhof, M. Vajtersic, editor, *Parallel Computation: Proc. of ACPC'99*, Lecture Notes in Computer Science, pages 13–27, Berlin, 1999. Springer-Verlag.

A Local Refinement Algorithm for Data Partitioning

Jarmo Rantakokko

Uppsala University, Information Technology, Dept. of Scientific Computing
Box 120, SE-751 04 Uppsala, Sweden
jarmo@tdb.uu.se

Abstract. A local refinement method for data partitioning has been constructed. The method balances the workload and minimizes locally the number of edge-cuts. The arithmetic complexity of the algorithm is low. The method is well suited for refinement in multilevel partitioning where the intermediate partitions are near optimal but slightly unbalanced. It is also useful for improvement of global partitioning methods and repartitioning in dynamic problems where the workload changes slightly. The algorithm has been compared with corresponding methods in *Chaco* and *Metis* in the context of multilevel partitioning. The cost of carrying out the partitioning with our method is lower than in *Chaco* and of the same order as in *Metis*, and still the quality of the partitioning is comparable and in some cases even better.

1 Introduction

The partitioning of a connected set of data is a primary issue in parallel computing. For a good parallel efficiency the data set must be partitioned in equally weighted parts with minimal dependencies between the partitions. In scientific computing with partial differential equations the data set corresponds to the computational grid, i.e. the discretized computational domain. The individual data items can be the grid points as in unstructured grid applications or they can be large data blocks as in domain decomposition applications [8]. For real three-dimensional problems, for example in flow calculations, the data sets can be very large. An optimal data partitioning can not be found with an exhaustive search, it is too expensive. A number of heuristics have been developed based on graph partitioning. A graph, corresponding to the data set and the data connections, is partitioned with an approximate solution method and a near optimal partitioning is found. A popular but rather complex method is the *Recursive Spectral Bisection* method. This method usually gives very good partitions but is computationally demanding. Global partitioning methods such as the Recursive Spectral Bisection method often give suboptimal partitions in the fine details. The partitioning can then be improved with a local refinement algorithm. One such algorithm that has been found to give good results is the *Kernighan–Lin/Fiduccia–Mattheyses* algorithm [1]. The algorithm can be used

T. Sørevik et al. (Eds.): PARA 2000, LNCS 1947, pp. 140–148, 2001.

to refine the partitions in the bisection steps of the Recursive Spectral Bisection method. Local refinement is also used in multilevel partitioning schemes [3, 6]. With a multilevel scheme the complexity of the partitioning can be reduced considerably. A multilevel scheme coarsens the graph in a number of steps, the coarsest graph is then partitioned and the graph is uncoarsened. In each uncoarsening step the partitioning is refined locally. Moreover, local refinement is also useful in dynamic partitioning. If the work load changes during the computations data must be repartitioned dynamically. Usually the work load changes are relatively small and localized. Then it is sufficient to refine the previous partitioning and not do a complete repartitioning.

We have developed a local refinement algorithm with a low complexity that both minimizes the number of edge-cuts and balances the arithmetic work load. The algorithm is described in Section 2 and experimental results with comparisons to related methods can be found in Section 3.

2   The Refinement Algorithm

The *Kernighan–Lin/Fiduccia–Mattheyses* refinement algorithm is currently considered as the state of the art. However, this algorithm is relatively time consuming. The KL/FM algorithm iterates until no better partition is found. In each iteration a loop is performed where vertices are swapped between two partitions. Preventing an infinite loop, the vertices are immediately looked in their new positions for the remainder of the pass. After each swap, the resulting partition is evaluated and the best one is recorded to be a new initial partition for the next iteration. The complexity of the inner loop is of order $O(E)$, where E is the number of edges in the adjacency graph, and the algorithm usually converges in a small number of iterations for a good initial partitioning. The KL/FM algorithm can be generalized for multipartition refinement [2]. The complexity then becomes $O(p^2 \cdot E)$, where p is the number of partitions. For a large number of partitions and vertices the partitioning work becomes high with this algorithm. In [5] a related local refinement algorithm, *Greedy Refinement*, with a low complexity is given. Here, only boundary vertices are considered and moved if they result in a better partitioning. Experimental results in [5] show that the simpler algorithm produces partitionings that have similar edge-cuts as the more complex variants of the KL/FM algorithm but with significantly less work. A drawback of the Greedy Refinement method is that it has no lookahead properties and can not climb out of local minima. This becomes a problem if the initial partitioning has a bad load balance but a low number of edge-cuts. Then, the algorithm is not able to smooth out the load balance and the initial partitioning is more or less retained, see Figure 1.

We have developed a variant of the algorithm that also considers the load balance without significantly increasing the complexity of the algorithm. Our refinement algorithm has two Greedy-like steps, see Figure 2. In the first pass, all gains of the boundary points are put into a list. The gain of a vertex v to

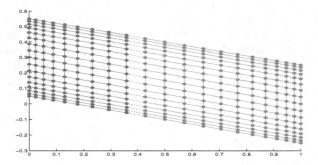

Fig. 1. Initial partitioning with a minimum number of edge-cuts but with a poor load balance. The Greedy Refinement method will fail to improve the partitioning. All possible moves results in negative gains, thus no vertices are moved.

move from partition A to partition B, (A and B neighbors), is defined as

$$g(v, A, B) = \sum_{w \in B} E_w - \sum_{w \in A} I_w,$$

where E_w (external degree) is the edge-weight between vertex v in partition A and vertex w in partition B. I_w (internal degree) is the edge-weight between vertex v and vertex w, both in partition A. This list of gains is used for selection of vertices. The algorithm iterates by moving vertices. In each iteration the best possible vertex move in terms of edge-cut and load balance is selected. The edge-cut gain is allowed to be negative if the load balance can be improved. The selected vertex is moved between partitions and then removed from the list, preventing repeated movement of the same vertex. All adjacent vertices and the list of gains are also updated. A removed vertex can then in succeeding iterations be reinserted in the list if one its neighbors is moved. The algorithm iterates until no further improvements are made. In contrast to the KL/FM algorithm no outer iterations are made. The second pass in our algorithm is the original *Greedy Refinement* algorithm. At this stage we have a good load balance and need only to consider the edge-cuts. Once again we iterate until no further improvements are made. In each iteration we go through all boundary points and select the vertex to move that gives the best positive gain with acceptable load balance. After the move we update the vertex and all its neighbors. Note, in the first pass vertices are removed from the list but in the second pass the partitioning is improved by considering *all* boundary points in each iteration. Since we only move vertices with a positive gain we do not need to worry about repeated movement of the same vertex and as we consider all the boundary points we can further improve the partitioning from pass 1. The complexity of the algorithm is very low as we only consider partition boundary points.

The key to a successful refinement is how to choose the vertex to move in each iteration. We presume that the initial partitioning is close to the optimal partitioning. Then, we must try to keep the shape of the partitions and only

Sketch of the local refinement algorithm

```
! Pass 1, smooth the load balance
Create a list of gains
do until (no allowed moves)
   S1:[select vertex to move]
   if (allowed move) then
      Perform move
      Remove vertex from gainlist
      Update status for all neighbors
      Update gainlist
   end if
end do

! Pass 2, decrease the number of edge cuts
do until (no allowed moves)
   S2:[select vertex to move]
   if (allowed move) then
      Perform move
      Update status for all neighbors
   end if
end do
```

Fig. 2. Sketch of the *Balanced Greedy Refinement* algorithm. In the first pass, the arithmetic work load is balanced. The transfered nodes are chosen with selection criteria $S1$, described in Figure 3. In the second pass, the edge-cut is locally minimized. The transfered nodes are here chosen with selection criteria $S2$, see Figure 4.

very carefully move vertices, i.e. we choose an *allowed* move with the best gain. In pass 1, see Figure 3, we allow a move only if the workload after the move is within certain limits, W_{max} for the partition the vertex is moved to and W_{min} for the partition the vertex is moved from. Experimentally the values $W_{max} = W_{ave} + \max_i w_i$ and $W_{min} = W_{ave} - \max_i w_i$ gives good results, where $W_{ave} = \sum_{i=1}^{V} w_i / N_p$ is the average workload and w_i are the weights of the vertices. The second criterion is that the gain must be positive or the load balance be improved. Then, we choose the move that gives the best gain and if several moves give the same gain we choose the one that gives the best load balance. In summary, the load balance is improved by excluding moves with a positive gain that do not decrease the load imbalance and by allowing moves with the least negative gain that decrease the load imbalance.

In the second pass we drop the W_{min} criterion and allow only moves with a positive gain or a gain equal to zero but with an improved load balance. The upper workload limit, W_{max}, can be chosen after any required load balance criterion trading the workload balance against the number of edge-cuts. Allowing for a larger load imbalance in the partitions we can in some cases further decrease

```
Sketch of the pass 1 selection criteria

   for all items in list
      if (work(part_to)+weight(vertex)<=wmax and
         work(part_from)-weight(vertex)>=wmin) then
         if (better_gain or equal_gain but better_load) then
            if (gain>0 or improved load_balance) then
               best_vertex=current_vertex
               gain=gain(vertex)
            end if
         end if
      end if
   end for
```

Fig. 3. Selection criteria in pass 1. A move is allowed if the work load remains within predefined limits. Then the node with the best gain is chosen. The gain can be negative only if the load balance is improved.

```
Sketch of the pass 2 selection criteria

   for all partition boundary vertices
      for all neighbor_partitions
         if (work(part_to)+weight(vertex)<=wmax) then
            if (better_gain) then
               if (gain>0 or gain=0 & better_load) then
                  best_vertex=current_vertex
                  gain=gain(vertex)
               end if
            end if
         end if
      end for
   end for
```

Fig. 4. Selection criteria in pass 2. A move is allowed if the load balance remains acceptable. Then the node with the best positive gain is chosen. Only if the load balance can be improved a vertex with a zero gain can be chosen.

the number of edge-cuts, compare for example the results for *Greedy* and *Mod Greedy* in Table 2. The selection algorithm is sketched in Figure 4.

In [11] a similar refinement algorithm is developed. This algorithm has three phases, (i) shape optimization, (ii) load balancing, and (iii) edge-cut minimization. Experimental results show that the quality is good compared to other methods but the load balancing phase is reported to suffer from slow convergence.

3   Experimental Partitioning Results

We have evaluated our refinement method on two different test cases, an unstructured grid and a structured multi-block grid. In the first test case, we partition an unstructured grid around a multi-element airfoil (see Figure 5) with the standard RSB method and refine the computed partitions with our refinement method. This reduces the edge-cut with up to 20%, see Table 1. We also compare the results with Chaco using RSB and Kernighan–Lin refinement. The more advanced Kernighan-Lin refinement method gives consistently better partitions.

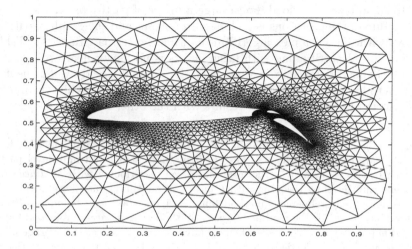

Fig. 5. Test case 1, an unstructured grid around a multi-element airfoil

Table 1. Edge-cuts for partitioning of a finite element grid, multi-element airfoil with 4720 vertices and 13722 edges. Three methods are compared, the standard *RSB* method without any refinement, post refinement of RSB with our *Greedy* method, and partitioning with *Chaco*. The Chaco partitioning is computed with RSB and Kernighan-Lin refinement in each bisection step.

np	RSB	Greedy	Chaco
2	117	113	90
4	260	250	245
8	469	454	395
16	752	723	652
32	1216	1171	1097
64	1909	1818	1742

As a second example, we use our refinement method in a multilevel context on a structured multiblock grid, see Figure 6(a). The results are very good and the number of edge-cuts is decreased with up to 45% compared to the standard RSB method. The results are also better than those computed with corresponding multilevel methods implemented in Chaco [4], Figure 6(b), and in Metis [7], Figure 6(c). Our multilevel algorithm [9,10] starts with large structured blocks and distributes these. The blocks are split in the refinement steps and redistributed until the block size is one grid point. Our algorithm can better utilize the structure of the composite grid than the pure graph partitioning algorithms. For structured grids it is advantageous to also have structured partitions. The initial partitioning on the coarsest level in our multilevel method is completely structured and is only marginally modified in the succeeding refinement steps in order to smooth the workload. For a large number of partitions on a small grid, the structure of the partitions is no longer important and the graph partitioning packages with more complex algorithms can find better partitions, see Table 2. Also, our initial partitioning with large blocks is no longer near optimal. If we compare the load balance we can see that the original Greedy Refinement method fails to improve work load within our multilevel method while the modified Greedy method gives comparable results with the other methods.

Table 2. Edge-cut and load imbalance for partitioning of the multiblock grid. Five approaches are compared, the standard *RSB* method, structured multilevel partitioning with *Greedy* refinement, structured multilevel partitioning with *Modified Greedy* refinement, multilevel partitioning with *Chaco*, and multilevel partitioning with *Metis*. The original Greedy refinement method fails to give a good load balance within our multilevel method while the modified Greedy method gives an acceptable load balance. The load imbalance of 1.076 arises when we have two vertices over the average number of vertices in the most loaded partition. The load imbalance is defined as the ratio between the maximum work load and the average work load in the partitions.

np	RSB	$Greedy$	$Mod\ Greedy$	$Chaco$	$Metis$
2	54 1.001	38 1.001	38 1.001	38 1.001	38 1.001
4	131 1.002	71 1.002	69 1.002	72 1.002	74 1.002
8	185 1.004	151 1.035	153 1.004	158 1.004	162 1.004
16	277 1.004	259 1.049	266 1.012	267 1.004	295 1.004
32	450 1.004	412 1.147	421 1.022	424 1.004	434 1.004
64	687 1.004	606 1.291	641 1.076	635 1.004	648 1.076

4   Conclusions

We have derived a local refinement method for data partitioning. The method balances the work load and minimizes the number of edge-cuts locally. The arithmetic complexity of the algorithm is low. The proposed refinement method is useful in the context of multilevel partitioning, dynamic partitioning, and refinement of global partitioning methods, i.e. in cases where we have an initial partitioning which we want to improve due to load balance or due to edge-cut. The algorithm tries to improve a previous partitioning by moving data

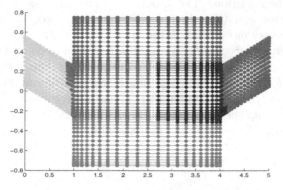

(a) Our structured Multilevel method

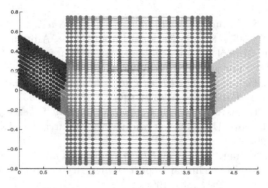

(b) Chaco, Multilevel partitioning

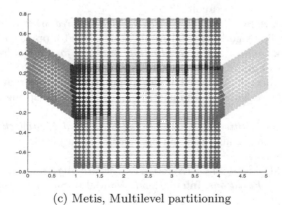

(c) Metis, Multilevel partitioning

Fig. 6. Multilevel partitioning of a structured multiblock grid (6 partitions).

items between the partitions. The data items which are moved are selected very carefully. It is essential to pick the right data items to get a fast algorithm avoiding unnecessary movements. If the initial partitioning is nearly optimal our algorithm converges very rapidly, i.e. only a few data items are moved.

We have compared our method with corresponding refinement methods both in *Chaco* and *Metis* in the context of multilevel partitioning and refinement of the Recursive Spectral Bisection method. The experiments show that the quality is comparable or even better in some cases with our method. The strength in our algorithm is the low complexity and the ability to balance the work load.

References

1. C.M. Fiduccia, R.M. Mattheyses, *A Linear-Time Heuristic for Improving Network Partitions*, in Proceedings of 19th IEEE Design Automation Conference, IEEE, pp. 175-181, 1982.
2. B. Hendrickson, R. Leland, *A Multidimensional Spectral Load Balancing*, Technical Report SAND 93-0074, Sandia National Laboratories, Albuquerque, NM, 1993.
3. B. Hendrickson, R. Leland, *A Multilevel Algorithm for Partitioning Graphs*, in Proceedings of Supercomputing '95, 1995.
4. B. Hendrickson, R. Leland, *The Chaco User's Guide Version 2.0*, Technical Report, Sandia National Laboratories, Albuquerque, NM, July 1995.
5. G. Karypis, V. Kumar, *Multilevel k-way Partitioning Scheme for Irregular Graphs*, Technical Report, University of Minnesota, Department of Computer Science, Minneapolis, 1996.
6. G. Karypis, V. Kumar, *A Fast and High Quality Multilevel Scheme for Partitioning Irregular Graphs*, Technical Report, University of Minnesota, Department of Computer Science, Minneapolis, 1995.
7. G. Karypis, V. Kumar, *Metis: Unstructured Graph Partitioning and Sparse Matrix Ordering System*, Technical Report, University of Minnesota, Department of Computer Science, Minneapolis, 1995.
8. J. Rantakokko, *A framework for partitioning structured grids with inhomogeneous workload*, Parallel Algorithms and Applications, Vol 13, pp:135-151, 1998.
9. J. Rantakokko, *Comparison of partitioning strategies for PDE solvers on multiblock grids*, in Proceedings of PARA'98, B. Kågström, J. Dongarra, E. Elmroth and J. Wasniewski (editors). Applied Parallel Computing, 4th International Workshop, PARA'98, Lecture Notes in Computer Science, No. 1541, Springer-Verlag, Berlin, 1998.
10. J. Rantakokko, *Partitioning Strategies for Structured Multiblock Grids*, accepted for publication in Parallel Computing, special issue on Graph Partitioning and Parallel Computing.
11. C. Walshaw, M. Cross, M.G. Everett, *A Localised Algorithm for Optimising Unstructured Mesh Partitions*, International Journal of Supercomputer Applications 9(4):280-295, 1995.

Feedback Guided Scheduling of Nested Loops

T.L. Freeman[1], D.J. Hancock[1], J.M. Bull[2], and R.W. Ford[1]

[1] Centre for Novel Computing, University of Manchester, Manchester, M13 9PL,
U.K.
[2] Edinburgh Parallel Computing Centre, University of Edinburgh, Edinburgh,
EH9 3JZ, U.K.

Abstract. In earlier papers ([2], [3], [6]) feedback guided loop scheduling algorithms have been shown to be very effective for certain loop
scheduling problems which involve a sequential outer loop and a parallel
inner loop and for which the workload of the parallel loop changes only
slowly from one execution to the next. In this paper the extension of
these ideas the case of nested parallel loops is investigated. We describe
four feedback guided algorithms for scheduling nested loops and evaluate
the performances of the algorithms on a set of synthetic benchmarks.

1 Introduction

Loops are a rich source of parallelism for many applications. As a result, a variety
of loop scheduling algorithms that aim to schedule parallel loop iterations to
processors of a shared-memory machine in an almost optimal way have been
suggested. In a number of recent papers, Bull, Ford and co-workers ([2], [3],
[6]) have shown that a Feedback Guided Dynamic Loop Scheduling (FGDLS)
algorithm performs well for certain parallel single loops. Parallel nested loops
are an even richer source of parallelism; in this paper we extend these feedback
guided scheduling ideas to the more general case of nested loops, and present
numerical results to evaluate the performance of the resulting algorithms.

2 Loop Scheduling

Many existing loop scheduling algorithms are based on either (variants of) guided self-scheduling (Polychronopoulos and Kuck [11], Eager and Zahorjan [5],
Hummel et al. [7], Lucco [8], Tzen and Ni [13]), or (variants of) affinity scheduling algorithms (Markatos and LeBlanc [9], Subramaniam and Eager [12], Yan et
al. [14]). The underlying assumption of these algorithms is that each execution
of a parallel loop is independent of previous executions of the same loop and
therefore has to be rescheduled from 'scratch'.

In a sequence of papers, Bull et al. [3] and Bull [2] (see also Ford et al. [6])
propose a substantially different loop scheduling algorithm, termed Feedback
Guided Dynamic Loop Scheduling. The algorithm assumes that the workload is
changing only slowly from one execution of a loop to the next, so that observed
timing information from the current execution of the loop can, and *should*, be
used to guide the scheduling of the next execution of the same loop. In this way

T. Sørevik et al. (Eds.): PARA 2000, LNCS 1947, pp. 149–159, 2001.
© Springer-Verlag Berlin Heidelberg 2001

it is possible to limit the number of chunks into which the loop iterations is divided to be equal to the number of processors, and thereby it is possible to limit the overheads (loss of data locality, synchronisation costs, etc.) associated with guided self-scheduling and affinity scheduling algorithms.

Note that almost all scheduling algorithms are designed to deal with the case of a single loop. The application of the algorithms to parallel loops usually proceeds by either (a) treating only the outermost loop as parallel (see Section 4.4), or (b) coalescing the loops into a single loop (see Section 4.2). In Section 4.1 we introduce a scheduling algorithm that treats the multidimensionality of the iteration space corresponding to the nested loops directly.

3   Feedback Guided Loop Scheduling

The feedback guided algorithms described in [3] and [2] are designed to deal with the case of the scheduling, across p processors, of a single loop of the form:

```
DO SEQUENTIAL J = 1, NSTEPS
    DO PARALLEL K = 1, NPOINTS
        CALL LOOP_BODY(K)
    END DO
END DO
```

We assume that the scheduling algorithm has defined the lower and upper loop iteration bounds, $l_j^t, u_j^t \in \mathbb{N}$, $j = 1, 2, \ldots, p$, on outer iteration step t, and that the corresponding measured execution times are given by $T_j^t, j = 1, 2, \ldots, p$. This execution time data enables a piecewise constant approximation to the observed workload to be defined, and the iteration bounds for iteration $t + 1$, $l_j^{t+1}, u_j^{t+1} \in \mathbb{N}$, $j = 1, 2, \ldots, p$, are then chosen to approximately equi-partition this piecewise constant function.

This feedback guided algorithm has been shown to be particularly effective for problems where the workload of the parallel loop changes slowly from one execution to the next (see [2], [3]) — this is a situation that occurs in many applications. In situations where the workload is more rapidly changing, guided self-scheduling or affinity scheduling algorithms are likely to be more efficient.

In this paper we consider the extension of feedback guided loop scheduling algorithms to the case of nested parallel loops:

```
DO SEQUENTIAL J = 1, NSTEPS
    DO PARALLEL K1 = 1, NPOINTS1
            .
            .
            .
        DO PARALLEL KM = 1, NPOINTSM
            CALL LOOP_BODY(K1,...,KM)
        END DO
            .
            .
            .
    END DO
END DO
```

As in the single loop case, the issues that arise in the scheduling of nested loops are load balancing and communication minimisation. In higher dimensions, the problem of minimising the amount of communication implied by the schedule assumes greater significance — there are two aspects to this problem: temporal locality and spatial locality. Temporal locality is a measure of the communication implied by the scheduling of different parts of the iteration space to different processors on successive iterations of the outer sequential loop. Spatial locality is a measure of the communication implied by the computation at each point of the iteration space.

4   Scheduling Nested Loops

Of the approaches described in this section, the first, described in Section 4.1, addresses the multi-dimensional problem directly by defining an explicit recursive partition of the M-dimensional iteration space.

The next two approaches are based on the transformation of the nested loops into an equivalent single loop, followed by the application of the one-dimensional Feedback Guided Loop Scheduling algorithm to this single loop; of these one-dimensional approaches, the first, described in Section 4.2, is based on coalescing the nested loops into a single loop; the second, described in Section 4.3, uses a space-filling curve technique to define a one-dimensional traversal of the M-dimensional iteration space.

The final approach, described in Section 4.4, is based on the application of the one-dimensional Feedback Guided Loop Scheduling algorithm to the outermost parallel loop only.

4.1   Recursive Partitioning

This algorithm generalises the algorithm described in Section 3 to multiple dimensions. We assume that the scheduling algorithm has defined a partitioning into p disjoint subregions, S_j^t, of the M-dimensional iteration space on the outer (sequential) iteration t:

$$S_j^t = [l_{1j}^t, u_{1j}^t] \times [l_{2j}^t, u_{2j}^t] \times \cdots \times [l_{Mj}^t, u_{Mj}^t], \ j = 1, 2, \ldots, p,$$

so that $\displaystyle\bigcup_{j=1}^{p} S_j^t = [1, NPOINTS1] \times [1, NPOINTS2] \times \cdots \times [1, NPOINTSM]$

(i.e. so that the union of the disjoint subregions is the complete iteration space). We also assume that the corresponding measured execution times, T_j^t, $j = 1, 2, \ldots, p$, are known. Based on this information, an M-dimensional piecewise constant approximation $\hat{W}^t$ to the actual workload at iteration t can be formed:

$$\hat{W}^t (x_1, x_2, \ldots, x_M) = \frac{T_j^t}{(u_{1j}^t - l_{1j}^t + 1)(u_{2j}^t - l_{2j}^t + 1) \ldots (u_{Mj}^t - l_{Mj}^t + 1)},$$

where $x_i \in \left[l_{ij}^t, u_{ij}^t\right]$, $i = 1, 2, \ldots, M$, $j = 1, 2, \ldots, p$. The recursive partitioning algorithm then defines new disjoint subregions S_j^{t+1}, $j = 1, 2, \ldots, p$, for scheduling the next outer iteration $t+1$, so that $\bigcup_{j=1}^{p} S_j^{t+1}$ is the complete iteration space and so that the piecewise constant workload function $\hat{W}^t$ is approximately equi-partitioned across these subregions. This is achieved by recursively partitioning the dimensions of the iteration space — in our current implementation, the dimensions are treated from outermost loop to innermost loop, although it would be easy to develop other, more sophisticated, strategies (to take account of loop length, for example). To give a specific example, if p is a power of 2, then the first partitioning (splitting) point is such that the piecewise constant workload function $\hat{W}^t$ is approximately bisected in its first dimension; the second partitioning points are in the second dimension and are such that the two piecewise constant workloads resulting from the first partitioning are approximately bisected in the second dimension; subsequent partitioning points further approximately bisect $\hat{W}^t$ in the other dimensions. It is also straightforward to deal with the case when p is not a power of 2; for example, if p is odd, the first partitioning point is such that the piecewise constant workload function $\hat{W}^t$ is approximately divided in the ratio $\lfloor p/2 \rfloor : \lceil p/2 \rceil$ in its first dimension ($\lfloor r \rfloor$ is the largest integer less than r and $\lceil r \rceil$ is the smallest integer greater than r); subsequent partitioning points divide $\hat{W}^t$ in the appropriate ratios in the other dimensions.

4.2   Loop Coalescing

A standard way of reducing a loop nest into a single loop is by loop coalescing (see [1] or [11] for details). The M parallel loops can be coalesced into a single loop of length NPOINTS1*NPOINTS2*· · ·*NPOINTSM. For example, for the case $M = 4$ the coalesced loop is given by:

```
DO SEQUENTIAL J=1,NSTEPS
    DO PARALLEL K=1,NPOINTS1*NPOINTS2*NPOINTS3*NPOINTS4
        K1 = ((K-1)/(NPOINTS4*NPOINTS3*NPOINTS2)) + 1
        K2 = MOD((K-1)/(NPOINTS4*NPOINTS3),NPOINTS2) + 1
        K3 = MOD((K-1)/NPOINTS4,NPOINTS3) + 1
        K4 = MOD(K-1,NPOINTS4) + 1
        CALL LOOP_BODY(K1,K2,K3,K4)
    END DO
END DO
```

Having coalesced the loops we can use any one-dimensional algorithm to schedule the resulting single loop. The numerical results of [2], [3], show that the feedback guided scheduling algorithms are very effective for problems where the workload of a (one-dimensional) parallel loop changes slowly from one execution to the next. This is the situation in our case and thus we use the one-dimensional Feedback Guided Loop Scheduling algorithm described in Section 3 to schedule the coalesced loop.

4.3   Space-Filling Traversal

This algorithm is based on the use of space-filling curves to traverse the iteration space. Such curves visit each point in the space exactly once, and have the property that any point on the curve is spatially adjacent to its neighbouring points. Again, any one-dimensional scheduling algorithm can be used to partition the resulting one-parameter curve into p pieces (for the results of the next section the Feedback Guided Loop Scheduling algorithm described in Section 3 is used). This algorithm readily extends to M-dimensions, although our current implementation is restricted to $M = 2$.

A first order Hilbert curve (see [4]) is used to fill the 2-dimensional iteration space, and the mapping of two-dimensional coordinates to curve distances is performed using a series of bitwise operations. One drawback of the use of such a space-filling curve is that the iteration space must have sides of equal lengths, and the length must be a power of 2.

4.4   Outer Loop Parallelisation

In this algorithm, only the outermost, K1, loop is treated as a parallel loop, the inner, K2, ..., KM, loops are treated as sequential loops; the one-dimensional Feedback Guided Loop Scheduling algorithm of Section 3 is used to schedule this single parallel loop. This approach is often recommended because of its simplicity and because it exploits parallelism at the coarsest level.

5   Numerical Results and Conclusions

This section reports the results of a set of experiments that simulate the performance of the four algorithms described in Section 4 under synthetic load conditions. Simulation has been used for three reasons;

- for accuracy (reproducible timing results are notoriously hard to achieve on large HPC systems (see Appendix A of Mukhopadhyay [10])),
- so that the instrumenting of the code does not affect performance, and
- to allow large values of p to be tested.

Statistics were recorded as each of the algorithms attempted to load-balance a number of computations with different load distributions. At the end of each iteration, the load imbalance was computed as $(T_{max} - T_{mean})\ /\ T_{max}$, where T_{max} was the maximum execution time of any processor for the given iteration and T_{mean} was the corresponding mean execution time over all processors. The results presented in Figures 1–4 show average load imbalance over all iterations versus number of processors; we plot average load imbalance to show more clearly the performance of the algorithms in response to a dynamic load.

The degree of data reuse between successive iterations was calculated as follows. For each point in the iteration space, a record was kept of the processor to which it was allocated in the previous iteration. For a given processor, the reuse was calculated as the percentage of points allocated to that processor that were also allocated to it in the previous iteration. As with load imbalance, the results presented in Figures 5–8 show reuse averaged over all iterations plotted against number of processors.

Four different workload configurations have been used in the experiments:

Gauss A Gaussian workload where the centre of the workload rotates about the centre of the iteration space. Parameters control the width of the Gaussian curve, and the radius and time period of the rotation.

Pond A constant workload within a circle centred at the centre of the iteration space where the radius of the circle oscillates sinusoidally; the workload outside the circle is zero. Parameters control the initial radius of the circle and the amplitude and period of the radial oscillations.

Ridge A workload described by a ridge with Gaussian cross-section that traverses the iteration space. Parameters control the orientation of the ridge, the width of the Gaussian cross-section and the period of transit across the iteration space.

Linear A workload that varies linearly across the iteration space, but is static.

Only the linear workload is static, the other three workloads are dynamic.

The computation is assumed to be executed on an iteration space that is 512×512. Each of the algorithms described in Section 4 was used to partition the loads for numbers of processors varying between 2 and 256. The results displayed are for 400 iterations of the loop scheduling algorithm (for the dynamic workloads, this corresponds to 2 complete cycles of the workload period). The algorithms were also run for 100 and 200 iterations and very similar results were obtained.

The four algorithms are labelled as *rcb* (recursive partitioning or recursive co-ordinate bisection algorithm), *slice* (loop coalescing algorithm), *sfc* (space-filling traversal or space-filling curve algorithm), *outer* (outer loop parallelisation).

We first consider the load balancing performance summarised in the graphs of Figures 1–4. For two of the dynamic loads (Gauss, Pond, Figures 1 and 2 respectively), there is little to choose between the four algorithms. For the third dynamic load (Ridge, Figure 3) *slice* and *outer* perform better than *rcb* and *sfc*; *slice* and *outer* traverse the iteration space in a line by line fashion and this is a very effective strategy for this particular workload. For the static load (Linear, Figure 4), *slice* and *sfc* outperform *rcb* and *outer*, particularly for larger numbers of processors; we note that *slice* and *sfc* achieve essentially perfect load balance. These results can be explained by considering the structures of the four algorithms. *Slice* and *sfc* are one-dimensional algorithms adapted to solve the two-dimensional test problems; as such they adopt a relatively fine-grain approach to loop-scheduling and this should lead to good load balance properties. In contrast *rcb* is a multi-dimensional algorithm and, in its initial stages (the first few partitions), it adopts a relatively coarse-grain approach to loop scheduling. For example, the first partition approximately divides the iteration space into two parts and any imbalance implied by this partition is inherited by subsequent recursive partitions. *Outer*, which schedules only the outermost loop, also adopts a coarse-grain approach to the scheduling problem. Given this, the load balance properties of *rcb* and *outer* for the dynamic workload problems are surprisingly good and are often better than either *slice* or *sfc*.

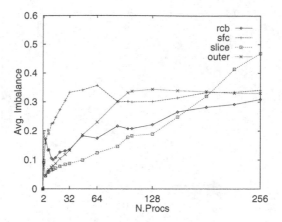

Fig. 1. Load Balance for Gauss Load

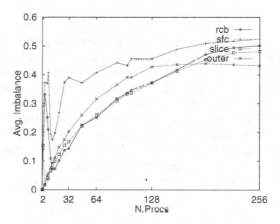

Fig. 2. Load Balance for Pond Load

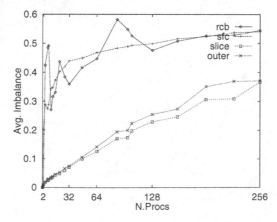

Fig. 3. Load Balance for Ridge Load

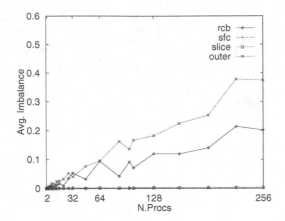

Fig. 4. Load Balance for Linear Load

We would expect the structure of *rcb* to be advantageous when we consider the data reuse achieved by the algorithms and this is borne out by the graphs in Figures 5–8 which show the levels of intra-iteration data reuse achieved. Note that data re-use is a measure of the temporal locality achieved by the different scheduling algorithms. In terms of intra-iteration data reuse, *rcb* clearly outperforms *slice*, *sfc* and *outer* for all the dynamic load types and for nearly all numbers of processors. In particular for larger numbers of processors *rcb* substantially outperforms the other algorithms. Again the explanation for these results lies in the structure of the algorithms; in particular, the fact that *rcb* treats the multi-dimensional iteration space directly is advantageous for data reuse.

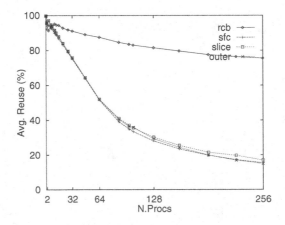

Fig. 5. Reuse for Gauss Load

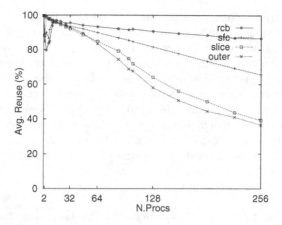

Fig. 6. Reuse for Pond Load

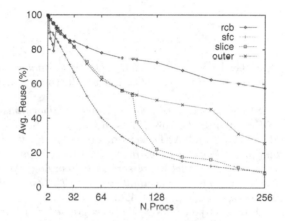

Fig. 7. Reuse for Ridge Load

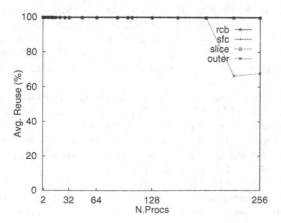

Fig. 8. Reuse for Linear Load

In broad terms we observe that the load balance properties of the four algorithms are very similar and it is not possible to identify one of the algorithms as significantly, and uniformly, better than the others. In contrast *rcb* is substantially better than the other algorithms in terms of intra-iteration data reuse, particularly for larger numbers of processors. How these observed performance differences manifest themselves on an actual parallel machine will depend on a number of factors, including: the relative performances of computation and communication on the machine; the granularity of the computations in the parallel nested loops; the rate at which the workload changes and the extent of the imbalance of the work across the iteration space. A comprehensive set of further experiments to undertake this evaluation of the algorithms on a parallel machine will be reported in a forthcoming paper.

Acknowledgements. The authors acknowledge the support of the EPSRC research grant GR/K85704, "Feedback Guided Affinity Loop Scheduling for Multiprocessors".

References

1. D. F. Bacon, S. L. Graham and O. J. Sharp, (1994) *Compiler Transformations for High-Performance Computing*, ACM Computing Surveys, vol. 26, no. 4, pp. 345–420.
2. J. M. Bull, (1998) *Feedback Guided Loop Scheduling: Algorithms and Experiments*, Proceedings of Euro-Par'98, Lecture Notes in Computer Science, Springer-Verlag.
3. J. M. Bull, R. W. Ford and A. Dickinson, (1996) *A Feedback Based Load Balance Algorithm for Physics Routines in NWP*, Proceedings of Seventh ECMWF Workshop on the Use of Parallel Processors in Meteorology, World Scientific.
4. A. R. Butz, (1971) *Alternative Algorithm for Hilbert's Space-Filling Curve*, IEEE Trans. on Computers, vol. 20, pp. 424–426.
5. D. L. Eager and J. Zahorjan, (1992) *Adaptive Guided Self-Scheduling*, Technical Report 92-01-01, Department of Computer Science and Engineering, University of Washington.
6. R. W. Ford, D. F. Snelling and A. Dickinson, (1994) *Controlling Load Balance, Cache Use and Vector Length in the UM*, Proceedings of Sixth ECMWF Workshop on the Use of Parallel Processors in Meteorology, World Scientific.
7. S. F. Hummel, E. Schonberg and L. E. Flynn, (1992) *Factoring: A Practical and Robust Method for Scheduling Parallel Loops*, Communications of the ACM, vol. 35, no. 8, pp. 90–101.
8. S. Lucco, (1992) *A Dynamic Scheduling Method for Irregular Parallel Programs*, in Proceedings of ACM SIGPLAN '92 Conference on Programming Language Design and Implementation, San Francisco, CA, June 1992, pp. 200–211.
9. E. P. Markatos and T. J. LeBlanc, (1994) *Using Processor Affinity in Loop Scheduling on Shared Memory Multiprocessors*, IEEE Transactions on Parallel and Distributed Systems, vol. 5, no. 4, pp. 379–400.
10. N. Mukhopadhyay, (1999) *On the Effectiveness of Feedback-guided Parallelisation*, PhD Thesis, Department of Computer Science, University of Manchester.
11. C. D. Polychronopoulos and D. J. Kuck, (1987) *Guided Self-Scheduling: A Practical Scheduling Scheme for Parallel Supercomputers*, IEEE Transactions on Computers, C-36(12), pp. 1425–1439.

12. S. Subramaniam and D. L. Eager, (1994) *Affinity Scheduling of Unbalanced Workloads*, in Proceedings of Supercomputing'94, IEEE Comp. Soc. Press, pp. 214–226.
13. T. H. Tzen and L. M. Ni, (1993) *Trapezoid Self-Scheduling Scheme for Parallel Computers*, IEEE Trans. on Parallel and Distributed Systems, vol. 4, no. 1, pp. 87–98.
14. Y. Yan, C. Jin and X. Zhang, (1997) *Adaptively Scheduling Parallel Loops in Distributed Shared-Memory Systems*, IEEE Trans. on Parallel and Distributed Systems, vol. 8, no. 1, pp. 70–81.

A Comparison of Partitioning Schemes for Blockwise Parallel SAMR Algorithms*

Johan Steensland, Stefan Söderberg, and Michael Thuné

Uppsala University, Information Technology, Dept. of Scientific Computing, Box 120, SE-751 04 Uppsala, Sweden. {johans, stefans, michael}@tdb.uu.se

Abstract. This paper presents an experimental comparison of dynamic partitioning techniques for blockwise parallel structured adaptive mesh refinement applications. A new partitioning technique, G-MISP, is described. Policies for the automatic selection of partitioner based on application and system state are outlined. Adaptive methods for the numerical solution to partial differential equations yield highly advantageous ratios for cost/accuracy compared to methods based upon static uniform approximations. Distributed implementations offer the potential for accurate solution of physically realistic models of important applications. They also lead to interesting challenges in dynamic resource allocation, e.g. dynamic load balancing. The results show that G-MISP is preferable for communication dominated cases where the block graph has high granularity. Recommendations for appropriate partitioning techniques, given application and system state, are given. It was found that our classification model needs to be extended to accurately capture the behavior of the cases studied.

Keywords: SAMR, dynamic load balancing, partitioning, inverse space filling curve, diffusion, RSB

1   Introduction

This paper presents an experimental comparison of dynamic partitioning and load-balancing techniques for blockwise parallel structured adaptive mesh refinement (SAMR) applications. The overall goal of this research is to support the formulation of policies required to drive a *dynamically adaptive* meta-partitioner for SAMR applications capable of decreasing overall execution time by selecting the most appropriate partitioning strategy (from a family of available partitioners) at runtime, based on current application and system state.

   The simulation of complex phenomena, described by partial differential equations (PDEs), requires adaptive numerical methods [19]. These methods yield highly advantageous ratios for cost/accuracy compared to methods based upon static uniform approximations. Distributed implementations of these methods offer the potential for accurate solution of physically realistic models of important physical systems. The next generation simulations of complex physical

---

* This research was supported by the Swedish Foundation for Strategic Research via the program of Industrial Computational Mathematics.

T. Sørevik et al. (Eds.): PARA 2000, LNCS 1947, pp. 160–169, 2001.

phenomenon will be built using such dynamically adaptive techniques executing in distributed computer environments, and will provide dramatic insights into complex systems such as interacting black holes and neutron stars, formations of galaxies, oil reservoirs and aquifers, and seismic models of entire planets. Distributed implementations also lead to interesting challenges in dynamic resource allocation, data-distribution and load balancing, communications and coordination, and resource management.

Our primary motivation for introducing the adaptive meta-partitioner is the observation that *no single partitioning scheme performs the best* for all types of applications and computer systems. Even for a given application, the most suitable partitioning technique may depend on runtime state [11,16].

In this paper, we present an experimental comparison of four partitioners, using a suite of four blockwise SAMR applications. The partitioners studied constitute a selection from software tools, viz. the partitioners in ParMetis[5] and Vampire[17]. We use the experimentation to characterize the behavior of the applications, and to determine partitioning requirements corresponding to the different application states. The experimental results will then be used to characterize the behavior of the partitioners using execution times as the ultimate metric, and to associate the partitioner(s) with application and system states. The paper makes four contributions: (1) It evaluates partitioners for blockwise SAMR applications by comparing gains/losses in (key parts of) execution time on two different parallel computers. (2) It presents a new domain-based partitioning/load-balancing technique (G-MISP) for general SAMR grid hierarchies. (3) It presents an experimental characterization of the new and existing partitioning/load-balancing techniques, using four different blockwise SAMR applications. (4) It outlines policies for the selection of partitioners for blockwise SAMR applications based on application and system state.

2   Structured AMR - Overview and Related Work

Dynamically adaptive numerical techniques for solving differential equations provide a means for concentrating computational effort to computationally demanding regions. General hierarchal AMR and load balancing for such methods can be studied in e.g. [3,2,18,15,8].

The present paper considers a blockwise approach to parallel SAMR. The initial, coarse grid is divided into $n_1 \times n_2$ blocks (n_1 and n_2 are parameters given by the user). Subsequent refinements are carried out with respect to entire blocks. That is, if some points in a block are flagged for refinement, then the whole block is refined. The blockwise approach is illustrated in Fig 1. This is *one* of several approaches to SAMR, and has been studied from a mathematical/numerical point of view in, e.g., [4,6,10].

In the partitioning phase, the blocks are considered to be atomic units. That is, load balance is achieved via migration of complete blocks between processors. Thus, the load balancer operates on the graph of blocks, where each node is a block, and the edges represent adjacency. We consider the situation where there

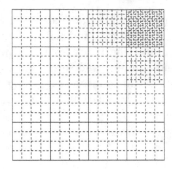

Fig. 1. In the blockwise approach, the domain is divided into $nx * ny$ blocks. A block is refined in its entirety, or not at all. This example shows a 4*4 block decomposition, and two levels of refinement

are more blocks than processors, so a partition will in general consist of a set of blocks. Since the blocks are organized in an $n_1 \times n_2$ grid (the "block graph"), this repartitioning problem is analogous to the repartitioning problem for a (small) structured grid with heterogeneous work load.

3   Requirements for Efficient Partitioning

The partitioning requirements for adaptive applications depend on the current state of the application and the computing environment. Therefore, it is of little importance to discuss the absolute "goodness" of a certain partitioning technique. In general, partitioning techniques tend to optimize some metric(s) at the expense of other. As a consequence, we base our characterization of partitioning behavior on the tuple (partitioner, application, computer system), (PAC). The goal is to capture the overall runtime behavior implied by the partitioner.

Classifying the state of the application and computer system at hand according to the *quadrant approach* (see Fig 2), enables the selection of the appropriate partitioning technique. The quadrant approach is a means to make a broad classification of applications and computer systems with respect to requirement for appropriate partitioning techniques. Classification is made with respect to *(a)* amount of communication (communication dominated vs. computation dominated) and *(b)* the distribution of the mesh refinements (localized vs. scattered). Thus, the classification space has four quadrants corresponding to different combinations of properties *(a)* and *(b)*. Applications may very well start off in one quadrant, and, as execution progresses, move to different quadrants. The same kind of dynamic behavior is exhibited by parallel computers, if the user does not have exclusive access to memory, communication bandwidth and CPU. This, too, will imply change of quadrants.

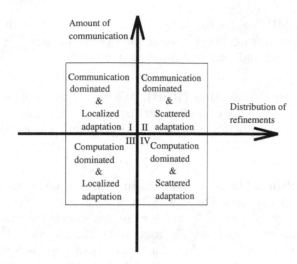

Fig. 2. According to the quadrant approach, applications and computer systems are classified with respect to distribution of refinements and amount of communication. Applications may very well start off in one quadrant, and, as execution progresses, move to any other quadrant. Quadrant changes may also be caused by the dynamic behavior of the computer system

4   Experimental Characterization of Partitioning Schemes

The overall goal of the presented research is to construct a dynamic meta-partitioner that can adapt to (and optimize for) fully dynamic PACs. In this paper, we move towards this goal by characterizing (P)artitioners for different (C)omputer systems and (states of) (A)pplications. The outline of the experiments is as follows. Partitioning quality is evaluated by measuring key parts of execution time for each PAC, each with different application behavior according to the quadrant approach. For each of the four partitioning schemes, we measure key parts of execution time on two different parallel computers, for the cases eight (DEC), and eight and 16 processors (SUN).

4.1   Partitioning Strategies

Wavefront diffusion (WD) based on global work load. The wavefront diffusion based on global work load is one of many partitioning algorithms included in the library *ParMetis* [5]. The justification for this choice is that this algorithm specializes in repartitioning graphs where the refinements are scattered. This objective is totally lacking in the other techniques. We want to investigate whether this technique is advantageous for applications residing mainly in quadrants corresponding to scattered refinements.

The WD algorithm was reported to perform well in [14]. It is in general more expensive than G-MISP. However, it can be expected to be competitive for blockwise SAMR, leading to small graphs to partition. In addition, it can be expected to give better quality partitions than G-MISP, since it explicitly takes both arithmetic load and communication into account.

Recursive Spectral Bisection (RSB). We include for comparison recursive spectral bisection [1], a high-quality, but expensive partitioner. For the small block graphs generated by the SAMR algorithm under consideration, it may be competitive.

Geometric multilevel inverse space-filling curve partitioning (G-MISP) with variable grain size. The G-MISP scheme is part of the parallel SAMR partitioning library *Vampire* [17], which combines structured and unstructured partitioning techniques in a fashion introduced by Rantakokko [11]. It is a multilevel algorithm that starts out by viewing a matrix of workloads as a one-vertex graph. The G-MISP scheme successively refines the graph where needed, and the parameter *atomic unit* acts as an upper bound of the granularity. Therefore, the resulting partitioning granularity is a function of both the atomic unit and the current state of the grid. The structured part is defined in terms of split and merge operations. Each cut is made at the geometrical mid point. The unstructured part is based on the Hilbert space-filling curve (SFC) [12,13]. The G-MISP algorithm strives to create an internal list of blocks, each having approximately the same weight. Load balance is thereafter targeted by trivially assigning the same number of blocks to each processor.

This approach favors speed at the expense of load-balance. Inverse SFC partitioning (ISP) has given promising results in other studies [7,9,8], even though its theoretical foundation could be stronger. In particular, the ISP schemes completely ignore graph edges, and can thus be expected to give a higher edge-cut than algorithms expertizing in minimizing this particular metric.

Stripes. The grid is partitioned once and for all by rectilinear cuts, into logical "stripes", one per processor. This partitioning is kept constant during the entire simulation. Thus, this scheme is not "dynamic". However, since it minimizes number of neighbors, it might be competitive for cases when latency dominates communication (and communication dominates the overall run-time).

4.2   The SAMR Solver and the Parallel Computers

In our tests, we use a 2D parallel SAMR code based on the approach described in section 2. On each block, the PDE problem is solved by means of a finite volume method. There is need for exchange of information along the interfaces between adjacent blocks. This leads to communication for blocks assigned to different processors, and to copying between blocks assigned to the same processor.

The structured block graph can be represented as an $n_1 \times n_2$ matrix, where the matrix elements are integers representing the arithmetic work loads for the blocks. This is the input format used by the G-MISP algorithm. For ParMetis, we use its required graph format [5].

The experiments were carried out on two different parallel computers. The first is a parallel platform consisting of Dec Alpha Server SMPs connected via a Memory Channel for which we used eight processors and 16*16 blocks. The second is SUN's new parallel platform based on Sun Enterprise SMPs and the Wildfire interconnect for which we used eight and 16 processors, and 16*16 as well as 20*20 blocks.

On one hand, moving from eight to 16 processors, affects our current position in the space of quadrants. The more processor one uses, the more relative communication one gets. As a consequence, the shift from the eight-processor DEC to the 16-processor SUN shifts all applications a bit *upwards* (towards more communication) in the space of quadrants.

On the other hand, on the SUN, all 16 processors are hosted within the same "box", i.e. there is no communication with processors on another SMP. On the DEC, we use two four-processor SMPs, hence introducing the need for communication with "the other box". Therefore, the same shift from the eight-processor DEC to the 16-processor SUN would imply a shift *downwards* (towards less communication) in the square of quadrants.

To avoid confusion, we make sure that our four applications, on each of the two platforms, clearly end up in their respective quadrants. That is, we create exactly one case per quadrant for each of the parallel computers.

4.3   The SAMR Applications

We have studied two test applications. One is the linearized Euler equations in a channel, where the refinements are scattered. The other is a scalar convection problem, where the refinements are strongly localized. These applications are communication dominated. For the experiments on the DEC, we used the same applications as a basis, and simulated computation dominated cases by adding arithmetic operations, to achieve an arithmetic load corresponding to a nonlinear Euler solver. For the experiments on the SUN, we used our new implementation of a real nonlinear Euler solver. Thus, for both parallel computers, we cover all four combinations according to the quadrant approach.

5   Experimental Results

Complete experimental results from all quadrants and partitioners are shown for two cases only, due to lack of space. Figure 3 shows the results for 16*16 blocks with 16 processors on the SUN. Stripes works best for the communication dominated cases, but worst for the computation dominated cases. In Figure 4 we have increased the block graph to 20*20 blocks and decreased the number of processors to eight. For this case, G-MISP works best for the communication dominated cases. Here follow the general characteristics of each partitioning scheme derived from our experiments.

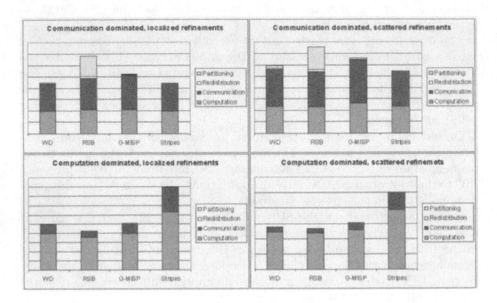

Fig. 3. Execution times for 16*16 blocks with 16 processors on the SUN. Stripes works best for the communication dominated cases, but worst otherwise

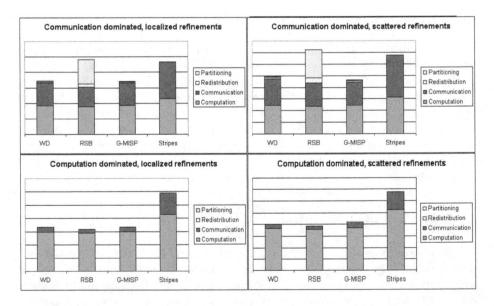

Fig. 4. Execution times for 20*20 blocks with 8 processors on the SUN. The G-MISP scheme works best for the communication dominated cases. For the computation dominated cases, RSB is best

WD. The quality of the partitions generated by WD is often close to (sometimes better) than that of RSB. Computation time is often a little better optimized than communication time. In general, WD performed better for the scattered cases when execution time was dominated by computation. No clear trend could be seen for the communication dominated cases.

RSB. The RSB technique gives the best quality partitionings in all computation dominated cases, and roughly half of the other cases as well. However, the time spent in repartitioning and redistribution is much longer than for the other schemes. The trend is that RSB gives shorter computation time, and even shorter communication time, than the other schemes.

G-MISP. The G-MISP scheme is by far the fastest. When the average number of blocks per processor is sufficiently large (about 32 for our experimentation), it performs the best of all schemes for the communication dominated cases. Both computation and communication time is greater than for the other schemes. However, its speed and its ability to optimize for redistribution balances that for these cases.

Stripes. Since stripes is a static method, no time is spent in repartitioning or in redistribution. In communication dominated cases where equally sized stripes could be assigned to the processors, it yields the shortest run-time.

6   Discussion and Conclusions

The results in this paper constitute the first part of an ongoing study. The aim is to use the categorization provided by the quadrant approach to create an *adaptive* meta-partitioner. By providing the load balancer with relevant parameters, it would be able to classify the application and computer system, and select the load balancing algorithm accordingly. (Note that the classification may change in the course of the simulation, calling for *dynamic* adaptation of the dynamic load balancer.)

An attempt to establish a mapping from quadrants onto partitioning schemes is shown if Fig 5. The mapping from quadrant I and II is identical. If the number of processors divide the number of rows or columns, i.e. it is possible to assign equally sized stripes to all processors, stripes is the best choice. Minimizing number of neighbors (2 for all processors) seems to be sufficient. Otherwise, if granularity is fine enough, say more than 32 blocks per processor in average, G-MISP is the best choice. Its ability to maintain locality and its speed give it the edge. The WD algorithm always gives decent results, and if none of the special cases above apply, it is a safe bet. The mapping from quadrant III and IV does not differ much from each other. The RSB algorithm is a good choice for both quadrants. It is interesting (and in line with earlier results [14]) to see that WD could be used for quadrant IV, since it is expertizing in cases of scattered refinements. Clearly, for cases III and IV it pays off to invest more time in the partitioning phase.

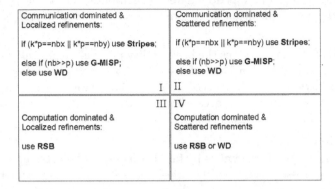

Fig. 5. The figure shows an attempt to establish a mapping from quadrants onto partitioning schemes. The complexity indicates that the quadrant approach is not powerful enough to fully capture the behavior of parallel SAMR applications

From the results, it is clear that a mapping from quadrants onto partitioning schemes is non-trivial, or even impossible, to achieve. The quadrant approach is not powerful enough to capture the nature of blockwise SAMR applications executed on parallel computers. These experiments show that parameters like graph granularity and whether number of processors divide number of rows or columns, must be included in the model.

In future work, we will extend the classification model to include the issues discussed above. Moreover, the high-quality results of RSB lead us to believe that better implementations, or even other scratch-remap techniques, could be competitive for many, or all, cases. We feel that this is important, since the block graphs are generally relatively small. Also, the partitioning library Vampire hosts a family of dedicated SAMR partitioners, which would be natural to include in the experiments.

In this paper, four partitioning schemes have been evaluated for four different blockwise SAMR applications on two parallel computers. The results in Fig 5 is intended to constitute a first step towards the dynamic meta-partitioner. It gives a mapping from quadrants onto partitioning schemes, i.e. a means of letting the current state of the application and computer system determine the partitioning requirements.

Acknowledgments. The authors would like to thank Manish Parashar for great collaboration and fruitful discussions regarding partitioning of general grid hierarchies. We also would like to thank Sumir Chandra, Andreas Kähäri, Jarmo Rantakokko, Pia Johansson and everyone at TASSL.

References

1. Stephen T. Barnard and Horst D. Simon. A fast multilevel implementation of re-cursive spectral bisection for partitioning unstructured problems. Technical report, NAS Systems Division, NASA Ames Research Center, 1994.
2. M. J. Berger and P. Colella. Local adaptive mesh refinement for shock hydrody-namics. *Journal of Computational Physics*, 82, 1989.
3. Marsha J. Berger and Joseph Oliger. Adaptive mesh refinement for hyperbolic partial differential equations. *Journal of Computational Physics*, 53, 1983.
4. L. Ferm and P. Lötstedt. Blockwise adaptive grids with multigrid acceleration for compressible flow. *AIAA J.*, 37:121–123, 1999.
5. G. Karypis, K. Schloegel, and V. Kumar. PARMETIS - parallel graph partitioning and sparse matrix ordering library, version 2.0. Univ. of Minnesota, Minneapolis, MN, 1998.
6. P. Lötstedt and S. Söderberg. Parallel solution of hyperbolic PDEs with space-time adaptivity. In D. Hänel R. Vilsmeier, F. Benkhaldour, editor, *Finite Volumes for Complex Applications II*, pages 769–776, Paris, 1999. Hermes Science.
7. Chao-Wei Ou and Sanjay Ranka. Parallel remapping algorithms for adaptive pro-blems. Technical report, Center for Research on Parallel Computation, Rice Uni-versity, 1994.
8. M. Parashar, et al. A common data management infrastructure for adaptive algo-rithms for PDE solutions. Technical Paper at Supercomputing '97, 1997.
9. John R. Pilkington and Scott B. Baden. Partitioning with spacefilling curves. Technical Report CS94-34, CSE, UCSD, March 1994.
10. K. G. Powell, et al. A solution-adaptive upwind scheme for ideal magnetohydro-dynamics. *J. Comput. Phys.*, 154:284–309, 1999.
11. Jarmo Rantakokko. *Data Partitioning Methods and Parallel Block-Oriented PDE Solvers*. PhD thesis, Uppsala University, 1998.
12. Hans Sagan. Space-filling curves. *Springer-Verlag*, 1994.
13. H. Samet. The design and analysis of spatial data structure. *Addison-Wesley, Reading, Massachusetts*, 1989.
14. K. Schloegel, et al. A performance study of diffusive vs. remapped load balancing schemes. Technical Report 98-018, Dept. of Computer Science, Univ. of Minnesota, Minneapolis, MN, 1998.
15. Mausumi Shee, Samip Bhavsar, and Manish Parashar. Characterizing the per-formance of dynamic distribution and load-balancing techniques for adaptive grid hierarchies. 1999.
16. Johan Steensland. http://www.tdb.uu.se/~johans/. Work in progress, 1999.
17. Johan Steensland. Vampire homepage, http://www.caip.rutgers.edu/~johans/vampire/. 2000.
18. Johan Steensland, Sumir Chandra, Manish Parashar, and Michael Thuné. Charac-terization of domain-based partitioners for parallel samr applications. Submitted to PDCS2000, 2000.
19. Erlendur Steinthorsson and David Modiano. Advanced methodology for simulation of complex flows using structured grid systems. *ICOMP*, 28, 1995.

Parallelizing an Adaptive Dynamical Grid Generator in a Climatological Trace Gas Transport Application

Jörn Behrens

Technische Universität München,
D-80290 München, Germany,
Internet: www-m3.ma.tum.de/m3/behrens/,
Email: behrens@mathematik.tu-muenchen.de

Abstract. Two disjoint paradigms in scientific computing, namely adaptivity and parallelism respectively, have been combined, in order to achieve very high local resolution for climatological simulations. The basis for the simulation is a parallel adaptive unstructured triangular grid generator. Load balancing is achieved using a space-filling curve approach. The space-filling curve data partitioning shows a very desirable feature: It keeps data local even in a time-dependent simulation. Moreover, it is very fast and easily parallelizable.

1 Introduction

The simulation of atmospheric trace gas transport requires very high local resolution, when fine spatial structures are to be revealed. In a simulation on a serial workstation type computer, a local resolution of 5 km could be achieved [1]. This experiment used a serial adaptive grid generator for atmospheric and oceanic applications, called amatos. A snapshot of the simulation is presented in figure 1.

In this study we combine adaptive mesh generation with parallelization in order to achieve the highest possible local resolution. These two paradigms of scientific computing often seem mutually exclusive, because adaptivity changes the data layout at runtime. Adaptive grid generation requires a fast and parallel dynamic load balancing process and an optimized data exchange between processors. We focus our attention on the load balancing strategy in this paper.

Dynamic load balancing has been studied for example by Diekmann et al., Jones and Plassman, Schloegel et al., and Walshaw et al. with good results [3, 6,9,10]. All these approaches operate on the graph of the grid. However, we adopt a different approach, based on a space-filling curve (SFC), because of its simplicity and its good characteristics regarding to edge length between partition boundaries (edge-cut when thinking in terms of a graph) and data locality. The SFC operates on the geometry of the mesh. A SFC approach has been mentioned by Roberts et al. [8]. Griebel and Zumbusch introduced SFC as a method of

T. Sørevik et al. (Eds.): PARA 2000, LNCS 1947, pp. 170–176, 2001.
© Springer-Verlag Berlin Heidelberg 2001

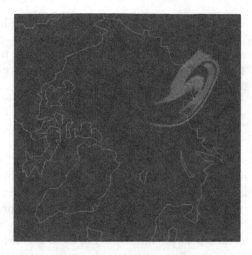

Fig. 1. Fine grain structures (filaments) can be observed in a trace gas transport simulation due to high spatial resolution. Continental outlines are given for orientation.

ordering sparse (rectangular) grids [4]. Details of the techniques used in our studies are given in [2].

The next section gives a brief overview on the parallel mesh generator **pamatos**. Section 3 demonstrates some of the characteristics of **pamatos**. Finally we draw conclusions from our experiments in section 4.

2   Parallel Mesh Generation with Space-Filling Curves

In this section we will give a brief introduction to **pamatos**, a parallel adaptive mesh generator for atmospheric and oceanic simulations. The programming interface (API) to **pamatos** consists of a small number of elementary routines. As **pamatos** is written in Fortran 90, the interface is a Fortran 90 module that allows access to the grid data structures and that contains methods for creating, refining, coarsening and destroying a grid.

2.1   Structure of the Parallel Program

One paradigm behind the design of **pamatos** is that all numerical operations are performed on arrays of consecutive data, while all mesh manipulations operate on the (object oriented) grid items. Thus, in each time step a gather-step collects data from grid items into arrays, and a scatter-step stores results back to the mesh. This paradigm allows for an almost unchanged API for the parallel and the serial version of the grid generator.

Parallelization is hidden away from the user of **pamatos**. Each process hosts a partition of the grid. However, the partitioning is performed within the grid generation library transparently. The utilization of lists allows for an optimized data communication within the grid generator. Data are collected in lists, and sent in large chunks, in order to avoid communication overhead. Several levels of abstraction (a communications primitives layer building on top of the communication library MPI, a high level communication layer on top of the primitives and a list layer on top of that) allow to handle the complexity of parallelizing the dynamic adaptive grid generation.

2.2 Partitioning with a Space-Filling Curve

A SFC, originally introduced by Peano and Hilbert [5], can be used to project a multidimensional (discrete) domain to a (discrete) unit interval. The SFC used for the triangular grid generation in **pamatos** is given in figure 2. The SFC can be refined arbitrarily by definition. Note that the SFC's nodes are chosen fine enough to cover each triangle in the finest level of local grid refinement. The SFC can be easily calculated for arbitrary triangular grids by a recursive algorithm. The algorithm resembles the refinement procedure of the triangulation, in our case a bisection algorithm. Details are given in [2].

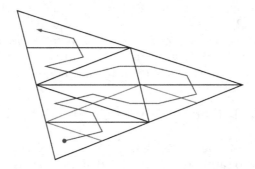

Fig. 2. Space-filling curve in a locally refined triangular grid

The mesh partitioning is straight forward. The curve's nodes can be numbered consecutively, inducing an ordering of all mesh elements. Now, distributing the consecutive list of elements evenly among the processes, yields an optimally load balanced partition of the grid. Interestingly this partition also has nice properties, when looking at the edge-cut or connectedness of the partitions (data locality). This is due to the neighborhood preserving character of the SFC.

Parallelization of the SFC partitioning is easy. The recursive indexing of SFC-nodes is embarrassingly parallel and requires no communication at all. Ordering of the global index set requires communication, but it can be implemented efficiently by pre-ordering the local index sets in parallel. Determination of bounds of the local index sets is again achievable in parallel.

3   Results

We performed tests on a clustered eight-processor SGI Origin 200 configuration. Our timing results show good scalability of the SFC partitioning algorithm (see table 1). Note that the super-linear speedup results from caching effects.

Table 1. Execution times and relative speedup (normalized to two processors) for the space-filling curve indexing (4342 elements).

no. of processors	2	4	8
time [ms]	68.7	26.5	14.9
speedup	1	2.6	4.6

The edge-cut is small and load balancing is almost optimal. Table 2 gives the load balancing parameter and the relative edge-cut. The *load balancing parameter* l is given by $l = e_{max}/e_{min}$, where e_{max} and e_{min} is the maximum and minimum number of elements on a single processor respectively. The *edge-cut* is the number of edges, that belong to more than one processor. This corresponds to the number of edges cut by a partition boundary in the dual graph (see [9]). The *relative edge-cut* is the edge-cut as a percentage of total edges. The number of edges differs for different numbers of processors, because edges can occur multiple times on different processors. This is the reason, why optimal load balancing can only be achieved, when looking at the elements per processor. However, the load imbalance with respect to edges or nodes is small (about 15%).

Table 2. Load balancing and relative edge-cut for the SFC-based algorithm on the model problem.

no. of processors	2	4	8
load balancing parameter	1.000	1.002	1.004
relative edge cut	0.03	0.04	0.06

Using the SFC partitioning algorithm yields well behaved partitions in an adaptive trace gas transport simulation. Especially, when looking at time-dependent simulations, the SFC partitioning yields higher locality of data from time step to time step. This is visualized in figure 3. We compare the partitions given by Metis [7] to those given by the SFC algorithm for two consecutive time steps. Note that the SFC based partition remains almost unchanged, while Metis repartitions the complete mesh, due to changes in the graph of the mesh.

Finally, the SFC based algorithm is fast. In our case, a serial version of the SFC based mesh partitioning is a factor of 5 faster then Metis on the same mesh.

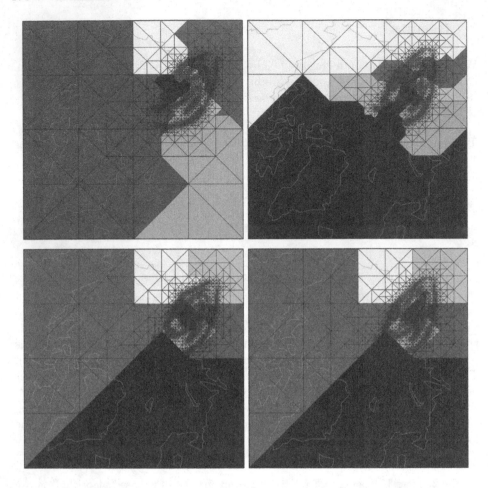

Fig. 3. Two consecutive meshes from a trace gas transport application. Metis redistributes almost every single element (upper row) while the SFC-based algorithm leaves most elements on the original processor (lower row).

This results in an acceleration of the complete adaptation step by a factor of 2.5.

4 Conclusions

We described briefly the parallelization of a dynamic adaptive grid generator in an atmospheric trace gas transport simulation application. The SFC approach for dynamical load balancing shows good performance and yields nicely connected partitions.

In the future **pamatos** can be used to implement different kinds of atmosphere and ocean models. The software is especially suited for semi-Lagrangian type algorithms, because it contains interpolation methods of different orders for calculating values between grid points. **pamatos** can also be easily extended to implement finite element based solvers for implicit fluid dynamics algorithms.

The parallel mesh generator **pamatos** has several sisters. A serial version (called **amatos**) has been implemented, and at the moment work is carried out on a version for spherical geometries (**samatos**).

The SFC based partitioning shows promising results and will be investigated further. It can be extended to three-dimensional tetrahedral meshes. Its property to operate on the geometry of the mesh makes it suitable for time-dependent calculations.

Acknowledgments. We would like to thank Jens Zimmermann (Ludwig-Maximilians-Universität, Munich) for implementing the recursive SFC algorithm. Furthermore we are indebted to Klaus Dethloff and Annette Rinke (both Alfred-Wegener-Institute for Polar and Marine Research, Potsdam) for the fruitful collaboration on arctic atmospheric tracer transport simulation.

References

1. J. Behrens, K. Dethloff, W. Hiller, and A. Rinke. Evolution of small scale filaments in an adaptive advection model for idealized tracer transport. Report TUM-M9812, Technische Universität München, Technische Universität München, Fakultät für Mathematik, D-80290 München, October 1998.
2. J. Behrens and J. Zimmermann. Parallelizing an unstructured grid generator with a space-filling curve approach. Technical Report TUM-M0002, Munich University of Technology, Center of Math. Sciences, D-80290 Munich, Germany, 2000. http://www-lit.ma.tum.de/veroeff/html/000.65001.html.
3. R. Diekmann, D. Meyer, and B. Monien. Parallel decomposition of unstructured FEM-meshes. In *Proceedings of IRREGULAR 95*, volume 980 of *Lecture Notes in Computer Science*, pages 199–215. Springer-Verlag, 1995.
4. M. Griebel and G. Zumbusch. Hash-storage techniques for adaptive multilevel solvers and their domain decomposition parallelization. *Contemporary Mathematics*, 218:279–286, 1998.
5. D. Hilbert. Über die stetige Abbildung einer Linie auf ein Flächenstück. *Math. Ann.*, 38:459–460, 1891.
6. M. T. Jones and P. E. Plassmann. Parallel algorithms for the adaptive refinement and partitioning of unstructured meshes. In *Proceedings of the Scalable High Performance Computing Conference*, pages 478–485. IEEE Computer Society Press, 1994.
7. G. Karypis and V. Kumar. *Metis – A Software Package for Partitioning Unstructured Graphs, Partitioning Meshes, and Computing Fill-Reducing Orderings of Sparse Matrices*. University of Minnesota, Dept. of Computer Science/ Army HPC Research Center, Minneapolis, MN 55455, 1998. Version 4.0.
8. S. Roberts, S. Kalyanasundaram, M. Cardew-Hall, and W. Clarke. A key based parallel adaptive refinement technique for finite element methods. Technical report, Australian National University, Canberra, ACT 0200, Australia, 1997.

9. K. Schloegel, G. Karypis, and V. Kumar. Multilevel diffusion schemes for repartitioning of adaptive meshes. *J. Par. Distr. Comp.*, 47:109–124, 1997.
10. C. Walshaw, M. Cross, M. Everett, and S. Johnson. A parallelizable algorithm for
    partitioning unstructured meshes. In A. Ferreira and J. D. P. Rolim, editors, *Par-
    allel Algorithms for Irregular Problems: State of the Art*, pages 25–46, Dordrecht,
    1995. Kluwer Academic Publishers.

Optimal Parameter Values for a Parallel Structured Adaptive Mesh Refinement Algorithm*

Michael Thuné and Stefan Söderberg

Uppsala University, Information Technology, Dept. of Scientific Computing, Box 120
SE-751 04 Uppsala, Sweden
{michael, stefans}@tdb.uu.se

Abstract. A blockwise approach to parallel structured adaptive mesh refinement is considered. The initial, coarse grid is divided into $n_1 \times n_2$ blocks. Subsequent refinements are carried out with respect to entire blocks.

The paper addresses the issue of choosing n_1 and n_2 optimally. A theoretical model for the execution time is formulated. Subsequently, it is suggested how to minimize the execution time with respect to the number of blocks. The approach is validated for test cases, where it successfully predicts the optimal choice of granularity. Finally, it is discussed how this can be automatized and integrated into the SAMR code.

Keywords: SAMR, parallel, granularity, optimization

1 Introduction

The simulation of complex phenomena, described by partial differential equations (PDEs), requires adaptive numerical methods [10] and parallel computers. The general idea is to begin computing on a coarse grid. Then, the pointwise errors in the computed solution are estimated, and points where the accuracy is too low (relative to a user-defined threshold) are flagged. Subsequently, the grid is refined in areas around the flagged points. This procedure is iteratively repeated, so that the grid continues to adapt to the evolving solution.

The goal for adaptive mesh refinement procedures is to attain a desired accuracy without using an unnecessarily large number of grid points. There is then a trade-off between the gain in reducing the number of grid points, and the loss in terms of overhead costs for the grid adaption. For adaptive mesh refinement in connection to *structured* grids, the Berger-Colella approach is often used [1]. There, the refinement consists of three steps: flagging (as explained above), clustering of flagged points, and grid fitting around such clusters. The refinements are thus inserted as new grid patches on top of the underlying coarser grid. The result is a hierarchical composite, structured grid [5].

* The research was supported by the Swedish Foundation for Strategic Research via the programme Industrial Computational Mathematics, and by Uppsala University via a faculty grant.

The present paper considers an alternative, *blockwise* approach to parallel structured adaptive mesh refinement (SAMR), with the purpose of avoiding the clustering and grid fitting steps. The initial, coarse grid is divided into $n_1 \times n_2$ blocks. Subsequent refinements are carried out with respect to *entire* blocks. That is, if some points in a block are flagged for refinement, then the whole block is refined. This approach to SAMR has been studied from a mathematical/numerical point of view in [3], [4]. A similar approach was used in [6]. As an example, Figure 1 shows a model problem and the corresponding blockwise refined grid. In this case $n_1 = n_2 = 8$. Many blocks remain coarse, others have been refined, some of them twice.

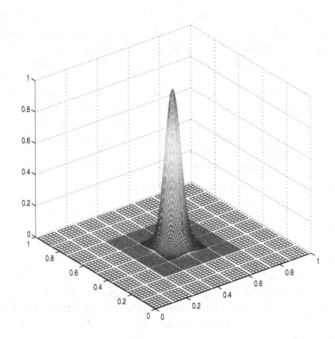

Fig. 1. A a model problem and the correspoding grid structure generated by the blockwise SAMR code. In this example, many blocks remain coarse, others have been refined, some of them twice.

The blockwise approach is appealing, since it avoids the complication of clustering/grid fitting. However, the choice of granularity (number of blocks) is crucial for the execution time of the blockwise adaptive algorithm. With $n_1 = n_2 = 1$, i.e., only one block, there is no gain at all, since the entire grid becomes refined. On the other hand, with a very large number of blocks, although the number of unnecessary grid points is strongly reduced, there is a significant overhead in terms of copying and interpolation of data between block boundaries. Figure 2 shows the execution time as a function of granularity $\sqrt{n_b}$ (here, $n_1 =$

$n_2 = \sqrt{n_b}$) for a 2D fluid flow simulation, solving the Euler equations on a square grid. Evidently, the granularity has a significant impact on the execution time.

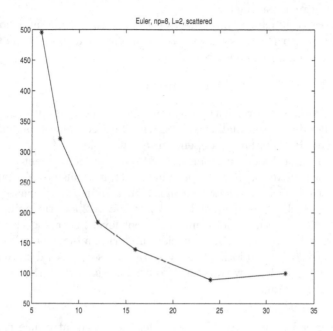

Fig. 2. The execution time in seconds as a function of granularity $\sqrt{n_b}$ for a 2D fluid flow simulation, solving the Euler equations on a square grid. This illustrates that the granularity has a significant impact on the execution time.

Thus, for the success of the blockwise approach, it is important to have a mechanism for selecting appropriate values of the parameters n_1 and n_2. In the present version of our SAMR code, these are input values given by the user. A better alternative would be to equip the code with an algorithm for automatic selection of optimal values of n_1 and n_2. The present paper presents a first step in that direction.

We introduce a theoretical model for the execution time of the parallel SAMR algorithm. Subsequently, we suggest a way of minimizing the execution time with respect to the number of blocks. The approach is validated for test cases, where it successfully predicts the optimal choice of granularity. Finally, it is discussed how this idea can be automatized and integrated into the SAMR code.

2   A Theoretical Model for the Execution Time

Let $n_b = n_1 n_2$ denote the number of blocks. In the following, we assume that $n_1 = n_2 = \sqrt{n_b}$. This can be generalized by introducing the aspect ratio of the grid as an additional parameter.

The number of blocks mainly affects the simulation time, and not the time for load balancing. It is thus sufficient to consider the execution time for a single *simulation step*. This can be expressed as:

$$t_s(\mathcal{P}) = t_a(\mathcal{P}) + t_c(\mathcal{P}). \tag{1}$$

Here, t_a is the time spent on arithmetic operations, and t_c is the communication time (in a general sense, including also copying between blocks belonging to the same partition). Both t_a and t_c depend on the partition $\mathcal{P}$.

Since the original blocks are maintained, and since refinement always affects a complete block (changing its level of refinement), it seems appropriate to count the load in terms of *coarse block equivalents*. If a coarse block has $k_1 \times k_2$ grid points, then a block on refinement level l has $2^l k_1 \times 2^l k_2$ grid points. Thus, in terms of arithmetic work, a block on refinement level l corresponds to 4^l coarse blocks. In terms of copying between block boundaries, a block on refinement level l corresponds to 2^l coarse blocks. For a given processor, let c_a denote its number of coarse block equivalents with respect to arithmetic, and c_c the corresponding with respect to copying.

The time for arithmetic work. In modelling t_a, we introduce the following parameters, in addition to those already mentioned:

- n_f, the number of floating point operations per grid point;
- t_f, the average time for a floating point operation;
- N the total number of coarse grid points (i.e., N is the size of the initial grid, before any refinement has taken place).

Note, that the number of grid points per coarse block is N/n_b. Consequently, the number of grid points in a particular processor, given its value of c_a, is $c_a N/n_b$. For the individual processor, the expression for t_a thus becomes:

$$t_a = c_a \frac{N}{n_b} n_f t_f. \tag{2}$$

The time for communication during simulation. There is both communication *between* processors, and communication between blocks *within* processors. (In practice, the latter will be implemented as copying of data from one block boundary to another.) In the present context, the external communication time can be disregarded, since it does not depend significantly on n_b. (The communication can be collected into one message for each neighbouring processor.)

In modelling the internal communication time, we introduce the following parameters:

– t_{copy}, the average time for copying the data associated to one grid point (this quantity depends on the computer architecture, and on the number of bytes to be copied for each point on the block boundary);
– m, the number of communication phases per simulation step;
– σ, the number of sides of a single block, times the number of lines of ghost cells along each side.

Since the side length of a coarse block is proportional to $\sqrt{N/n_b}$, the total amount of data to be copied in a processor, given its value of c_c, is proportional to $c_c\sqrt{N/n_b}$. For the individual processor, the internal communication time for a simulation step can then be expressed as:

$$t_c^{(i)} = \sigma m c_c \sqrt{\frac{N}{n_b}} t_{copy}. \tag{3}$$

3   An Approach to Finding the Optimal Number of Blocks

3.1   General Considerations

The optimal choice of $\sqrt{n_b}$ is thus the one that minimizes

$$g = \max\left(t_a + t_c^{(i)}\right),$$

where the maximum is taken over all processors. The quantities t_a and $t_c^{(i)}$ depend on n_b both explicitly, and via the parameters c_a and c_c. The other parameters involved are constants.

Since $\sqrt{n_b}$ has to be an integer value of modest size, it is appropriate to address this optimization problem with a search algorithm. However, the function g has local minima, since, for a given number of processors, certain values of $\sqrt{n_b}$ are more advantageous from the point of view of load balance. Thus, a straightforward golden section search [2] is not applicable.

A closer look at (2) and (3) reveals that $t_a = \mathcal{O}(1)$ and $t_c^{(i)} = \mathcal{O}(\sqrt{n_b})$. (The observation follows from the fact that both c_a and c_c are $\mathcal{O}(n_b)$.) Case studies confirm this. Fig. 3 shows the typical behaviour. As $\sqrt{n_b}$ increases, t_a rapidly approaches a constant value, whereas $t_c^{(i)}$ grows linearly.

Now, let η^* denote the optimal choice of $\sqrt{n_b}$, i.e., g assumes its minimum for $\sqrt{n_b} = \eta^*$. Moreover, let $\bar{\eta}$ denote the point where t_a ceases to decrease and begins to behave as an (almost) constant function. Clearly,

$$\eta^* \leq \bar{\eta}, \tag{4}$$

since for $\sqrt{n_b} > \bar{\eta}$, t_a is constant whereas $t_c^{(i)}$ grows linearly.

The relation (4) gives an upper bound for the search interval that the optimization algorithm has to consider. A trivial lower bound is $\sqrt{n_b} \geq \sqrt{p}$ since we want to have at least one block per processor.

In our case studies, we have observed that $\bar{\eta}$ is a relatively small value. Thus, we have come to the following *straightforward optimization algorithm*:

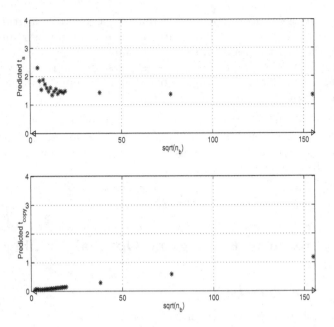

Fig. 3. The typical behaviour of t_a (top) and $t_c^{(i)}$ (bottom) as a function of $\sqrt{n_b}$. As $\sqrt{n_b}$ increases, t_a rapidly approaches a constant value, whereas $t_c^{(i)}$ grows linearly

1. Choose a value η_0, such that $\eta_0 > \bar{\eta}$.
2. Carry out a linear search for the optimum, by evaluting g for all integer values of $\sqrt{n_b}$ in the interval $[\sqrt{p}, \eta_0]$.

In the case studies reported below, this procedure has been adequate. The final search interval is small enough that a linear search is sufficiently efficient.

So far, we have chosen η_0 heuristically. In all our case studies, it has turned out to be sufficient with a value of $\eta_0 = 40$. (In fact, an even smaller value would have been enough.) Finding a more stringent way of choosing η_0 is part of our continued investigations.

3.2 Computing the Coarse Block Equivalents

Having found an appropriate optimization algorithm, the remaining issue to resolve is how to compute c_a and c_c. The coarse grid equivalents depend on n_b itself, on the number of processors p, and on the refinement pattern that develops during the simulation. For a given processor, they also depend on the partition. Consequently, they can not be known a priori, but must be estimated.

The crucial part is the refinement pattern. Let there be $L + 1$ refinement levels, from the initial coarse grid (level 0) to the finest refinement (level L). Moreover, let n_l denote the *total* number of blocks on refinement level l (on all processors together).

Now, assume that there is an *estimated* refinement pattern for the case $\sqrt{n_b} = \eta_0$, i.e., the upper limit of the initial search interval. Let this pattern be given as a matrix B ($\eta_0 \times \eta_0$), where $B_{i,j}$ is an integer denoting the refinement level of the corresponding block. (*How* to make such an estimate is discussed below.)

For a value of $\sqrt{n_b}$ in the search interval, we can now estimate the refinement pattern by dividing B into $\sqrt{n_b} \times \sqrt{n_b}$ boxes (as equally sized as possible). The refinement level for each box will be the maximum of the corresponding elements of B. Given the refinement pattern, it is straightforward to compute $n_l, l = 0, \ldots, L$.

Next, we theoretically distribute these blocks onto the p processors. In our case studies below, we used a bin-packing algorithm for this. Thus, we get local values of n_l on each processor. The local values of c_a and c_c for the processor will then be

$$c_a = \Sigma_{l=0}^{L} 4^l n_l$$

and

$$c_c = \Sigma_{l=0}^{L} 2^l n_l,$$

respectively. These values are used for computing $g(\sqrt{n_b})$.

4   Experimental Validation

4.1   Background

For the experimental validation of the approach suggested above, we used a Fortran 90 code, which implements the blockwise SAMR algorithm for problems in two space dimensions. The code is parallel of SPMD type, and uses MPI for the communication between processors.

The tests were set up with the purpose to cover a range of cases with different properties:

a. A first series of tests involved the 2D scalar convection equation on the unit square. It was solved by means of a second order finite volume approximation in space, and a two-stage Runge-Kutta method in time. This leads to a relatively small *computation-to-communication* ratio.

   In a second series of tests we solved the nonlinear Euler equations, in 2D, on the same domain as above and with the same numerical approach. Here, the computations clearly dominate over the computations.

b. In each of the two test series, we varied the refinement pattern by using different initial conditions. For the convection equation, the initial condition

$$u(0, x, y) = f(x, y, 0.5, 0.5),$$

   where

$$f(x, y, x_0, y_0) = \exp\left(-300 * \left((x - x_0)^2 + (y - y_0)^2\right)\right),$$

generates a case with strongly localized refinements, whereas the condition

$$u(0, x, y) = f(x, y, 0.25, 0.25) + f(x, y, 0.25, 0.75) + f(x, y, 0.75, 0.25) +$$
$$f(x, y, 0.75, 0.75) + f(x, y, 0.5, 0.5)$$

leads to more scattered refinements. Similar initial conditions were used for the Euler equations.

c. Moreover, for each combination of equation and initial condition, we varied the number of refinement levels, by varying the accuracy requirements.

d. Finally, each of the cases obtained from different combinations of (a)–(c) was executed on different sizes of processor sets.

4.2   Tests with a Posteriori Data

Since the purpose is to apply the optimization algorithm as a preprocessing step to the parallel SAMR algorithm, the input data to the algorithm have to be available a priori, i.e., before the parallel SAMR algorithm has been executed. This means in particular that the refinement pattern that develops during the execution of the SAMR algorithm must be estimated a priori.

However, in a first series of tests, we used a posteriori input data to the optimization algorithm. The purpose was to find out if the algorithm was at all useful. (If the algorithm would fail in *these* tests, then it would certainly not be of any value for realistic a priori data.)

The a posteriori input data were generated in the following way. We first ran the simulation, using a large number of blocks. The resulting refinement pattern was registered, and used as input to the optimization algorithm (the matrix B). Then, we calculated a theoretically optimal value of $\sqrt{n_b}$. This was compared to the experimentally observed optimal value of $\sqrt{n_b}$.

The parallel computer platform used in these experiments was a cluster of three Dec Alpha Server 8200 (each with four Alpha EV5 processors), connected via the Memory Channel. The code was compiled using the DIGITAL Fortran 90 V5.1 compiler.

In this series of tests, we solved the convection problem, with localized refinements. We ran two different cases, with different numbers of refinement levels, $L := 1$ and $L = 2$, respectively. Each test was carried out on two different processor configurations, with $p = 4$ and $p = 8$ processors, respectively.

The results were very encouraging: *In all four tests, the theoretically computed optimal value of $\sqrt{n_b}$ agreed with the one experimentally observed.*

4.3   Tests with a Priori Data

The results for cases with a posteriori data established that the optimization algorithm is basically sound. It was thus meaningful to go on with a series of tests with a priori data, the way the optimization algorithm is intended to be used in practice.

In these tests, we used a Sun Ultra Enterprise 6000, with 16 UltraSparcII processors, and the Sun WorkShop Compiler Fortran 90 2.0 for Solaris 7.

The a priori data were collected in the following way. The PDE problem (to be solved by the parallel SAMR algorithm) was solved in a preliminary, *very inexpensive* run on a very coarse grid (40×40 grid points). The error estimator was activated, but *no adaption was carried out*. The results of the preliminary simulation were used to generate estimates of the refinement pattern.

Using the input data generated in this way, the optimization algorithm predicted an optimal value for $\sqrt{n_b}$. The experimental optimum for comparison was found by carrying out full parallel blockwise SAMR simulations, with an inital coarse grid of 384×384 grid points. Our experimental code assumes that n_b is a divisor of the number of grid points. Thus, the experiments were repeated for the following values of $\sqrt{n_b}$: 6, 8, 12, 16, 24, 32 (whereas the optimization algorithm could generate *any* integer value in the interval $[\sqrt{p}, \eta_0]$).

The results of the experiments with a priori data were similar for different numbers of processors. The results for $p = 8$ are summarized in Table 1. The rows represent the communication dominated convection equation, and the computation dominated Euler equations. The columns represent the different refinement patterns, localized and scattered, respectively. For each of these four quadrants, there are two cases, corresponding to different numbers of refinement levels L.

Table 1. Predicted/observed optimal value for $\sqrt{n_b}$, for the different test cases described in the text. The number of blocks recommended by our optimization algorithm comes close to the experimentally observed optimum in most of the cases.

	Localized refinements		Scattered refinements	
	$L = 1$	$L = 2$	$L = 1$	$L = 2$
Convection	5/16	16/16	14/16	24/16
Euler	10/24	28/24	24/24	21/24

The number of blocks recommended by our optimization algorithm comes close to the experimentally observed optimum in all but three cases. Moreover, in the case of the convection equation, with scattered refinements, and $L = 2$, although the predicted optimal value of $\sqrt{n_b}$ (24) is far from the observed one (16), the *execution time* for the predicted granularity is close to the optimal exection time. Thus, there are only *two* cases where our algorithm failed, in that it recommended a value of $\sqrt{n_b}$ which led to an unnecessarily large execution time. These two cases are the ones with localized refinements, and $L = 1$, i.e., cases with a relatively low computation-to-communication ratio. This suggests that we may need to revise the model for cases with this property. This is part of our continued research.

5   Conclusions and Discussion

The conclusion of the tests reported above is that the suggested optimization approach is very promising, and consequently worth pursuing further. To get feed-back for improving the details of the algorithm, we will conduct further experiments, using test problems with different characteristics, and varying the number of processors.

Moreover, we will take the steps needed for automatizing the approach. The key issue in that context is how to find the initial estimate of the refinement pattern a priori. The idea to run an inexpensive preliminary simulation on a very coarse grid worked well in the tests above, and will consequently be our main alternative.

In some situations, the user may have some additional insight into the application, that makes it possible to predict the refinement pattern. Then, this can be used as an alternative way of generating the necessary input.

Finally, we note that a problem similar to the one treated here, is how to choose the granularity of the blocks in the block-based partitioning approach of [7], [8], [9]. The approach suggested above will be applicable in that context as well, but then the theoretical model for the execution time needs to be reformulated. This is another line of continued research.

References

1. M. J. Berger and P. Colella. Local adaptive mesh refinement for shock hydrodynamics. *J. Comp. Phys.*, 82:64–84, 1989.
2. W. Cheney and D. Kincaid. *Numerical Mathematics and Computing*. Brooks/Cole Publishing Company, Pacific Grove, CA, fourth edition, 1999.
3. L. Ferm and P. Lötstedt. Blockwise adaptive grids with multigrid acceleration for compressible flow. *AIAA J.*, 37:121–123, 1999.
4. P. Lötstedt and S. Söderberg. Parallel solution of hyperbolic pdes with space-time adaptivity. In D. Hänel R. Vilsmeier, F. Benkhaldour, editor, *Finite Volumes for Complex Applications II*, pages 769–776, Paris, 1999. Hermes Science.
5. M. Parashar, et al. A common data management infrastructure for adaptive algorithms for PDE solutions. Technical Paper at Supercomputing '97, 1997.
6. K. G. Powell, et al. A solution-adaptive upwind scheme for ideal magnetohydrodynamics. *J. Comput. Phys.*, 154:284–309, 1999.
7. J. Rantakokko. Partitioning strategies for structured multiblock grids. Accepted for publication in Parallel Computing, special issue on Graph Partitioning and Parallel Computing.
8. J. Rantakokko. Comparison of partitioning strategies for PDE solvers on multiblock grids. In B. Kågström, J. Dongarra, E. Elmroth, and J. Wasniewski, editors, *Applied Parallel Computing, 4th International Workshop, PARA'98, Lecture Notes in Computer Science, No. 1541*, Berlin, 1998. Springer-Verlag.
9. J. Rantakokko. A framework for partitioning structured grids with inhomogeneous workload. *Parallel Algorithms and Applications*, 13:135–152, 1998.
10. E. Steinthorsson and D. Modiano. Advanced methodology for simulation of complex flows using structured grid systems. Technical Report 95-28, ICOMP, NASA Lewis Research Center, Cleveland, OH, 1995.

Partition of Unstructured Finite Element Meshes by a Multilevel Approach

Noureddine Bouhmala[1] and Xing Cai[2]

[1] Vestfold College, Norway
noureddine.bouhmala@hive.no
[2] Department of Informatics, University of Oslo. P.O. Box 1080, Blindern,
N-0316 Oslo, Norway
xingca@ifi.uio.no

Abstract. We consider the topic of partitioning unstructured finite element meshes by a class of multilevel graph partitioning algorithms. Two issues are studied, where the first issue concerns the coarsening phase in such multilevel graph partitioning algorithms. In particular, we propose a new heuristic for matching the vertices of a graph during the coarsening phase. We compare our heuristic with two other known matching heuristics in respect of matching ratio and quality of the final partition. As the second issue of the paper, we look at the relation between the parallel efficiency of finite element computation and different aspects of the partition quality.

Keywords: parallel finite element computation, unstructured meshes, multilevel graph partitioning algorithms.

1 Introduction

The finite element (FE) method is used extensively in the numerical solution of partial differential equations (PDEs). Such FE computations often arise from discretizations that are carried out on unstructured FE meshes, which typically consist of triangles in 2D and tetrahedra in 3D. If parallel FE computations are to be run on unstructured meshes in the style of domain decomposition (see e.g. [6]), the original large global mesh needs to be explicitly partitioned into a set of small sub-meshes, which are to be hosted by different processors. The quality of the mesh partition determines, to a great extent, the achievable parallel efficiency. Not only should the partition be well-balanced in terms of computation volume per processor, but the volume of the resulting inter-processor communication should be kept as small as possible.

2 Parallel FE Computation

In the present paper, we restrict our attention to non-overlapping partition of unstructured FE meshes. If we denote an FE mesh by $\mathcal{M}$, then the result of such

T. Sørevik et al. (Eds.): PARA 2000, LNCS 1947, pp. 187–195, 2001.

a non-overlapping partition is a set of sub-meshes $\mathcal{M} = \cup_{i=1}^{p} \mathcal{M}_i$. The partition of the elements of $\mathcal{M}$ is non-overlapping in that each element of $\mathcal{M}$ belongs to only one sub-mesh $\mathcal{M}_j$. Mesh nodes lying on internal boundaries between sub-meshes, however, have multiple owners. Computation associated with those nodes needs contributions from all of their owner sub-meshes. So communication between processors is necessary. To limit the CPU-time overhead spent on the inter-processor communication, it is desirable to keep both the number of inter-processor messages and the message size as small as possible. The following terms will be used in Sect. 5 to quantify the inter-processor communication volume:

- $N_{i,j}$ denotes the number of nodes belonging to both $\mathcal{M}_i$ and $\mathcal{M}_j$, indicating the size of the message to be exchanged between processor number i and j.
- $N = \sum_{i \neq j} \sum_{j=1}^{p} N_{i,j}$ indicates the entire volume of communication.
- D_i denotes the number of neighbors for sub-mesh $\mathcal{M}_i$. More precisely, D_i is the size of an index set $\mathcal{J}_i$ such that $j \in \mathcal{J}_i \Rightarrow N_{i,j} > 0$, $j \neq i$. The average number of neighbors is denoted by $D = \sum_i D_i / p$.

3   Formulation of Graph Partitioning Problem

The above-mentioned problem for a non-overlapping partition of an unstructured FE mesh can be formulated as a graph partitioning problem. The FE mesh is to be associated with a graph $G(V, E)$ that consists of vertices and edges connecting them, where every vertex v has weight 1 and corresponds to an element of the FE mesh. An edge between two vertices v and u indicates that the two corresponding elements are neighbors. There are two criteria for determining whether two elements of an FE mesh are neighbors:

1. Two elements of an FE mesh are considered neighbors if they share at least one common nodes. An edge thus connecting the two corresponding graph vertices will be assigned an integer weight that is the number of shared nodes between the two elements. This criterion will be used for the present study.
2. Two elements of an FE mesh are considered neighbors in the same way above for 1D FE meshes. However, two neighboring elements are required to share a common side in 3D or a common edge in 2D. All edges of the resulting graph will be assigned weight 1. We note that a graph arising from using this criterion will have fewer edges compared with that using the above criterion.

The objective of partitioning a graph $G(V, E)$ is to obtain a set of nonempty and disjointed sub-graphs $\{G_{s_i}(V_i, E_i)\}_{i=1}^{p}$, where $V = \cup_i V_i$, $V_i \cap V_j = \emptyset$ for $i \neq j$. The sum of weights of all the vertices $|V_i|$ of sub-graph G_{s_i} should be as close to $w_i |V|$ as possible, where $w_i > 0$ are prescribed partition weights and we have $\sum_i^{p} w_i = 1$. We note that for perfect load balance we have $w_i = 1/p$. In addition, the sum of weights of all the edges whose incident vertices belong to different sub-graphs should be as small as possible. This quantity is normally referred to as the *edge-cut* of a partition, we hereby denote it by EC. To measure the quality of load balance, we introduce two more quantities:

$$\gamma_1 = \frac{\max_i \frac{|V_i|}{w_i}}{|V|}, \quad \gamma_2 = \frac{\max_i \frac{|V_i|}{w_i} - \min_i \frac{|V_i|}{w_i}}{|V|},$$

where $\gamma_1 \geq 1$ and $\gamma_2 \geq 0$ and both quantities are desired to have as small values as possible.

4   Multilevel Graph Partition

The graph partitioning problem is NP-complete. Recent research activity in this field has been focused on finding efficient algorithms that are able to produce reasonably good partitioning result. One particular class of partitioning algorithms is the so-called *multilevel* algorithms; see e.g. [3] and the references therein. Multilevel graph partitioning algorithms are made up of the following three phases:

Coarsening phase. The coarsening phase aims at building a sequence of smaller graphs from the original graph using a coarsening scheme such that $G \equiv G^0$ and $|G^0| > |G^1| > |G^2| > \ldots > |G^m|$. Given a graph G^j, a coarser graph G^{j+1} is obtained by merging two and two adjacent vertices of G^j, where the edge linking the two vertices is collapsed and a supervertex is formed. The supervertices are weighted by the sum of the weights of the merged vertices. In order to preserve the connectivity information, a superedge between two supervertices u and v is the union of all the edges connecting the supervertices u and v.

Initial partitioning phase. The hitherto coarsest graph G^m is partitioned according to a load-balancing criterion in respect of w_i.

Uncoarsening and refining phase. The partition of G^m is projected backward level by level. When the partition on level G^{j+1} is projected back to level G^j, the partition is refined on G^j to reduce the edge-cut and, if possible, to improve the quality of load balance.

4.1   Different Matching Heuristics

In the coarsening phase, the vertices of a graph G is traversed in a random order. If vertex v is not yet matched, we try to match it with one of its unmatched neighboring vertices. One of the objectives of the present paper is to study the effect of different matching heuristics. We hereby propose a so-called *gain vertex matching* heuristic and study it together with two known heuristics. The definitions of the three heuristics are as follows.

1. Random matching (RDM) - a random unmatched neighbor u is picked to match v.
2. Heavy edge matching (HEM) - an unmatched neighboring vertex u is chosen, if the edge (v, u) has the heaviest weight among all the unmatched neighbors.
3. Gain vertex matching (GVM) - an unmatched neighboring vertex u is chosen to minimize the total weight of edges that will come out from the new supervertex formed by v and u.

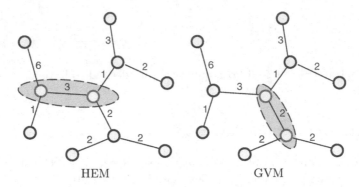

Fig. 1. A comparison between the matching heuristics HEM and GVM, where the two encircled vertices are to form a supervertex

As a comparison between the matching heuristics HEM and GVM, we refer to Fig. 1 where the vertex in the center is matched with one of its three neighbors. Numerical experiments in Sect. 5 will show that GVM is able to reduce the number of vertices during the coarsening phase more rapidly than both RDM and HEM. This can be intuitively explained by considering a graph where all the edges have equal weight. In this case, it is easy to see that GVM will match a vertex with a neighbor that has the fewest neighbors. Vertices with small numbers of neighbors are thus matched first, leaving vertices with more neighbors to be matched later. This is advantageous in respect of obtaining a large matching ratio. We also mention that GVM can be very efficiently implemented. Before going randomly through the vertices, every vertex i is first assigned with an integer se_i, which is equal to the summed weight of all the edges coming out from vertex i. Then the summed weight of all the edges coming from a supervertex, which is formed by merging vertices u and v, can be easily obtained as $se_u + se_v - 2e_{uv}$.

4.2   Details of the Two Other Phases

The size of the coarsest graph G^m is normally very small, so we can afford to use any of the sophisticated graph partitioning algorithms for the initial partition. For the present paper we have used a recursive bisection procedure for the initial partition of G^m, which typically has only a couple of hundreds vertices. After each bisection during the initial partition procedure, the edge-cut between the two sub-graphs is reduced by using the Kernighan-Lin heuristic [4].

During the uncoarsening and refining phase, when the partition is projected from G^{j+1} to G^j, we traverse all the vertices of G^j in a random order. Suppose vertex v is located on the internal boundary, i.e., at least one of v's neighbors belongs to a different sub-graph than that of v. We then allow v to be moved to one of its neighboring sub-graphs, see Fig. 2, provided that the move does not cause the edge-cut to increase and one of the following conditions is fulfilled:

1. The load balance in form of $\gamma_1 + \gamma_2$ is improved;
2. The new values of γ_1 and γ_2 are both within some prescribed thresholds.

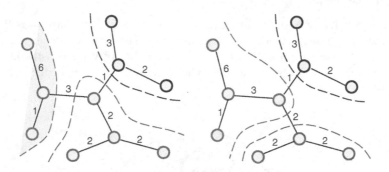

Fig. 2. An example of moving a vertex from one sub-graph to another one for reducing the edge-cut

Figure 3 shows one example of how the uncoarsening and refining phase is able to reduce the edge-cut. In this example the edge cut is reduced from 5,628,483 on the coarsest graph level to 1,529,995 on the finest graph level.

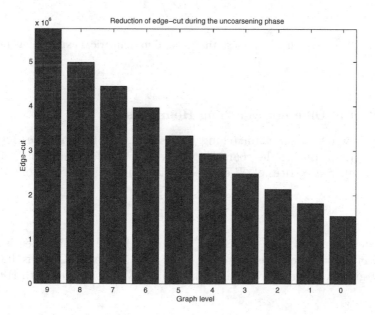

Fig. 3. An example of reducing the edge reduction by the uncoarsening and refining phase

5   Some Numerical Experiments

We study two test cases of partitioning unstructured FE meshes. The concerned graphs are as follows:

- Graph G1 arises from a 2D FE mesh; see Fig. 4. The graph has 260,064 vertices and 1,564,129 edges, where the weight of each edge is either 1 or 2.
- Graph G2 arises from a 3D FE mesh, its number of vertices is 320,048 and the number of edges is 11,540,788, where the weight of each edge is between 1 and 3.

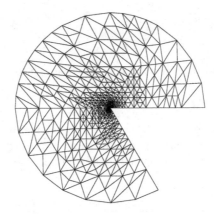

Fig. 4. The 2D unstructured FE mesh that is used for numerical experiments in Sect. 5

5.1   Effect of Different Matching Heuristics

To study how fast different matching heuristics are able to reduce the size of coarsened graphs during the coarsening phase, we define a matching ratio ϱ as half the number of vertices of G^j divided by the number of vertices of G^{j+1}, i.e.,

$$\varrho = 0.5 \cdot \frac{|G^j|}{|G^{j+1}|}.$$

The developments of ϱ associated with the three matching heuristics are depicted in Fig. 5. The larger matching ratio associated with GVM is due to the fact that graphs get smaller more quickly with GVM compared with HEM and RDM.

5.2   Parallel Efficiency of FE Computation

To see how the parallel efficiency of FE computation depends on the quality of partition, we choose to run 500 conjugate gradient (CG) iterations on the

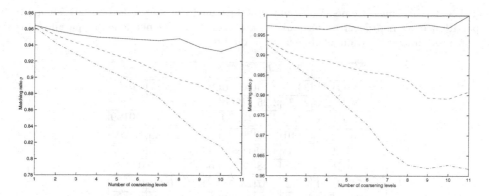

Fig. 5. Effect of different matching heuristics on the coarsening phase. The left diagram depicts the matching ratio developments associated with G1; whereas the right diagram is associated with G2. In both diagrams, the GVM matching heuristic is represented by the whole line, RDM by the dashed line and HEM by the dash-dotted line

linear system of equations arising from discretizing the Poisson equation on the concerned FE meshes. The FE simulation code is written in Diffpack (see [2,5]).

The measurements are obtained on a cluster of 24 PC nodes running the Linux operating system. The total number of processors is 48. More specifically, the Linux cluster consists of 24 dual Pentium-III computing nodes, where every processor runs at 500MHz. The computing nodes are inter-connected with a 100Mbit/s ethernet network, through a 26-port switch. Such clusters have relatively slow communication, therefore keeping the average number of neighboring sub-meshes small is critical for achieving good parallel efficiency of FE computation. Take for instance the 3D FE computation running on 48 processors, see Table 2. If the RDM matching heuristic is used, the average number of neighbors D immediately after the initial partition is 19.04, whereas it decreases to 10.63 after the uncoarsening and refining phase. The impact of this decrease of D is such that the CPU time is reduced from 1882.06 seconds to 9.24 seconds!

6   Concluding Remarks

Although different coarsening heuristics tend to reduce the graph size at different speeds, as Fig. 5 shows, they do not have a decisive effect on the final partition result, let alone the parallel efficiency of FE computation. This is because the objective of the uncoarsening and refining phase is to minimize the edge-cut, which does not completely reflect the communication overhead of parallel FE computation, as is pointed out in [1]. We can observe from Tables 1 and 2 that for achieving high efficiency of parallel FE computation, especially when the parallel computer is utilized at its full capacity, it is very important for the partitioning algorithm to keep the average number of neighbors D as small as possible. Thereafter small values of EC, γ_1 and γ_1 are also advantageous.

Table 1. CPU-measurements of running 500 parallel CG iterations on a 2D FE mesh partitioned by multilevel graph partitioning algorithms. Dependency of parallel efficiency on different aspects of the partition quality

		CPU	D	EC	N	γ_1	γ_2
$p = 8$	GVM	23.05	2.75	10560	2427	1.0087	0.0199
$p = 8$	HEM	23.35	3.25	11966	2776	1.0077	0.0200
$p = 8$	RDM	22.76	3.50	12290	2841	1.0100	0.0199
$p = 16$	GVM	12.01	3.88	20694	4792	1.0065	0.0204
$p = 16$	HEM	12.45	4.00	18906	4368	1.0111	0.0253
$p = 16$	RDM	12.46	4.38	23707	6562	1.0284	0.0587
$p = 24$	GVM	9.73	4.67	23414	5455	1.0234	0.0486
$p = 24$	HEM	8.98	4.42	22395	5180	1.0321	0.0586
$p = 24$	RDM	8.55	4.08	24848	5811	1.0286	0.0569
$p = 48$	GVM	6.57	5.21	46652	12396	1.0460	0.1178
$p = 48$	HEM	6.09	4.96	36875	9435	1.0371	0.0842
$p = 48$	RDM	6.32	5.17	38776	9046	1.0397	0.1000

Table 2. CPU-measurements of running 500 parallel CG iterations on a 3D FE mesh partitioned by multilevel graph partitioning algorithms. Dependency of parallel efficiency on different aspects of the partition quality

		CPU	D	EC	N	γ_1	γ_2
$p = 8$	GVM	20.17	5.75	636583	11430	1.0082	0.0209
$p = 8$	HEM	19.88	5.50	622199	10969	1.0106	0.0241
$p = 8$	RDM	20.04	5.50	631833	11286	1.0111	0.0233
$p = 16$	GVM	12.56	7.88	940900	16965	1.0341	0.0931
$p = 16$	HEM	12.21	7.63	896595	15844	1.0086	0.0209
$p = 16$	RDM	12.36	7.38	903281	16052	1.0239	0.0600
$p = 24$	GVM	9.38	8.92	1131259	20969	1.0289	0.0846
$p = 24$	HEM	9.25	8.58	1119779	19816	1.0259	0.0613
$p = 24$	RDM	9.30	8.75	1166809	21063	1.0197	0.0502
$p = 48$	GVM	8.58	10.13	1529995	27945	1.0443	0.1098
$p = 48$	HEM	7.88	9.92	1488135	26786	1.0413	0.0829
$p = 48$	RDM	9.24	10.63	1515539	27721	1.0567	0.1350

References

1. E. Elmroth, *On grid partitioning for a high performance groundwater simulation software.* In B. Engquist et al (eds), Simulation and Visualization on the Grid. Springer-Verlag, Berlin. Lecture Notes in Computational Science and Engineering, No. 13, 2000, pp 221-233.
2. Diffpack Home Page, *http://www.nobjects.com/Products/Diffpack.*
3. G. Karypis, V. Kumar, *Multilevel k-way partitioning scheme for irregular graphs.* J. Parallel Distrb. Comput. 48 (1998), pp. 96–129.
4. B. W. Kernighan, S. Lin, *An efficient heuristic procedure for partitioning graphs.* The Bell System Technical Journal 49 (1970), pp. 291–307.

5. H. P. Langtangen. *Computational Partial Differential Equations – Numerical Methods and Diffpack Programming.* Springer-Verlag, 1999.
6. B. F. Smith, P. E. Bjørstad, W. D. Gropp, *Domain Decomposition, Parallel Multilevel Methods for Elliptic Partial Differential Equations.* Cambridge University Press, 1996.

GRISK: An Internet Based Search for K-Optimal Lattice Rules

Tor Sørevik* and Jan Frode Myklebust**

Dept. of Informatics,
University of Bergen,
NORWAY

Abstract. This paper describe the implementation and underlying philosophie of a large scale distributed computation of K-optimal lattice rules. The computation is huge corresponding to the equivalent of 36 years computation on a single workstation. In this paper we describe our implementation, how we have built in fault tolerence and our strategy for administrating a world wide computation on computer we have no account on.

We also provide a brief description of the problem, that of computing K-optimal lattice rules, and give some statistics to describe the extent of the search.

1 Introduction

Using idle cycles on Internet connected workstations for large scale computation is an appealing idea. The combined power of hundreds or thousands of workstations equals the computational power of a top-of-the-line supercomputer. Moreover, utilizing the idle cycles on networked workstations is an essential free-of-cost supercomputer, compared to the multi million dollar price tag on a high end HPC-system.

Nevertheless only a few heroic examples on successful large scale computation on Internet connected systems exists. The most noticeable being, the factorization of RSA 129 [AGLL94] and the record breaking Traveling salesman computation by Bixby et. al. [ABCC98].

Given todays hype about GRID-computing [FK] this fact may seems strange. There are however, good reasons for the situation. First of all the Internet has order of magnitude lower bandwidth and higher latency than the internal network in todays parallel computers, thus communication intensive parallel applications don't scale at all on WAN (Wide Area Network) connected computers. Secondly, one can't assume that each computational node in a NOW (Network of Workstations) has equal power. This might be due to difference in hardware as well a computational load imposed by other usage. Consequently one need a dynamical load balancing scheme which can adapt to available resources of the NOW and

---

* http://www.ii.uib.no/~tors
** http://www.ii.uib.no/~janfrode

T. Sørevik et al. (Eds.): PARA 2000, LNCS 1947, pp. 196–205, 2001.
© Springer-Verlag Berlin Heidelberg 2001

the demand of the application. Most existing parallel program do apply static load balancing, and can for that reason not be expected to run well on NOWs.

These two reasons alone exclude most of the problems traditionally solved on parallel computers. There is however still important subclasses of large scale problems which possess a course grain parallelism and lends itself naturally to a dynamic scheduling approach. But only few of these problems are solved by this style of wide area, Internet base, computing. Again there is good reason for this. One is the problem of having heterogeneous platforms working together. Another is the administration problem; how to get your program running on far away computers where you don't have an account and without disturbing the interactive computation going on? Reliability and fault tolerance is also a problem which have to be dealt with. With hundreds of system running for weeks and even months, you can't expect them all to run without any interference. Thus you need a survival strategy in case of system crash on any of these

In this paper we report on our experiences with Internet based computation on a problem that satisfy the *Problem requirements:* (communication"thin" and dynamic load balancing). We explain our solution on how to deal with heterogeneous platforms, the administration problem and fault tolerance.

This paper is organized as follows. In section 2 we briefly describe our problem, that of finding K-optimal lattice rules in dimension 5. The underlying algorithm and how it is parallelized is explained in section 3. In section 4 we describe our implementation and our approach to heterogeneity, administration and fault tolerance. In section 5 we present statistics of our search to give a flavor of the size and involvement. In section 6 we give some references to related work, while we in section 7 sum up our experiences and indicate the direction in which we intend to develop GRISK in near future.

2   K-Optimal Lattice Rules

Lattice rules are uniform weight, numerical integration rules for approximating multidimensional integrales over the s-dimensional unit cube. It takes its abscissaes as those points of an integration lattice, Λ, which lies within $[0, 1)^s$.

$$Qf = \sum_{\forall x \in \Lambda \cap [0,1)^s} f(x) \qquad (1)$$

An integration lattice is a lattice which contains all integer points in the s-dimensional space. This makes its reciprocal lattice an integer lattice. The truncation error of a lattice rule is conveniently described by its error functional

$$Ef = If - Qf = \sum_{\forall r \in \Lambda^\perp} a_r(\hat{f}) \qquad (2)$$

Where $a_r(f)$ is the Fourier coefficients of the integrand and $\Lambda^\perp$ is the reciprocal lattice. Since the Fourier coefficients decay rapidly for large $| r |$ for natural periodic integrands, we seek lattice rules, Q, for which $\Lambda^\perp$ has as few points as few points as possible for small $|\, r\, |$.

A standard way of meassuring the quality of an integration rule for periodic intergrands is the trigonometric degree. For lattice rules this correspond to The 1-norm of the point on $\Lambda^\perp$ closest to the origin, δ, defines the trigonometric degree, d, as $d = \delta - 1$.

A cubature rule needs not only be accurate, a good rule is also economical. A convenient measure of the cost of the rule is $N(Q)$, the number of function evaluation needed to carry out the calculation. This depends directly on the density of the points in the lattice $\Lambda^\perp$.

An optimal rule of degree, d, is optimal if its abscissas count is less or equal to the abscissa count of any other rule of degree d. Optimal rules are known for all degrees in dimension 1 and 2. For higher dimension optimal rules are only known for the degree 0,1,2 and 3. Since the number of lattice rules with $N(Q)$ less or equal a fixed number is finite. It is, in theory possible to do a computer search through all these lattices to find optimal lattices. The number of potenial interesting lattice rules is however huge and increases exponentially with the dimension. Thus to conduct such a search in practice the search needs to be restricted and even then lots of computing resources is needed.

Cools and Lyness [CL99] have recently conducted a computer search for good lattice rules in dimension 3 and 4. They define a class of lattices $K(s,\delta)$ and restrict their search to either this class, or well defined subclass of this. There are good reasons to believe that any optimal lattice rule belongs to class $K(s,\delta)$, but this has not been proved. They have termed the rules found in their search K-optimal.

We have modified their code so it can do the search in any dimension and parallelized it in a master slave style. The parallelism is very coarse grained. A typical task takes anything from 1 to 100 hours to do, and the amount of data exchange between the master and a slave is very small.

3   Brief Outline of the Algorithm

A set of s linearly independent vectors provides a basis for an s-dimensional lattice. If we arrange these vectors as rows in an $s \times s$ matrix. This matrix is called a generator matrix for the lattice. There is a beautiful theory [Lyn89] stating that a lattice is an integration lattice if and only if the generator matrix of its reciprocal lattice is an integer matrix. The idea behind the set $K(s,\delta)$ is that lattices of this class have a generator matrix for which the 1-norm of each row is exactly δ. We can than carry out a search by creating all such lattices, checking their abscissas count, and if it is a potential good one, compute its degree. To speed up the search a number of short cuts are taken. These are described in the paper by Cools and Lyness [CL99].

Think about the entire search as s nested loops. In each we loop over all possible $\mathbf{b}_i$, such that $\|\, \mathbf{b}_i\, \|_1 = \delta$. This will generate a huge, but finite number of lattices with degree at most $\delta - 1$. We could now compute the degree of each of these lattices as well as the abscicca number of the associated lattice rule. This is in principal what we do, though in pratice a number of short cuts designed to speed up the computation are taken. These are described in the paper by Cools and Lyness [CL99].

3.1 Parallelization

In the naive search described above the computation for each lattice is totally independent. The speed-up techniques do however try to eliminate computation in the $s - 2$ inner loops. To take full advantage of the speed-up techniques we can't parallelize over these loops. However, the iteration of the two outer loops remains totally indepent and thus we are parallelizing over these. A parallel 'task' is identified by the indeces of the 2 outer loops.

Solving a 'task' is equivalent to creating all possible combination of the $s - 2$ last rows and do the above described computation. The time needed to to compute a task shows huge differences. In the simplest case some simple checksums proves that none lattices with the two upper rows of the specified task as basis vector can have a degree equal $\delta - 1$ thus we can return immidiately. In the computaion reported in section 5 as much as 40 % of the tasks belong to this class of "quick-return". This computation is done within milliscconds. If the task is not in the class of "quick-return" it typically might need hours. But also within this class we find huge differences.

The computational work of the different tasks is impossible to predict in advance. Thus to efficiently load balancing the computation, the distribution of tasks needs to be done dynamically. In our case this is quite trivial. The different tasks are totally independent, thus as soon as a client return with an answer the next task in line is forward to that client.

4 Implementation

4.1 The Distributed Search

A central server is keeping the pool of tasks, keeping track of what's done and not done, and updates the current best solution. Each client is running a small java program that connects to the server, retrieves a task and start the calculations.

The initiative is in the hand of the clients. The server only reponds to request. A task consists of a task id which is used to construct the two basis vector constituting a task, and an integer vector, specifying problem parameters such as δ and upper and lower bounds for the abscissa count. The client-Java program then calls a fortran library which does the actual calculations of the task.

4.2 Implementation

Java is an ideal language when you program for an unknown number of different system. Being platform independent, Java program runs on whatever platform you happen to have. Moreover, there should be no possibility for nasty format mismatch when two Java program communicate, even if they run on totally different systems running totally different OS. In our case we've found that this holds not only in theory, but in practice as well.

However, of reasons such that history, laziness and computational speed, the core numerical number crunching is carried out in Fortran. The Fortran code has of course to be compiled for the different OS it is supposed to run under. We have compiled Fortran library for all the systems we have had access to, which is a wide variety of Unix-system. We have experienced some problem with the binary fortran-code on some system, even if it is compiled on the approriate version of the particular OS and the same compiler release. The problem is due to differences in the local installation of the runtime libraries. This is without any comparison the main barrier to seemless portability of our code. First it imposes a significant amount of extra work in moving the code around, compiling it for N different systems and keeping it updated with the latest OS and compiler version, secondly it is still no foolproof guaranty for portability.

By using RMI (Remote Method Invocation), we quickly had a simple and elegant method for communicating distributed java objects. One possible enhancement of our program might be to instead of using a native fortran library, we could have the routine for solving the problem included in the distributed objects. Then we would have a platform and problem independent distributed solver.

4.3   Administration

The administration burden we have distributed to our Internet-friends. What we do is to make a compressed tar-file available at our web-page http://www.ii.uib.no/grisk, which anyone can download, unzip, untar and kick-off on all systems that NSF-mounts the catalogue with the downloaded executable. This takes you about 10 minutes of which eight is needed to read the instruction. The run-script assure that the code runs with lowest possible priority, and since the code is pretty slim memory wise (less than 10 MB) you hardly notice that it runs on your desktop when you're processing your daily computing there. (Try it!!)

The administration duties left to us is those related to maintaining the web-page and keeping the server on the road. The server collects all the incoming data which, including statistics on the search. This add up to no more than 20 MB. The workload on the server is low. We run the server program on an O2-workstation which also process a full load of standard workstation tasks. The extra load imposed by our server program has been unnoticeable, even when more than 100 clients have been actively engaged in the search.

4.4   Fault Tolerance

More than 250 systems have been taking turns as clients. Only few, if any, have been active all the time. The owner must have full freedom to abort the client-program any time he likes. If that happens the computation of that particular task is lost and have to be recomputed. The server keeps track of, not only which task he has sent out, but also which he receives answer to. After having waited for an answer a couple of days there is a high possibility that the client is down. But it might of course well be that the system has been occupied with more pressing computation, and only few cycles has been allocated to the low priority grisk-job. There is however no need to any immediate action. Thus recomputation of unfinished tasks are postponed to the very end of the computation. This is perfectly all right as there is no reason to believe that the order in which the tasks are completed has any significant influence on the overall running time, and it has definitely no influence on the computational results. For $\delta = 11$ there is 50505 tasks of these 5.6% (3030 tasks) where redistributed. The 5.6% provides an upper bound for the extra computational cost due to recomputation. We find this an acceptable cost, and our fallback scheme simple and reliable.

Cases of permanently network failure will be treated as if the client is aborted. Temporarily network failure are dealt with by letting the client resubmit its message to the server every 5. minute. If the server fails, it will not stop active clients from working, but when they finish with their current task they will have no place to deliver their answer, nor any possibility to get new ones. In the first place this will be treated by the client as a network failure. It simply resubmit the sending of results every 5. minute. If the server recover, we are back in business. To recover a server process we do however rely on manual supervision. Keeping in mind that the server runs 24 hour a day, 7 days a week, through holidays and vacations this is not foolproof. Even the grisk-administrator takes occational breaks. We had, however, no serious problems with the search in the 6 months it took place.

In case of a total disaster at the server, like serious hardware failure, the server might be replaced by another piece of hardware. The address for the clients message is just an IP-address and it does not care who is processing it at this address. Backup of the results are taken once every night, so in case of disk crash the latest results will have to be recomputed.

5 Some Statistics on the Search

A test version of the code was used to compute k-optimal rules for $\delta = 10$. Some bugs were fixed and improvements made for the final version which was used for the $\delta = 11$. The great search had kick-off November 11 1999 and the last task was recieved at the server at May YY 2000. In this periode 263 systems[1] asked for tasks to compute and 218 from 13 different countries delivered one or more answers.

---

[1] More precisely; We have recorded request from 263 IP-adresses. We know that in some cases this is a firewall, and don't know how many systems which have been running grisk behind this firewall.

The top ten systems were:

Name processors	No. of tasks
parnass2.iam.uni-bonn.de 450 Mhz Pentium II 144 CPU linux cluster, University of Bonn	9423
cs.kotnet.kuleuven.ac.be Pentium II(Deschutes) 16 linux PC at a stuent lab at CS, K.U.Leuven	8531
lolligo.mi.uib.no 350 Mhz Pentium II A 10 CPU Linux cluster at Math. dept. UoB	4032
dontask-20.ii.uib.no 250 Mhz MIPS R10000 A 4 CPU Origin 2000 at Parallab, UoB	2178
dke.cs.kuleuven.ac.be Pentium III (Katmai) Single CPU linux PC at CS, K.U.Leuven	1066
cke.cs.kuleuven.ac.be Pentium III (Katmai) Single CPU linux PC at CS, K.U.Leuven	992
korkeik.ii.uib.no 300 Mhz Ultra Sparc-II A 4 CPU Sun Ultra 450 at Dept. of Inf., UoB	984
bke.cs.kuleuven.ac.be Pentium III (Katmai) Single CPU linux PC at CS, K.U.Leuven	719
madli.ut.ee ?? 'Something' at the University of Tartu, Estonia	706
pandora.cs.kuleuven.ac.be Pentium III (Katmai) Single CPU linux PC at CS, K.U.Leuven	695

The number listed for the multiprocessor system is the total number of CPU they have. Not all of these have been used for GRISK.

The multiprocessors systems were running multiple clients. No effort was made to parallelize the client program. Thus at peak in January there might have been as many as 300-400 CPUs simulatiously running grisk.

The server got 53595 requests of these 3030 never was completed. The total amount of CPU time used for those completed was: 1140638417 CPU-seconds $\approx$ 316844 CPU-hours $\approx$ 36 CPUyears. Which shows that there is no way this computation could be performed on a single workstation.

For comparison we note the May 8th 2000 annoucement from SGI were they claim that: "...The Los Alamos' Blue Mountain supercomputer recently set a world record by running 17.8 years of equivalent single-processor computing in just 72 hours. The Blue Mountain system, one of the world's fastest supercomputers with a peak speed of three trillion floating point operations per second (TFLOPS)..." Our computation is twice as big, has been performed without any additional hardware investment and moderate extra administration work.

The following table shows the distribution of tasks as a function of their completion time.

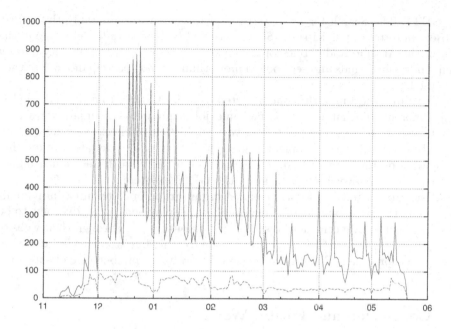

Fig. 1. The number of completed tasks pr. day in red and the number of contributing systems pr. day in green

	0-2 sec	2-10000 sec	10000-100000 sec	> 100000 sec	Sum
no of tasks:	20486	67	29542	761	50856
percentage	40.3	0.1	58.1	1.5	100
Total time	0	515982	1041182248	98940187	

The table shows that more than 40% of the tasks gave 'quick-return'. The fastest of the tasks, not included in the 'quick-return-class' used 6738 sec or almost 2 hours while the longest lasting task which was completed used 382383 sec or more than 100 CPU-hours. The huge bulk of the tasks use between 3 and 30 CPU-hours.

Note that there is no way to know the time consumption of a task in advance. Thus the table about provides a convincing arguement for the need of dynamic load balancing.

6   Related Work

A number of other groups are working on system for pooling together distributed resources. In this section we briefly mention some of them. A more comprehensive list of distributed computing project might be found at
http://distcomp.rynok.org/dcprojects.htm

SETI. The Search for ExtraTerrestrial Intelligence is probably the largest project in terms of participants. Signal received by largest radio telescope in the world, the Arecibo Radio Telescope, The signals are chopped up in very narrow frequency/time bands and sent out to more than 100 000 partcipants where they analysed.

Condor. The aim of the condor system is as for GRISK to utilize idle cycles. The difference is that it targets multiple jobs and works in many ways as a batch system. It does however have a very sophisticated check pointing and job migration system, which enable it migrate jobs from one workstation to another depending on the current load. This requires the server to have a much higher degree of control over its slave than in the GRISK system.

distributed.net is a loosely coupled organization of distributed individuals how share the common goal of "..development, deployment, and advocacy, to be pursued in the advancement of distributed computing...". Through their website they, very much like GRISK, invites everyone to contribute their cycles to solving large scale problems. They have successfully solve a number of code-cracking problems

7   Conclusion and Future Work

There is no way we could have carried out this computation on a traditional supercomputer. We don't have enough money to buy our own computer or sufficiently political influence to occupy enough computing time on a communitee system to accomplish the computation described above. We've found the results of this project very encouraging. Our problem seems tailored to large scale internet based computing. Using Java with RMI the implementation was rather easy and our strategy for fault tolerance and administration worked very well.

We will continue this work along two different axis. 1) Improvements on the core lattice-search program and 2) Improving the functionality of the internet search.

The lattice-search program will be extended to handle all different bases for k-optimal lattices rules. We also plan to implement a Java-only version of the entire system. Whether this will be used in next version depends on whether or not it can compete with FORTRAN in speed.

Planed improvements on the internet search include updating and improvements of the web-pages and better backup for the server program.

References

[ABCC98]   D. Applegate, R. E. Bixby, V. Chvatal, and W. Cook. On the solution of traveling salesman problems. *Documenta Mathematica*, ICM(III):645–656, 1998.

[AGLL94]   Derek Atkins, Michael Graff, Arjen K. Lenstra, and Paul C. Leyland. The magic words are squeamish ossifrage. In Josef Pieprzyk and Reihanah Safavi-Naini, editors, *Advances in Cryptology – ASIACRYPT '94*. Springer, 1994.

[CL99]   Ronald Cools and James Lyness. Lattice rules of moderate trigonome-
           tric degree. Technical report, Math. and CS. dept. Argonne Nat. Lab.,
           November 1999.
[FK]   Ian Foster and Carl Kesselman. *The Grid: Blueprint for a New Computing
           Infrastructure*. http://www.mkp.com/index.htm.
[Lyn89]   James N. Lyness. An introduction to lattice rules and their generator
           matrices. *IMA J. of Numerical Analysis*, 9:405–419, 1989.

Parallel and Distributed Document Overlap Detection on the Web

Krisztián Monostori, Arkady Zaslavsky, and Heinz Schmidt

School of Computer Science and Software Engineering
Monash University, Melbourne, Australia
{krisztian.monostori, arkady.zaslavsky, heinz.schmidt}@infotech.monash.edu.au

Abstract. Proliferation of digital libraries plus availability of electronic documents from the Internet have created new challenges for computer science researchers and professionals. Documents are easily copied and redistributed or used to create plagiarised assignments and conference papers. This paper presents a new, two-stage approach for identifying overlapping documents. The first stage is identifying a set of candidate documents that are compared in the second stage using a matching-engine. The algorithm of the matching-engine is based on suffix trees and it modifies the known matching statistics algorithm. Parallel and distributed approaches are discussed at both stages and performance reslults are presented.

Keywords: copy-detection, string matching, job-distribution, cluster

1. Introduction

Digital libraries and semi-structured text collections provide vast amounts of digitised information on-line. Preventing these documents from unauthorised copying and redistribution is a hard and challenging task, which often results in avoiding putting valuable documents on-line [7]. Copy-prevention mechanisms include distributing information on a separate disk, using special hardware or active documents [8]. One of the most current areas of copy-detection applications is detecting plagiarism. With the enormous growth of the information available on the Internet users have a handy tool for writing or compiling documents. With the numerous search engines users can easily find relevant articles and papers for their research. These documents are available in electronic form, which makes plagiarism feasible by cut-and-paste or drag-and-drop operations while tools to detect plagiarism are almost non-existent.

There are few systems built for plagiarism detection, namely, CHECK [18], Plagiarism.org [17], SCAM [9], Glatt [10], and the "shingling" approach of [3]. They are very similar in their approaches except for Glatt, which does not use computational power but rather assumes that everyone has a different style of writing and users more easily remember their own words rather than plagiarised sentences. The drawback of Glatt's approach is that users must be involved in the checking process, which makes this approach very cumbersome.

T. Sørevik et al. (Eds.): PARA 2000, LNCS 1947, pp. 206-214, 2001.
© Springer-Verlag Berlin Heidelberg 2001

Approaches using computational power include Plagiarism.org, SCAM, and the "shingling" approach. These approaches are common in the way that they are building a special purpose index on documents that reside either on the Internet or in a local repository. There are a few problems when we apply these indexing technics on large data sets. The first issue is finding an appropriate chunking granularity. It is impractical to index all possible chunks of a document. Suffix tree, which is used in this paper, is an index structure that holds all possible suffixes, therefore all possible chunks, of a document. However, suffix trees are too space-consuming to be used as an index [16], though they can be applied when comparing local documents.

Garcia-Molina et al. [9] study document overlap with different chunk sizes. Word chunking is obviously too fine granularity while sentence chunking is not obvious because sentence boundaries are not always easy to detect. Considering a number of words as a chunking primitive is easy to cheat because adding just one-word shifts chunk boundaries. Garcia-Molina et al. [9] propose a so-called hashed breakpoint chunking, which somehow overcomes this problem. The size of the index in the SCAM system is quite large: 30-60% of the size of the original documents depending on the chunking primitive. The "shingling" approach [3] selects chunks to index based on Rabin's fingerprints. Each ten-word chunk is considered but fingerprinting keeps only every 25th chunk. Using Rabin's fingerprinting method ensures that selection is random. The storage requirement of this method is more effective: for a 150 GByte data set it takes up only 10GB. The problem with this approach is the elimination of most of the possible chunks but larger chunks still have more chance to be caught because they participate in many chunks.

Because of the extensive space usage of indexes we focus on distributing the index and workload among nodes on the Internet and on a local cluster. Local clusters can be easily built of commodity workstations at low cost. Global clusters are also available by utilising the services provided by Globus [6] and Nimrod/G [1].

In the following sections we discuss our two-stage approach: during the first stage candidate documents are identified using an index and in the second stage candidate documents are compared using the matching engine[15]. Different parallel and distributed approaches of both stages are presented. In Section 2 we outline the algorithm underpinning the matching engine that identifies identical chunks of documents. Section 3 introduces our prototype system, MatchDetectReveal(MDR). Section 4 discusses how we can use the local cluster to analyse documents in the local repository. Section 5 gives a brief overview of the Globus system. Section 6 shows how we can use idle workstations and dedicated servers on the Internet to analyse Internet documents. In section 7 we summarise the work we have done and discuss future work.

2. Identifying Overlap in Documents

The basic problem we address could be summarised in the following way. If given a suspicious document, we want to identify all chunks that overlap with any document either in the local repository or on the Internet. To achieve our goal we use Chang's matching statistics algorithm [4]. Comparing document P to T the matching statistic value of ms(i) for P is the length of the longest chunk in T that matches the chunk in P starting at position i. It is obvious that we can identify the overall overlap percentage of document P if we have the matching statistics of P for each document.

The original algorithm builds a suffix tree for T and compares P to this suffix tree. We can use this algorithm in a reverse fashion by building a suffix tree only for P and compare all candidate documents to the suffix tree of P. It is important that the suffix tree is built only once because of potential time and cost saving. By storing some additional information on nodes we can even identify the starting position of a chunk matching another chunk in the suspicious document.

Suffix trees are very versatile [13] but they are very space-consuming data structures. In [16] we proposed a more space-efficient representation of suffix trees, which inserts only suffixes that start at the beginning of words because we are not interested in chunks starting in the middle of a word. Not only does this save space but also the building and matching statistics algorithms benefit from the space reduction towards performance objectives.

Figure 1 depicts the running time of the matching statistics algorithm on the modified suffix tree. We have a single document of 14K and we compare it to sets of documents with different total size. The figure bears out that the algorithm is linear.

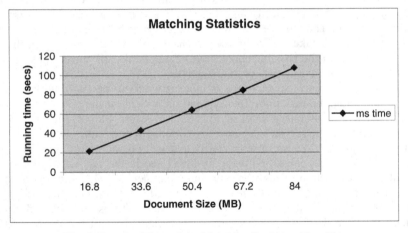

Fig. 1. Running Time of the Matching Statistics Algorithm

3. MatchDetectReveal(MDR) System Architecture

Figure 2 depicts the system architecture of our proposed prototype system MatchDetectReveal [15]. The core of the system is the Matching Engine, which uses the algorithms discussed in the previous section.

Texts must undergo some preprocessing before they can be presented to the Matching Engine. Not only does the Converter component convert documents in different formats (e.g. Adobe Portable Document Format - PDF, PostScript, MS Word etc.) into plain text but it also eliminates multiple whitespaces, converts each character to lowercase, and shifts some characters, so that the matching algorithm can work more efficiently.

The Matching Engine can only work on a limited number of documents in case of Internet documents because it needs complete documents. Therefore MDR uses a Search Engine, which builds a special-purpose index on the documents from the Internet and only candidate documents identified by the Search Engine are presented to the Matching Engine. Different indexing methods of search engines were discussed in Section 1.

Some additional components provide a user-friendly interface to the system. The Online Submission component enables users to submit documents via a simple Internet browser and results are presented to the user by the Visualiser component.

As depicted in *figure 2* we can use parallel and distributed processing of documents. The following sections discuss how to utilise a local cluster and after a short introduction to the Globus project [6], we propose different extended architectures to utilise idle nodes on a global scale.

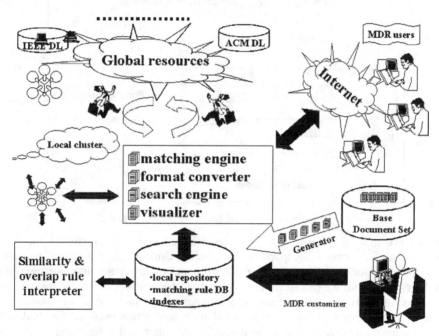

Fig. 2. MatchDetectReveal(MDR) System Architecture

4. Using a Local Cluster to Identify Overlapping Documents

There are several cluster systems built of commodity workstations for different applications including the Berkeley NOW project, the Beowulf project, and Solaris-MC [2]. Most of them provide a Single System Image (SSI), which hides the complexity of using several nodes. For our experiments we used the Monash Parallel Parametric Modelling Engine. It uses the Clustor tool [5] for transparent distribution and scheduling jobs. There are different ways of utilising a cluster of workstations.

One application is to use it as a final comparison engine in stage two. We identify candidate documents on the Internet, download them, and have different nodes compare them to the suspicious document. When using the Clustor tool we had to pay a time penalty of building a suffix tree for the suspicious document each time a new job is started. Batching candidate documents together, of course, can reduce this time penalty, and Clustor performs well in terms of job scheduling. We only have to provide a plan file, which provides a framework and Clustor does all distribution and scheduling of jobs. If we have one node that downloads documents and distributes them to other nodes using Clustor, the network can easily get congested, which is illustrated in *figure 3*. In this figure we compare the same batch of documents using different number of nodes. The figure shows that using more than 4 or 5 nodes does not add significant speed-up.

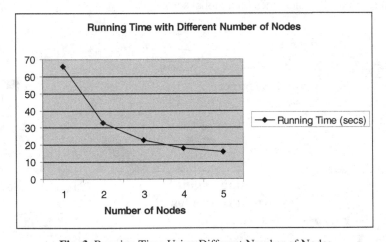

Fig. 3. Running Time Using Different Number of Nodes

If we use a local repository we can have a dedicated server storing the local repository and distribute comparison jobs using Clustor. In this case the same network-congestion problem comes up. If we bypass Clustor and use our own job-scheduling algorithm we can store the repository distributed on nodes. In this case we only have to distribute the suspicious document or its suffix tree and nodes can primarily compare documents on their local disks and when they are finished with local documents they can request documents from other nodes for comparison. No matter which approach we use we have to merge results from different nodes. Fortunately the matching statistics values are easy to merge: it is a simple maximum on each element of the matching statistics array.

5. The Globus Project

In this section we briefly introduce the Globus project [11]. The Globus project aims to provide a general metacomputing infrastructure that can connect heterogenous environments and act like a virtual supercomputer. Networked virtual supercomputers can both increase utilisation of resources and use unique capabilities of separate resources that otherwise could not be created effectively [6]. Three significant testbeds currently running include National Technology Grid by NPACI and NCSA,

NASA Information Power Grid (IPG), and the Globus Ubiquitous Supercomputing Testbed (GUSTO). The Monash Parallel Parametric Modelling Engine is part of the latter one. *Figure 4* shows the integrated grid architecture.

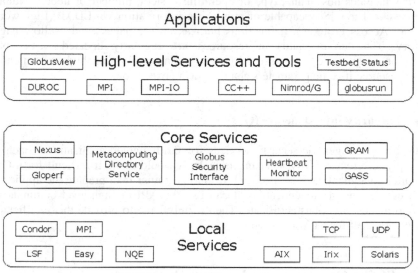

Fig. 4. Globus Architecture

At the lowest level there are local services, i.e. local supercomputers, clusters, or individual computers with their own operating systems, network connections, network libraries etc.

Core Grid services are built on top of local services to provide a uniform interface for higher-level components. Some important core services are detailed in the following subsections.

High-level services use core services to provide an interface for applications, which sit at the highest-level of the architecture. One of these high-level tools is Nimrod/G [1], which is a Clustor-like tool for the global grid rather than the local cluster. Nimrod/G uses a simple cost formula and allocates resources in a way that minimises cost while meeting time constraints. MPICH-G is the Globus version of MPI, which allows users to write applications for the global grid using the standard Message Passing Interface [12].

In the following subsection we briefly describe the three most important core services, i.e. Global Resource Allocation Manager, Grid Information Services (formerly known as Metacomputing Directory Service), and Grid Security Infrastructure.

5.1 Global Resource Allocation Manager (GRAM)

GRAM is a basic library service that enables remote job submission. There are specific application toolkit commands based on the top of this service. For example the `globus-job-run` command uses an RSL file (Resource Specification Language) to describe which resources to allocate and which jobs must be executed on those resources.

5.2 Grid Information Services (GIS)

The GIS stores information about the state of the grid. Information available about resources includes host name, type of operating system, number of nodes, available memory, etc. The GIS is capable of publishing information via LDAP (Lightweight Directory Access Protocol). It functions both as "white pages" and "yellow pages" directory. The current status of resources are regularly updated, so adaptive applications might for example use a different communication method or even choose another resource if some parameters change during execution.

5.3 Grid Security Infrastructure (GSI)

The GSI provides generic security services for Grid-enabled applications. Each site may apply different security mechanisms, which would make it difficult for a Grid-application to authenticate itself to each resource. Rather than prepare the application for different types of authentication, it can use the API or high-level commands of GSI, and request a Globus certificate. Then it only have to provide this certificate to authenticate itself for a given resource.

6. Identifying Overlap Using Internet Resources

In Section 1 we outlined the storage requirements of an appropriate index for overlap detection. At today's growth-rate of the Internet it is hard to store and browse such a large index. One option is to partition the index and store these partitions on a local cluster. Hence on comparison each node can analyse documents against their own index partitions and identify their own sets of candidate documents. These sets are then merged and candidate documents are analysed by the matching-engine. The matching process can be either distributed on the local cluster as discussed in the previous section, or we can use idle resources on the Internet to execute the final comparison task using the resource discovery capabilities of Globus.

Using GIS we can assign jobs to resources that are actually in the close proximity of the document to be analysed and downloading time can be reduced in this way.

Not only can our distributed index reside on a local cluster but it can also be distributed on the Internet. We can have servers in different subnets of the Internet, which are responsible for indexing their own subnets and comparison jobs can also be executed on those nodes. Results of these comparisons, that is sets of candidate documents, are collected and the final matching process can be either executed on the same nodes or scheduled to be run on different nodes.

There have been some efforts to use the Grid for data sharing [14], which may also be applied in our system. There could be a testbed set up for copy-detection. Each participating organisation (universities, digital libraries, etc.) would publish resources in this testbed for copy-detection jobs. A copy-detection task could be divided into jobs submitted to separate resources that would compare suspicious documents to their local documents.

7. Conclusion and Future Work

In this paper we discussed different overlap-detection schemes. We found that the storage and time requirement of identifying overlapping documents are large enough to consider parallel and distributed approaches. We introduced a two-stage process. In the first stage candidate documents are identified while in the second stage candidate documents are compared using our matching-engine. We showed how these two stages can be parallelised and distributed both on a local cluster and on the Internet.

Performance results are presented using the local cluster deployed in our school. They show that the Clustor tool looses efficiency when processing data-intensive jobs, so we are implementing our own job-distribution scheme to achieve better performance. We are also currently analysing the efficiency of distributed approaches and implementing a plagiarism-detection prototype named MatchDetectReveal (MDR).

Acknowledgment. Support from Distributed Systems Technology Centre (DSTC Pty Ltd) for this project is thankfully acknowledged.

References

1. Abramson, D., Giddy, J. and Kotler, L. High Performance Parametric Modeling with Nimrod/G: Killer Application for the Global Grid?. International Parallel and Distributed Processing Symposium (IPDPS), pp 520- 528, Cancun, Mexico, May 2000.
2. Baker M., Buyya R. Cluster Computing at a Glance in Buyya R. High Performance Cluster Computing. (Prentice Hall) pp. 3-47, 1999.
3. Broder A.Z., Glassman S.C., Manasse M.S. Syntatic Clustering of the Web. *Sixth International Web Conference*, Santa Clara, California USA. URL http://decweb.ethz.ch/WWW6/Technical/Paper205/paper205.html
4. Chang W.I., Lawler E.L. Sublinear Approximate String Matching and Biological Applications. *Algorithmica 12*. pp. 327-344, 1994.
5. Clustor Manual (1999). URL   http://hathor.cs.monash.edu.au/clustor/
6. Foster I., Kesselman C. Globus: A Metacomputing Infrastructure Toolkit. Intl J Supercomputer Applications 11(2), pp. 115-128, 1997.
7. Garcia-Molina H., Shivakumar N. (1995a). The SCAM Approach To Copy Detection in Digital Libraries. *D-lib Magazine*, November.
8. Garcia-Molina H., Shivakumar N. (1995b). SCAM: A Copy Detection Mechanism for Digital Documents. *Proceedings of 2nd International Conference in Theory and Practice of Digital Libraries (DL'95)*, June 11 - 13, Austin, Texas.
9. Garcia-Molina H., Shivakumar N. (1996a). Building a Scalable and Accurate Copy Detection Mechanism. *Proceedings of 1st ACM International Conference on Digital Libraries (DL'96) March, Bethesda Maryland.*
10. Glatt Plagiarism Screening Program. URL http://www.plagiarism.com/screen.id.htm, 1999.
11. Globus Quick Start Guide. URL http://www.globus.org/toolkit/documentation/QuickStart.pdf
12. Gropp W., Lusk E., Skjellum A. (1994). Using MPI. Portable Parallel Programming with the Message-Passing Interface. (The MIT Press)
13. Gusfield D. *Algorithms on Strings, Trees, and Sequences. Computer Science and Computational Biology.* (Cambridge University Press), 1997.

214 K. Monostori, A. Zaslavsky, and H. Schmidt

14. Kesselman C. Data Grids for Next Generation Problems in Science and Engineering. In this proceedings.
15. Monostori K., Zaslavsky A., Schmidt H. MatchDetectReveal: Finding Overlapping and Similar Digital Documents. Proceedings of IRMA International Conference, Anchorage, Alaska, 21-24 May, 2000.
16. Monostori K., Zaslavsky A., Schmidt H. Parallel Overlap and Similarity Detection in Semi-Structured Document Collections. *Proceedings of 6th Annual Australasian Conference on Parallel And Real-Time Systems (PART '99)*, Melbourne, Australia, 1999.
17. Plagiarism.org, the Internet plagiarism detection service for authors & education URL http://www.plagiarism.org, 1999.
18. Si A., Leong H.V., Lau R. W. H. CHECK: A Document Plagiarism Detection System. *Proceedings of ACM Symposium for Applied Computing*, pp.70-77, Feb. 1997.

A Parallel Implementation of a Job Shop Scheduling Heuristic*

U. Der[1] and K. Steinhöfel[1,2]

[1] GMD - Research Center for Information Technology, 12489 Berlin, Germany.
[2] ETH Zurich, Institute of Theoretical Computer Science, 8092 Zurich, Switzerland.

Abstract. In the paper, we present first experimental results of a parallel implementation for a simulated annealing-based heuristic. The heuristic has been developed for job shop scheduling problems that consist of l jobs where each job has to process exactly one task on each of the m machines. We utilize the disjunctive graph representation and the objective is to minimize the length of longest paths, i.e., the overall completion time of tasks. The heuristic has been implemented in a distributed computing environment. First computational experiments were performed on several benchmark problems using a cluster of 12 processors. We compare our computational experiments to sequential runs and show that stable results equal or close to optimum solutions are calculated by the parallel implementation.

1 Introduction

In the job shop scheduling problem, where the objective is to minimize the makespan, one considers a set of l jobs consisting of tasks and a set of m machines. In this problem setting each job consists of a chain of tasks. A schedule is an allocation of the tasks to time intervals on the machines. The goal is to find a schedule that minimizes the overall completion time which is called the makespan.

Finding the minimum makespan is a difficult combinatorial optimization problem. Even for small problem instances it appears to be very complex to solve (e.g., even in the case of a single machine). For example, an instance with ten jobs, ten machines and one hundred tasks which was introduced by Muth and Thompson [8] in 1963 remained unsolved until 1986. The job shop scheduling problem has been shown to be NP-hard (see [6]). It is also NP-hard to find a job shop schedule that is shorter than $5/4$ times the optimum [15].

We employ the disjunctive graph model to represent the problem of minimizing the makespan. The graph $G = (V, A, E, \mu)$ is define such that V contains a node for each task and an additional source (I) and sink (O). Each node in V has a weight $\mu(v)$ that is given by the processing time of the corresponding task $(\mu(I) = \mu(O) = 0)$. The arcs in A represent the given precedences between the tasks. The edges in E represent the machine capacity constraints, i.e., they connect tasks that require the same machine.

* Research partially supported by the HK-Germany Joint Research Scheme under Grant No. D/9800710, and by the RWCP under Grant No. D-00-026.

T. Sørevik et al. (Eds.): PARA 2000, LNCS 1947, pp. 215–222, 2001.

Feasible solutions correspond to acyclic directed graphs that are admitted by a disjunctive graph. The makespan of a feasible schedule is determined by the length of a longest path from I to O. The problem of minimizing the makespan therefore can be reduced to finding an orientation on E that minimizes the length of longest paths.

There are a few papers on parallel implementations for job shop scheduling problems. E. Taillard [12] uses a method that performs the calculation of longest paths in parallel for successor nodes at any step from the starting node (source) to the sink within the disjunctive graph model. The nodes v are ordered by the maximum $A(v)$ of predecessor nodes on the machine and within the job. The paper does not report computational experiments.

In [9], results from implementing two parallel branch-and-bound algorithms for job shop scheduling are presented. The implementations are running on a 16-processor system based on Intel i860 processors, each with 16 MB internal memory. The parallel algorithms are performing a parallel search for solutions, i.e., the branching operation of sequential algorithms is parallelized. For the FT10 problem, they obtain a speed-up of about 14.67 in runs finding the optimum solution. However, branch-and-bound methods tend to an exponential increase of branching steps.

Verhoeven et al. [14] suggest the parallel execution of independent sequential search algorithms. They analyze a theoretical model for predicting the speed-up of such parallel versions of local search algorithms.

In the paper, we present first computational results of our parallel implementation for the job shop scheduling heuristic. The underlying algorithmic method is simulated annealing and sequential versions of these algorithms have been studied earlier in [10]. The paper continues our research presented in [11] where we investigate a theoretical parallelization of our heuristic based on algorithms that employ $O(n^3)$ processors.

Since $O(n^3)$ is an extremely large number of processors for real world applications, the heuristic has been implemented in a distributed computing environment. Another reason for our distributed approach is the stochastic character of simulated annealing. The stochastic process of performing several runs in order to increase the probability of computing good solutions is the underlying idea of our parallel implementation. We are using the software package PROMISE that consists of multiple levels, including the language extension PROMOTER [3, 4]. The framework is used for a global distribution of our heuristics where each processor starts its own simulated annealing processes with an independently chosen initial schedule. In the present scheme, the sequential versions were running on 12 processors in parallel by using different time frames for the exchange of intermediate results. Our first computational experiments on FT10, LA36 – LA40, and YN1 - YN4 show that results equal or close to optimum solutions are calculated by the distributed implementation.

2   Our Simulated Annealing-Based Heuristic

Simulated annealing algorithms are acting within a configuration space in accordance with a certain neighborhood relation, where the particular transitions between adjacent elements of the configuration space are governed by an objective function. More information about simulated annealing and its application in diverse areas can be found in [1,2]. Throughout the paper, we follow the notations from [2].

The configuration space, i.e., the set of feasible solutions of a given problem instance, is denoted by $\mathcal{F}$. The neighborhood of a solution $S \in \mathcal{F}$ is defined by $\eta : \mathcal{F} \to \wp(\mathcal{F})$. We use the non-uniform neighborhood function η proposed in [10] that is based on changing the orientation of several arcs, i.e., reversing the orientation of a single machine subpath. The selection of the subpath depends on the number of longest paths in the current schedule.

Given a pair of feasible solutions $[S, S']$, $S' \in \eta(S)$, we denote by $G[S, S']$ the probability of generating S' from S and by $A[S, S']$ the probability of accepting S' once it has been generated from S. The acceptance probabilities $A[S, S']$, $S' \in \eta(S)$ are derived from the underlying analogy to thermodynamic systems is the following:

$$(1) \qquad A[S, S'] := \begin{cases} 1, & \text{if } \mathcal{Z}(S') - \mathcal{Z}(S) \leq 0, \\ e^{-\frac{\mathcal{Z}(S') - \mathcal{Z}(S)}{c}}, & \text{,otherwise,} \end{cases}$$

where c is a control parameter having the interpretation of a *temperature* in annealing procedures. Finally, the probability of performing the transition between S and S', $S, S' \in \mathcal{F}$, is defined by

$$(2) \qquad \mathbb{P}_{S \to S'} = \begin{cases} G[S, S'] \cdot A[S, S'], & \text{if } S' \neq S, \\ 1 - \sum_{Q \neq S} G[S, Q] \cdot A[S, Q], & \text{,otherwise.} \end{cases}$$

By definition, the probability $\mathbb{P}_{S \to S'}$ depends on the control parameter c.

For our cooling schedule, the control parameter $c(t)$ is decreased in accordance with the following *hyperbolic* function:

$$(3) \qquad c(t + 1) := \frac{c(t)}{1 + \varphi(p) \cdot c(t)} = \frac{c(0)}{1 + (t + 1) \cdot \varphi(p) \cdot c(0)},$$

where $\varphi(p)$ is defined by $\varphi(p) := \ln(1 + p)/(\mathcal{Z}^{\max} - \mathcal{Z}_{\min}) < 1$, where $\mathcal{Z}^{\max}$ and $\mathcal{Z}_{\min}$ are the maximum value, respective the minimum value, of the objective function.

3   Parallel Implementation of the Heuristic

Our simulated annealing-based heuristic was implemented in a distributed computing environment by using the software package PROMISE [3,4]. It is a high-level, massively parallel programming environment for data-parallel operations/programs. It comprises an extension (PROMOTER) of the programming languages C/C++ for SPMD programs (single program, multiple data). The processes created by such a program perform parallel computations locally. PROMISE allows the user to formulate a parallel program in terms of application-specific concepts for an abstract machine rather than for a particular architecture. It handles low-level issues such as optimal data distribution, communication and co-ordination of the parallel threads.

Table 1. Results on solved problems using a single processor without communication.

Instance	J × M	Optimum	λ				T_{av}
FT10	10 × 10	930	937	937	930	930	2472
LA38	15 × 15	1196	1196	1196	1202	1198	41985
LA40	15 × 15	1222	1234	1233	1238	1237	31644

We use PROMISE for a global distribution of our scheduling heuristics (written in C++), where each processor starts its own simulated annealing process with an independent initial schedule (SPMD). Since the annealing processes are individual for each processor, we use the (sequentially) optimized code without any changes except the communication strategy. This parameter determines when and how the processes exchange the best solution found so far.

Table 2. Results on solved problems using 12 processors.

Instance	25	T_{av}	λ	50	T_{av}	λ	75	T_{av}	λ	100	T_{av}	λ
FT10		241	930		261	930		260	930		250	930
LA38		3213	1196		3109	1196		3128	1201		3142	1196
LA40		3413	1228		3507	1234		3657	1228		3471	1233

Table 3. Results on solved problems with stopping criterion.

Instance	25	T_{av}	λ	50	T_{av}	λ	75	T_{av}	λ	100	T_{av}	λ
LA38		982	1209		1100	1210		1071	1210		783	1202
LA40		3413	1228		3507	1234		3657	1228		3471	1233

We present first results of our computational experiments. The individual calculations were interrupted when M times the temperature was lowered, where

$M = 25, 50, 75, 100$ was chosen, i.e., we made four runs with different communication strategies for each problem. The computations were performed on a Linux PC-Cluster with 12 Athlons of 550 MHz, 128 MB RAM, and by using Myrinet.

Table 4. Results of single processor runs compared to 12 processors.

Instance	J×M	Opt.	single(100)		50		75		100	
			λ	sec	λ	sec	λ	sec	λ	sec
FT10	10×10	930	930	199	930	234	930	232	930	234
LA36	15×15	1268	1268	8964	1268	4083	1268	6118	1268	5976
LA37	15×15	1397	1397	14442	1397	5613	1397	5337	1397	4616
LA38	15×15	1196	1196	10658	1196	6136	1196	6726	1196	6829
LA39	15×15	1238	1238	16082	1238	15698	1238	4961	1238	6795
LA40	15×15	1222	1233	18760	1228	18063	1224	18408	1224	18080

The problems LA38 and LA40 are believed to be the most challenging instances among the benchmarks introduced by S. Lawrence [7], we started with these two problems and the smaller instance FT10 due to [8]. We used the hyperbolic cooling schedule, where the parameter settings were taken from [10]. In Tab. 1 we provide our results of consecutive runs using a single processor. These computational experiments have been performed on a Sun Ultra 1/170 and without communication, i.e., the run was allowed to jump back to best found solutions. The stated makespan λ was computed within an average time T_{av} that is in seconds.

Table 5. Results on still unsolved problems YN.

Instance	J × M	LB	UB	$\lambda_{t \to \infty}$	λ	Time
YN1	20 × 20	846	888	886	894	70267
YN2	20 × 20	870	909	909	918	63605
YN3	20 × 20	840	893	895	904	61826
YN4	20 × 20	920	968	970	975	63279

As in the sequential experiment the distributed computation produced stable results equal to optimum solutions for FT10 and LA38 in almost all runs, see Tab. 1 and 2. The results for LA40 are close to the optimum. The run-time is relatively short in all runs, i.e., for our range of M there is no remarkable impact of this parameter. But the speed-up compared to our sequential run-time was about 3 by using 12 processors and our results for LA40 could be improved.

From Tab. 3 one can see that the run-time improves significantly when the condition to reach the optimum is slightly weakened. For LA38 the execution stopped when a makespan $\lambda < 1210$ was found and for LA40 when $\lambda < 1237$.

In Tab. 4 we compare results obtained by a run using a single processor to runs using 12 processors. In this experiment the run with using one processor only performs after 100 decreasing steps of the temperature a communication step (indicated by single(100)), i.e., it jumps back to its best found solution. All runs stopped as soon as they found the optimum solution. For all benchmark problems the runs using 12 processors obtained better makespans or found the optimum solution much faster.

In a second series of experiments we attacked larger 20×20 benchmark problems YN1 – YN4. Tab. 5 displays again our computational results of sequential runs without communication. We used a stopping criterion of finding a makespan λ less or equal to 1% of the known upper bound and $\lambda_{t \to \infty}$ indicates the makespans found without such a stopping criterion. The results denoted by $\lambda_{t \to \infty}$ were obtained after long runs, in general, with the tempering strategy. We succeeded in improving the upper bound for YN1.

Table 6. Results of single processor runs compared to 12 processors.

Instance	J×M	LB	UB	single(100)		50		75		100	
				λ	sec	λ	sec	λ	sec	λ	sec
YN1	20×20	846	888	900	13718	891	13895	891	14064	898	13916
YN1	20×20	846	888	890	48743	891	52719	891	52962	886	53076
YN2	20×20	870	909	918	23829	923	24295	919	24401	909	24004
YN2	20×20	870	909	911	53474	910	60075	910	59563	909	64421
YN3	20×20	840	893	906	22420	905	24871	906	22852	902	22427
YN4	20×20	920	968	983	24074	986	24169	975	23994	973	24397

In Tab. 6 we present our results of the distributed computation. All these four benchmark problems are still unsolved. In column LB the known lower bounds on the optimum of these problems are given as well as in column UB the known upper bounds, i.e., best know solutions. We compare again a run using a single processor but performing a communication step according to $M = 100$. Since the value of the optimum solution is unknown the runs stop when the final temperature determined by the parameters is reached. Therefore, it is naturally that the run-time of the single processor run is more or less the same as in the 12 processor runs.

Note that we found again the makespan 886 for YN1 that improves the previous upper bound. In terms of the speed-up we were able to obtain our results for YN1 and YN2 of the very long sequential runs (several day, i.e., between 12 and 14) within 15 to 16 hours for YN1 and within 7 hours for YN2.

The run-time given in Tab. 6 is in seconds and is the time when the last processor terminated.

Again, the parameter M that determines when individual calculations are interrupted does not seem to have a remarkable impact.

4   Concluding Remarks

We implemented a simulated annealing-based heuristic for job shop scheduling in a distributed computing environment of 12 processors. Each processor starts its own simulated annealing process with an independent initial schedule. A communication strategy is used where individual calculations are interrupted when M times the temperature was lowered. During the interruptions the processes exchange the best solution found so far.

Our computational experiments produced stable results equal or close to optimum solutions for the benchmark problems FT10, LA38, and LA40 in a relatively short time. The choice of the parameter M does not seem to have a remarkable impact of the results or the run-time. Compared to the sequential results we could improve our results for LA40 and obtained a speed-up of about 10. The speed-up for runs on larger unsolved benchmarks was even more significant. Our best results obtained after very long sequential runs are computed in parallel within a few hours.

In forthcoming research, we will increase the number of processors and apply the implementation to other large scale benchmark problems. Moreover, we will investigate other communication strategies that might have more impact than our parameter M.

References

1. E.H.L. Aarts. *Local Search in Combinatorial Optimization.* Wiley, NY, 1998.
2. E.H.L. Aarts and J.H.M. Korst. *Simulated Annealing and Boltzmann Machines: A Stochastic Approach.* Wiley, NY, 1989.
3. M. Besch, H. Bi, P. Enskonatus, G. Heber, M. Wilhelmi. High-level Data Parallel Programming in PROMOTER. *Proc. Intern. Workshop on High-level Parallel Programming Models and Supportive Environments*, pp. 47-54, 1997.
4. H. Bi, M. Kessler, H. W. Pohl, M. Tief. Promise - High-level Data-parallel Programming Environment. *Proc. Workshop on Advanced Parallel Processing Technologies*, pp. 207-211, Publishing House of Electronic Industry, 1999.
5. E. Dekel and S. Shani. A Parallel Matching Algorithm for Convex Bibartite Graphs and Applications to Scheduling. *Journal of Parallel and Distributed Computing*, 1:185 – 205, 1984.
6. M.R. Garey and D.S. Johnson. Complexity Results for Multiprocessor Scheduling under Resource Constraints. *SIAM Journal on Computing*, 4(4):397–411, 1975.
7. Lawrence, S., Resource Constrained Project Scheduling: An Experimental Investigation of Heuristic Scheduling Techniques (Supplement), Technical Report, Graduate School of Industrial Administration, Carnegie-Mellon University, Pittsburgh, Pennsylvania, 1984.

8. J.F. Muth, G.L. Thompson, and P.R. Winters, editors. *Industrial Scheduling.* Prentice-Hall, Englewood Cliffs, N.J., 1963.

9. M. Perregaard and J. Clausen. Parallel Branch-and-Bound Methods for the Job-Shop Scheduling Problem. *Annals of Operations Research,* 83:137 – 160, 1998.

10. K. Steinhöfel, A. Albrecht, and C.K. Wong. Two Simulated Annealing-Based Heuristics for the Job Shop Scheduling Problem. *European Journal of Operational Research,* 118(3):524 – 548, 1999.

11. K. Steinhöfel, A. Albrecht, and C.K. Wong. On Parallel Heuristics for the Job Shop Scheduling Problem. In: S.Q. Zheng, editor, *Proc. 11th International Conf. on Parallel and Distributed Computing and Systems,* pp. 806 – 811, MIT, Cambridge, Mass., November 3-6, 1999.

12. E. Taillard. Parallel Taboo Search Techniques for the Job-Shop Scheduling Problem. *ORSA Journal on Computing,* 6:108 – 117, 1994.

13. P.J.M. Van Laarhoven, E.H.L. Aarts, and J.K. Lenstra. Job Shop Scheduling by Simulated Annealing. *Operations Research,* 40(1):113–125, 1992.

14. M.G.A. Verhoeven, H.M.M. ten Eikelder, B.J.M. Aarts, E.H.L. Aarts. Sequential and Parallel Local Search Algorithms for Job Shop Scheduling. *Meta-Heuristics (Advances and Trends in Local Search Paradigms for Optimization),* pp. 359-371, 1999.

15. D.P. Williamson, L.A. Hall, J.A. Hoogeveen, C.A.J. Hurkens, J.K. Lenstra, S.V. Sevast'janov, and D.B. Shmoys. Short Shop Schedules. *Operations Research,* 45:288–294, 1997.

16. Yamada, T. and Nakano, R., A Genetic Algorithm Applicable to Large-Scale Job Shop Problems, In: R. Manner and B. Manderick (eds.), *Proc. 2nd International Conf. on Parallel Problem Solving from Nature,* North-Holland, Amsterdam, 1992, pp. 281-290.

Restructuring Irregular Computations for Distributed Systems Using Mobile Agents

Rocco Aversa, Beniamino Di Martino, and Nicola Mazzocca

Dipartimento di Ingegneria dell' Informazione - 2nd University of Naples - Italy
Real Casa dell'Annunziata - via Roma, 29
81031 Aversa (CE) - ITALY
{beniamino.dimartino,n.mazzocca}@unina.it
rocco.aversa@unina2.it

Abstract. One of the main problems arising when designing distributed High Performance programs, or when restructuring sequential programs for distributed systems, is workload unbalance. The Client-Server paradigm, which is usually adopted for distributed systems, is not flexible enough to implement effective dynamical workload balancing strategies. The Mobile Agent paradigm can increase the flexibility in the creation of distributed applications (and the restructuring of sequential applications for distributed systems), and can in particular provide with a robust framework for managing dynamical workload balancing. In this paper we show, through a case-study, how the restructuring of a sequential code implementing an irregular algorithm, with adoption of the mobile agent model, allows for yielding a load-balanced distributed version of the algorithm without completely rethinking its structure, and for reusing a great deal of the sequential code.

1 Introduction

Heterogeneuos distributed architectures are gaining more and more importance for High Performance Computing. One of the main problems arising when designing distributed high performance programs, or when restructuring sequential programs for distributed systems, is workload unbalance. This issue certainly arises with irregular computations on distributed architectures (and on homogeneous tightly coupled parallel architectures as well). But even regular computations could be affected by this problem when executed on heterogeneous loosely coupled distributed architectures. The Client-Server paradigm, which is usually adopted for distributed systems, is not flexible enough to implement effective dynamical workload balancing strategies.

Mobile Agents [6] are emerging as an effective alternative to Client/Server for programming distributed systems. Several application fields such as e-commerce, brokering, distributed information retrieval, telecommunication services, monitoring, information dissemination, etc. can benefit from mobile agent technology in many aspects, ranging from network load reduction and overcoming of network

T. Sørevik et al. (Eds.): PARA 2000, LNCS 1947, pp. 223–232, 2001.
© Springer-Verlag Berlin Heidelberg 2001

latency to asynchronicity and autonomy, heterogeneity, dynamic adaptivity, robustness and fault-tolerance [3].

Parallel and, in general, High Performance Computing can likewise benefit from mobile agent technology, expecially when targetted towards heterogeneous distributed architectures. Several characteristics of potential benefit for scientific distributed computing can be provided by the adoption of the mobile agent technology, as shown in the literature [1,2,5]: they range from fault-tolerance to portability to paradigm-oriented development.

We focus in this paper on two features the adoption of the mobile agent technology can provide for irregular numerical algorithms: ease of restructuring of sequential applications, and provision of a robust framework for managing dynamical workload balancing.

We show, through a case-study, how the restructuring of a sequential code implementing an irregular algorithm, with adoption of the mobile agent model, allows to yield a load-balanced distributed version of the algorithm without completely rethinking its structure, reusing a great deal of the sequential code and trying to exploit the heterogeneity of the target computing architecture. The chosen application solves the well-known N-body problem.

The paper is organized as follows: section (2) is devoted to the description of the case study under consideration. In section (3) the parallelization and workload balancing strategies devised are described, while in (4) experimental results are provided, which validate the strategy adopted for workload balancing. Section (5) provides conclusions and a glance on planned future work.

2   A Case-Study: The N-Body Algorithm

The chosen case-study is a sequential algorithm that solves the N-body problem by computing, during a fixed time interval, the positions of N bodies moving under their mutual attraction. The algorithm is based on a simple approximation: the force on each particle is computed by agglomerating distant particles into groups and using their total mass and centre of mass as a single particle. The program repeats, during the fixed time interval, its three main steps: to build an octal tree (*octree*) whose nodes represents groups of nearby bodies; to compute the forces aging on each particle through a visit in the octree; to update the velocities and the positions of the N particles.

The sequential algorithm can be summarized using the pseudo-code of fig.1, where p is the particle array, t is the *octree* array, N the total number of particles and *delt* the time step according to which evolves the algorithm. The procedure *tree()* , starting from the N particles, builds the *octree* storing its nodes in t array using a depth-first strategy. At each level in the *octree* all nodes have the same number of particles until each leaf remains with a single particle. The subsequent level in the *octree* can be obtained by repeatedly splitting in two halves the particles in each node on the basis of the three spatial coordinates (x,y,z). The splitting phase requires three consecutive activations of a *quicksort* routine, that orders the particles array according to each of the spatial coordinates. The tree is constructed *breadth first* but stored *depth first*. At the end of this stage a tree node represents a box in the space characterized by its position (the two distant corners of the box) and by the total mass and the center of mass of all the particles enclosed in it. A leaf of the tree obviously coincides with a box containing a single particle.

```
    /* main loop */
  while (time<end) {

    /* builds the octree */
    tree(p,t,N);

    /* computes the forces */
    forces(p,t,N);

    /* computes the minimal time step */
    delt=tstep(p,N);

    /* updates the positions and velocities of the particles */
    newv(p,N,delt);
    newx(p,N,delt);

    /* updates the simulation time */
    time=time+delt;
  }
```

Fig. 1. The N-body sequential algorithm.

The second stage of the algorithm (the procedure *force()*) computes the forces on each of the N particles in p by visiting the above built t tree. Each particle traverses the tree. If the nearest corner of the box representing the current node is sufficiently distant, the force due to all particles is computed from the total mass and centre of mass of the box. Otherwise the search comes along the lower levels of the tree, if necessary, till the leaves. Thus, distant particles contributions are taken in account by looking at nodes high in the octree; nearby particles contributions by looking at leaves. Once obtained the force components affecting the particles, it's possible to update their velocities and positions. The procedures *newv()* and *newx()* computes respectively the new velocities and the new positions of the N particles using a simple Euler step. The procedure *tstep()* returns the time step *delt* to be used for the updates. The time step is chosen so as to limit the velocity change of any particle (tstep   1/fmax).

A first analysis of the sequential algorithm suggests some preliminary considerations that can drive the parallelizzation strategy:

- The *tree()* procedure appears to be a typical *master-slave* computation: a master reads input data and starts building the first levels of the octal tree until the elaboration can proceed in parallel;
- The *force()* procedure can be classified as a *worker farm* computation that is an elaboration phase during which each node should receive a dynamically well-sized portion of the work (in terms of the number of assigned particles);
- Finally, the computation steps that updates the particles velocities and positions (*newv()* and *newx()*) can be recognized as a data-parallel portion of the code that can be parallelizzable in a trivial way.

3   A Parallelization Strategy Using the Mobile Agents

Before describing the detailed steps of the parallelization procedure, it is convenient to explain the main characteristics of the chosen programming environment: The Aglet mobile agent system. The Aglet Workbench is a framework for programming mobile networks agents in Java developed by IBM Japan research group [4].

An *aglet* (agile applet) is a lightweight Java object that can move to any remote host that supports the Java Virtual Machine. An Aglet server program *Tahiti* provides the agents with an execution environment that permits an *aglet* to be created and disposed, to be halted and dispatched to another host belonging to the computing environment, to be cloned, and, of course, to communicate with all the other *aglets*. For our aims we needed to add new facilities to the Aglet system by extending the scope of the *multicast* messages. Infact, the Aglet Workbench supports the *multicast* messages but only within a local context: an *aglet* that requires to receive a multicast message with a specific label needs to subscribe to that kind of *multicast* message. We extend this mechanism allowing a *multicast* message to reach both the local and remote *aglets* using a *multicast aglet server* running on each host. Such *aglet server* performs the following actions (fig. 3):

- subscribes to a specific *multicast* message labelled "remote multicast";
- captures in a local context a "remote multicast" message;
- dispatches a *multicast slave aglet* to each host of the computing environment with a copy of the *multicast* message.

Each *multicast slave aglet* on its arrival to destination:

- sends in a local context the *multicast* message but with a new label ("external multicast"), so allowing any *aglet* on that host to receive the message;
- disposes itself.

To summarize, an *aglet* that wishes to send a remote multicast, must label its message with the label "remote multicast", while an *aglet* interested to receive multicast messages from remote *aglets* has to subscribe to a *multicast* message labelled "external multicast". According to the considerations, described in the

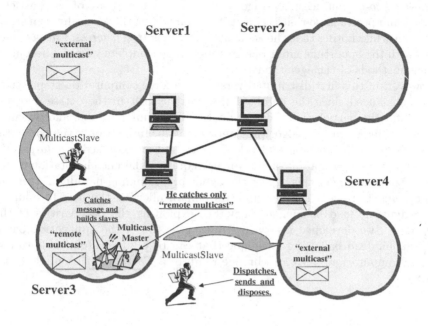

Fig. 2. The *multicast* extension to the Aglet system.

precedent section, about the parallel nature of the different phases of the sequential algorithm, a first distributed version can be easily obtained. The parallelization of the code relative to the construction of the *octree*, on the basis of the master-slave paradigm, require that the reading of the input data and the production of the first level in the octree is carried out by a single *master* agent. As soon as the nodes in the current level of the *octree* exceeds in number the computing nodes, the *master* agent cloned itself and dispatches the clones to each host making up the computing environment. Every agent, *master* and *slaves*, is responsible for the building of the subtree assigned to it, an operation that can be carried out in parallel. At the end of this stage every agent has filled up a slice of the complete octree data structure and can send a remote *multicast* message to all the other agents so that each of them is able to get a complete copy of the *octree*. At this point of the algorithm the force computation stage *forces()* and the steps that update the particles velocities and positions (*newv()* and *newx()*) can happen in a completely independent way. Each of the N particles, in fact, computes the force aging on it by traversing separately the *octree*, just as the updating phase of the particles velocities and the positions also performs the same computation on different data. So, an initial parallelization can be obtained by simply distributing the particles among the agents. We adopt a static workload distribution technique: the number of particles assigned to each agent is calculated taking in account the different computational characteristics of the available computing nodes. At the end of their work the agents, by means of a new *all-to-all* multicast message, obtain an updated copy of the particles velocities and positions so that the next iteration step of the algorithm can start. It is worth while noting that the sequential routines *tree()*, *forces() newv()* and *newx()* with the opportune input parameters can be completely reused becoming different methods of the agent code.

However, in this first distributed version, the force computation step results highly unbalanced since the depth of the particle visit in the octree depends by the data (the particles spatial coordinates) in an unforeseeable and dynamic way (it is different in each algorithm iteration step. In fact, only the computation of the particles position and velocity can be seen as a strict *data-parallel* code, while the computation of the forces requires different elaboration times for each particle. Under this aspect the agent programming paradigm, through the mechanism of agent cloning, appears helpful allowing the user to adopt a flexible dynamic load balancing policy. So, exploiting the peculiarities of the agent model, we developed a new distributed version of the application based on a dynamic load balancing technique that can be easily customizable to any irregular computation. Such a technique requires a *Coordinator* agent that (see fig. 3):

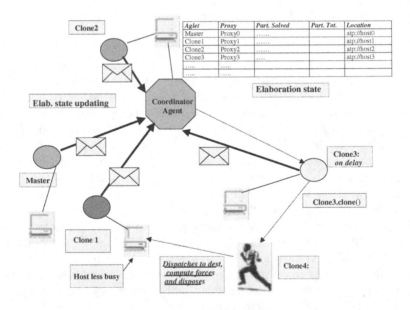

Fig. 3. Graphical description of the workload balancing strategy using a *Coordinator* agent.

- periodically collects information about the elaboration state of the agents;
- on the basis of a suitable load-balancing policy (a policy that, for example, can take in account the computing speed of the nodes, the availability of new computing resources, etc.) orders some agents to clone themselves and to dispatch the clones with a partial load towards less busy hosts.

To particularize this technique to our case-study, we developed a *Coordinator* agent that:

- periodically asks the agents for the number of particles they have already solved;
- tries to balance, on each node of the computing environment, the number of particles yet to process by distributing on the network some *miniclones* carrying a partial load of particles.

Obviously, the cloning and the dispatching operations will be performed only if the load imbalance exceeds a fixed threshold.

4 Experimental Results

The performance analysis of parallel application has been carried out using a distributed system made up of Windows PC with different hardware configurations: 1 PC with a CPU Intel Pentium II (400 Mhz) and 128MB of RAM memory (*Epicuro*); 2 biprocessor PC each with 2 Intel Pentium II (350 Mhz) and 256MB of RAM memory (*Platone* and *Hegel*).

We measured the execution times of the single algorithm phases (the building of the octree (*t_tree*) the computation of the forces aging on the particles (*t_forces*); the updating of the particles positions and velocities (*t_update*)) of a sequential Java version of the algorithm (see table n. 1) and of the parallel version running on three non homogeneous computing nodes (*Epicuro*, *Platone* and *Hegel*) (see table n. 2). The number of particles used in this test is fixed to 32768.

Table 1. Execution times (in milliseconds) of the different phases of the sequential version of the application (number of particles: 32768).

	t_tree	t_forces	t_update	t_total
Epicuro	36150	6210810	210	6247170
Hegel	47844	7750734	265	7798843
Platone	47485	7893180	266	7940969

As we expected, the extracted measures show that the force computation stage is the heaviest portion of the algorithm and that the workload distribution plays a crucial role on the overall performance of the parallel application since the collective communications, in fact, force all the application agents to synchronize

Table 2. Execution times (in milliseconds) of the different phases of the distributed version of the application without the Coordinator agent (number of particles: 32768).

	t_tree	t_forces	t_update	t_total
Epicuro	40700	2979100	170	3068140
Hegel	35562	86266	124	3060859
Platone	36344	104359	141	3057922

in different algorithm points. This implies that the completion time of a single simulation step of the algorithm will result strictly tied to the performance of the slowest node in the network. Furthermore, a static workload distribution technique assigning to each agent a number of particles proportional to the different computational characteristics of the nodes doesn't pay, considering the strong data dependence that can be found in this stage of the algorithm. In fact, the PC *Epicuro* that, according to the sequential tests, should be the fastest computing node, in the parallel version slows down the execution time since it probably received the most demanding portion of the particles load. The table n. 3 shows the beneficial effects introduced in the parallel version of the algorithm by the presence of the *Coordinator* agent that dinamically yields an almost perfect balance of the computational load among the nodes.

Table 3. Execution times (in milliseconds) of the different phases of the distributed version of the application with the Coordinator agent (number of particles: 32768).

	t_tree	t_forces	t_update	t_total
Epicuro	41360	2414030	50	2475440
Hegel	43594	2411563	156	245313
Platone	43422	2416625	140	2470203

Finally, we stressed the proposed load balancing technique on a hardware platform consisting of 3 computing nodes, where two of them (*Hegel* and *Platone*)were burdened with an extra computational load. This addictional load on the hosts was artificially created using 5 new agents that periodically move between the two hosts. In this way we can test the *Coordinator* agent balancing action against both an irregular application and an hardware platform characterized by dynamically changing available computing resources. The tables n. 4 and n. 5 respectively summarizes the experimental results of the test without and with the *Coordinator* agent, clearly showing how the load balancing technique is able to promptly react to a dinamically changing hardware/software computing environment.

Table 4. Execution times (in milliseconds) of the different phases of the distributed version of the application without Coordinator agent and in presence of a dynamical changing external workload (number of particles: 32768).

	t_tree	t_forces	t_update	t_total
Epicuro	42120	3061710	50	3239500
Hegel	34969	823751	140	3239406
Platone	97843	958438	457	323734

Table 5. Execution times (in milliseconds) of the different phases of the distributed version of the application with the Coordinator agent and in presence of a dynamical changing external workload (number of particles: 32768).

	t_tree	t_forces	t_update	t_total
Epicuro	114570	2578870	50	2723590
Hegel	106000	2579125	172	2715344
Platone	97890	2584938	140	2712968

5 Conclusions and Future Work

In this paper we have addressed the problem of an effective utilization of the mobile agent model for High Performance Distributed Computing, particularly focussing on the dynamical workload balancing problem in irregular applications. We have showed, through a case-study, how the design of a dynamical workload strategy, based on the cloning and migration feasibilities of the agent model, and coupled with coordination and reactiveness properties of the balancing algorithm, provides promising results over an heterogeneously distributed architecture. The workload dynamic balancing solution was compared with a static one, and was tested under dynamically changing external workload conditions for the computing hosts, and results were satisfactory both in term of overall efficiency, and reactiveness to dynamically changing external workload conditions. In addition the restructuring of the sequential code implementing the case, with adoption of the mobile agent model, was achieved by reusing a great deal of the sequential code's structure.

Our future work in this area is addressed towards defining Computational Paradigms expecially tailored to the Mobile Agent model, and implementing compilation techniques and run-time support for those. Paradigms or Skeletons are a promising technique for the high level development of parallel code; we wish to study how this technique could be exploited within the framework of mobile agent model, by taking particular care of provision of paradigm compositionality and performance issues related to the paradigm's selection and composition.

Acknowledgments. We wish to thank Salvatore Venticinque who contributed to the implementation of the technique, and performed the experimental measures.

References

1. T. Drashansky, E. Houstis, N. Ramakrishnan, J. Rice, "Networked Agents for Scientific Computing", Communications of the ACM, vol. 42, n. 3, March 1999.
2. Gray R., Kotz D., Nog S., Rus D., Cybenko G., "Mobile agents: the next generation in distributed computing" Proc. of Int. Symposium on Parallel Algorithms/Architecture Synthesis, 1997.
3. D. Lange and M. Oshima, *Programming and Deploying Java Mobile Agents with Aglets*, Addison-Wesley, Reading (MA), 1998.
4. D. Lange and M. Oshima, "Seven good reasons for Mobile Agents", Communications of the ACM, vol. 42, n. 3, March 1999.
5. H. Kuang, L.F. Bic, M. Dillencourt, "Paradigm-oriented distributed computing using mobile agents", Proc. of. 20th Int. Conf. on Distributed Computing Systems, 2000.
6. V. A. Pham, A. Karmouch, "Mobile software agents: an overview", IEEE Communications Magazine, Vol. 36(7), July 1998, pp. 26-37.

An Information System
for Long-Distance Cooperation in Medicine

Harald Kosch[1], Renata Słota[2], Lászlo Böszörményi[1],
Jacek Kitowski[2,3], and Janusz Otfinowski[4,5]

[1] Institute of Information Technology, University of Klagenfurt, Universitätsstrasse
65-67, 9020 Austria
[2] Institute of Computer Science, AGH, al. Mickiewicza 30, 30-059 Cracow, Poland
[3] ACC CYFRONET, ul. Nawojki 11, 30-950 Cracow, Poland
[4] Collegium Medicum Jagellonian University, ul. Św. Anny 12, Cracow, Poland
[5] Cracow Rehabilitation Centre, Thraumatology Department, ul. Modrzewiowa 22,
Cracow, Poland
email: kito@uci.agh.edu.pl

Abstract. In the paper we present an information system which is being
developed for medical purposes, especially for providing medical staff
with a large database of image and video data, distributed over several
medical centers in Europe. The aspects of the system are to develop
user-friendly and intelligent interfaces for database querying as well as
services for multimedia data management and archiving with quality of
service requirements. The aim of the system is to support medical diag-
nostics and teleeducation.

Keywords: multimedia hierarchical database system, telemedicine, teleeducation, video and storage servers.

1 Introduction

The principal goal of the presented system [1] is to take advantage of multimedia
telematics services, especially delivery of medical images and video sequences.
Another important aspect is the support of long-distance collaboration and cooperation of several medical centers and medical teleeducation.

The delivery of the video and image data is organized in a hierarchical form. It
is enabled by content-based retrieval which relies on descriptive information also
called indexes. The indexes are stored in a meta-database which is intended to
be implemented in a high-performance Database Management System (DBMS).
This database acts as a server for incoming video and image queries.

Medicine is a highly specialized field of science, which results in high concentration of knowledge and know-how. The communication between physicians of
different areas and/or at different geographical places is difficult, although extremely important. Telemedicine [2] might assist physicians to overcome the problem, enabling them to get substantial information from sites located elsewhere,
in an usable form even in time-critical cases. This is an essential contribution
to medical practice. Moreover, this possibility might bring entirely new ideas

T. Sørevik et al. (Eds.): PARA 2000, LNCS 1947, pp. 233–241, 2001.

into medical research, since it allows comparison of multimedia data in such extent which has not been possible previously. Last but not least, distributed telemedicine systems can be used in local and remote education [3].

The fields of applications of our system are usage and exchange of scientific medical information, as well as, teleeducation rather than standard use of databases for storing and retrieving patients' information. The usefulness of exchanging scientific video and audio data on demand is not limited to the medical field. Concepts and techniques of the PARMED Project can be easily adopted in many other application fields, where collaboration is required.

This paper presents the technical aspects of the PARMED multimedia architectures and points out its innovative aspects. Implementation, as far as they were already realized, are described. The project PARMED is now in the second year. In the next two sections we summarize the objectives and present the architecture of the system. Section 4 describes the advances in the storage, meta-database and teleeducation components, i.e., data acquisition, storage server and proxy-cache for videos, meta-database, supporting tool for teleeducation. Finally Section 5 concludes and points out future developments.

2   Objectives

There are several objectives defined for the system:

- medical cooperation, at the initial stage – to get knowledge and data for some medical fields, at the operational stage – the usage of system for knowledge and data acquisition and for teleeducation,
- active client software consisting of several modules – to browse rough medical video and image data, for data processing (like filtering, compression, pattern reconstruction for operation planning), teleeducation subsystem,
- meta-database concept – how to model the contents of the medical images and videos – implementation of the meta-database should be done in a way that complex queries can be executed efficiently in order to guarantee a desired throughput for multiple users,
- video and image archiving systems – to store and retrieve medical data,
- cooperative resource management – to deal with quality of service problems, avoiding large communication requirements.

3   Architecture of the System

There are several tasks of the system, which overall architecture is presented in Fig. 1:

1. insertion of raw medical data into the system,
2. annotation of data and storing of the indexes in the meta-database,
3. management of user's queries, consisting in data searching and delivering,
4. teleeducation.

Four kinds of sites are defined, each with LAN services. The first kind, *Site 1*, is in the most advanced state. It consists of the Video Server (VS) and the Storage Server (SS). *Site 4* has at the time a limited configuration.

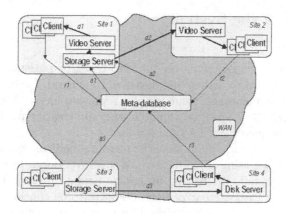

Fig. 1. System architecture

The meta-database is the central component of the system. It maps user queries onto the physical address of the required data. As users might access local data faster, it can be replicated over the sites.

Three levels of data hierarchies have been introduced. Raw data resides on the SS, which represents data vault for the system. One or more SS can coexist in the system, depending on the requirements. The VS has been introduced in order to get better Quality of Service (QoS) for multimedia streams (it serves as a proxy cache; see section 4.2). Thus, the base site configuration contains the VS holding the data which are requested locally by the users (e.g. *Site 2*). The VS is fed with data (i.e. *d2*) from the SS on request (*r2→a2*). The data stored in the VS can then be used with a guaranteed QoS. The similar case, but with higher data locality, is represented for *Site 1*, for which data is fed to the VS from the local SS (*r1→a1→d1*). Repeated VS feeding is possible when another data is required and replacing data which is no longer useful.

Guaranteed QoS from the SS to the client is not maintained, however data delivering is still possible with services represented by *Site 3* and *Site 4*. With *r3→a3→d3*, data is placed from the SS to the Disk Server (DS) for local clients, while in *Site 3* data from the SS can be used locally with the SS cache.

There are some specific tasks assigned to the sites. In order to introduce data into the system, the configuration typical for *Site 4* , is applied. Raw data is placed on the DS and then annotated locally. The annotation is moved to the meta-database, while the data is moved to the SS, which location is the most appropriate according to the data use. To present data with QoS configurations like *Site 1* or *Site 2*, the proxy cache functionality of the VS is used. This is suitable for teleeducation purposes, like regular courses, for which data are prepared in advance.

Since the objective elements are at different development stage at the moment we will focus in the following sections on selected aspects.

4   Elements of the System

4.1   Input Data

Our first experience concentrates on the databases of patients with diseases. Special interest is given to orthopedic diseases. In general, medical centers have large sets of Roentgen pictures. Some of them represent interesting medical cases from the past, others – unconventional history of treatment. Therefore it is worthwhile to introduce them into the system.

Another kind of data are video tapes concerning medical operation, like grafting a knee prosthesis using some modern or unconventional prosthesis models. This video data is mainly interesting for course definition. Lectures or seminars given to students or presented at conferences have to be considered as well.

The Roentgen pictures are introduced into the system using scanner facility with different kind of optical and digital filtering techniques, while video tapes are hardware compressed to MPEG1 or MPEG2 formats.

4.2   Video and Storage Server Utilities

There are high storage requirements concerning the medical data. Due to the huge amount of data it is necessary to make use of tertiary storage, like an automated tape library with suitable software management. In our implementation [4] we employ an ATL 4/52 tape library and the UniTree Central File Manager software for building the SS. The library itself is equipped with a disk cache in order to assist data streams. No QoS guarantee could be maintained with the SS, therefore a strategy of the VS and the SS cooperation is being developed [4, 5].

The VS behaves as a proxy cache between the SS and the clients in order to provide reasonable start-up times and smooth streaming of video data. Such a caching and buffering proxy is able to compensate fluctuating network bandwidth and to take advantage of interactive applications accessing repetitively the same set of video fragments.

In the sample application, we are currently using a simple video server, following the techniques employed in the Stony Brook VS [6,7], to store incoming video and to deliver them to the clients. We developed a mechanism to adapt the outgoing stream to the requirements of the clients and the average arrival video rate. In principle, it performs in the following way: after a specific amount of data is accumulated, the proxy begins to transmit the video data to the requesting client. Depending on the current network situation between the server of the corresponding video and the proxy, the proxy chooses the start-up latency for the transmission to the client. After transmission the video is cached further requests.

4.3   Meta-database and Client Connection

At present, two kinds of data are of interest in the project – digitized Roentgen pictures and MPEG files. We implemented a meta-database capturing the segment and high-level indexing part of the videos, as semantic information about the stored video streams. It contains the basic set of high-level content classes that are 'events', 'objects', 'persons' and 'locations'. These classes are subclasses of a general 'ContentObject' which may refer to a low-level motion object describing its spatio-temporal characteristics. The model provides means for segmenting video streams in different granularities, such as: 'shot', 'scene' and 'sequence'. A detailed description of the generic index model VIDEX can be found in [8]. We present here the implementation for orthopedic operations where the high-level indexing part of the VIDEX model was extended with a rich set of orthopedic-specific classes.

Consider as example the event of the Anesthesia in an operation video. This event occurs in some structural component of the video which can be either a shot or a scene: it involves surely the persons anesthetist and the patient and occurs before the Incision event.

We provide furthermore a JAVA-based client who provides means to annotate, query and browse video material using the graphical library Swing. The core components of the client include the annotator, querier and browser interfaces. The annotator allows the specification of video segments and high-level content objects. The querier follows a text-based, structured query specification technique. It enables the definition of video queries by specifying conditions on content objects and the specification of semantic and temporal relations to other content objects. The browser allows the navigation through the contents of the index database.

The client implementation using the index database contains an abstract class which defines all methods a database manager has to provide for the requirements of the client. Some of the methods are already implemented, as they are valid for all kinds of databases, many others have to be implemented by a special database manager. We provide at time an implementation for the *Oracle8i* DBMS.

Let us now detail the organization of the indexes in the database. The architecture of the meta-database is based on three medical operations: artificial hip joint replacement, lumbar interbody fusion, screwing in of the titanium block, for which the annotation has been developed in collaboration with orthopedic surgeons.

As mentioned beforehand, four types can be indexed in the database model [8]: `objects`, `events`, `persons`, `locations`.

The `object` is an *operation* or a *patient*. The patient is a passive subject of the operation. For the *patient* we need his name tag and his case history. Additional information like sex and age are also useful. The *operation* attributes are more informative: a patient name tag, the date and the method of operation, the part of the body that is operated, and some information about complications arising during and after operation. Name tags of medical staff participating in the operation are also included.

Table 1. Stages of the operation concerning the artificial hip joint replacement with events for stages 8 and 9 depicted

Stage	Description	Stage	Description
1	Anesthesia	8.2	The femoral shaft
2	Placement of the patient	8.2.1	Preparing the femoral canal
3	Ischemia	9	Implantation
4	Preparation of operating field	9.1	Inserting the acetabular component
5	Covering of the field	9.2	Inserting the femoral stem
6	Incison	10	Attaching the femoral head
7	Joint dislocation	11	Implant repositioning
8	Preparation of bony placenta	12	Insertion the drain
8.1	The hip joint acetabulum	13	Suturing
8.1.1	Removing the femoral head		
8.1.2	Reaming the acetabulum		

The **event** is an action which is undertaken during the operation stages (an example of operation stages is shown in Table 1). Definitions of the events have been introduced on the basis of available video material. In-deep knowledge of the operation methodology is required in order to set it properly – to be specific enough for categorization but also general enough to be used for different operations.

Table 2. Example of the annotation terms for incisions

Number	Description	Term examples
a	layer	*skin*
b	type of approach	*anterior*
		posterior
c	course of incision	*longitudinal*
		oblique
d	incision area	*hip joint*
e	tools	*knife*
		blunt hook

Incision, for example, is a part of each operation. In this case the following information is required: the name tag of the physician who made the cut, the type of approach, the course of incision, the incision area, the layer, the names of the tools that have been used, and (optional) the complications. Another example can be the preparation of the bony placenta, for which similar information are required.

At present, about thirty types of events are implemented. The current database model will be enhanced with new events for other types of operations.

The indexed **persons** are *physicians* or *nurses*. Some additional information is required for a **person**, like professional position, specialization and so on. The **location** can be a *hospital* with auxiliary data included, e.g. hospital address and name. In Table 2 we give an example of annotation terms (mentioned in *italics*) used only for one stage, namely incision.

Since the full scope of the information included into the database can be available for users with the highest privileges only, we work on methods of access privileges for different kinds of users, e.g. only restricted access to the name tags of patients.

4.4 Teleeducation

Although many standards concerning teleeducation and the virtual university have been developed [9] we are adopting a rather simple approach. There have been defined two functions of the teleeducation subsystem:

1. semi-automatic preparation of the lectures and courses to support the lecturer,
2. individual and group learning with verification of knowledge acquisition.

We have defined a lecture or a course as a set of presentation material consisting of multimedia information, logically organized by the lecturer, and presented electronically to the participants. The presentation is assembled with data already introduced into the system (and typically existing on the SS). Besides the physical data from medical cases, some written material like slides with text and other graphical objects are also considered.

The tool manages several windows. The main window is used for the lecture preparation while the rest for browsing the existing data. The user is offered the automatic links construction when he selects two items (called parent and child items) in order to organize the presentation flow. Typically, the parent item is a word or a phrase on the text screen presented in the main window, while the child item – selected from other windows – can be another text screen or its element (with other links defined subsequently), a picture (with links to other objects) or a movie.

Such an organization could be represented as a graph as well, in which nodes define lecture elements (atomic data) and edges – relations between them. Two versions are being developed: portable, which can be taken away to be installed on a local computer elsewhere and a stationary version making use of the VS.

The graph representation is also implemented for computer based learning, which is being adopted in PARMED based on [10]. The subsystem incorporates some learning methods like preliminary learning recognition or self pace [11, 12,13]. In the graph of knowledge representation the nodes determine atomic problems (knowledge elements), while edges – problem relations. The relations specify sequence of knowledge learning.

Since learning is associated with knowledge evaluation, two kinds of nodes have been introduced. The first one represents problems which have to be learned and answered by a student. He/she receives a qualitative feedback and a mark. Based on the mark and the problem relations, the system enters one of the second kind nodes – the transition node. Its role is to determine an action sequence for the student. It could be a next portion of the knowledge or request for material revision.

5 Conclusion and Future Developments

Innovative technical aspects, advances and experiences of the PARMED projects (at the moment in the second of its three years) have been presented. Some of its elements are already operational (e.g. video and storage servers). The meta-database and computer based training subsystems are being implemented and are tested currently.

The database schema and the concept of the multimedia-storage are supposed to be general enough to be used for a rather wide scope of knowledge from different fields. This must be, however, validated by a number of additional applications.

Current advances and experiences in the cooperation between the computer scientists and the medical doctors show that we can achieve a running subsystem at the end of the second project year. The exchange of researchers could be intensified, as the cooperation with the medical doctors become more and more active. The medical doctors operate not only as suppliers of raw data, i.e. Roentgen pictures and operation videos, but also as experts with valuable knowledge, e.g. in how to index correctly the multimedia data.

The planned development for the next months concerns: the completion of the interfacing of the meta-database with the storage system and the development of the control and status flow in the whole system (e.g. information for the client about the status of the demanded data). The following activities are concerned: refinement of the implementations of the meta-database and the storage server as well as development of further client's applications and concept of client's application framework.

Acknowledgments. We gratefully acknowledge the contribution of K. Breidler, M. Bubak, R. Krasowski, J. Mościński, D. Nikolow, B. Nitka, S. Podlipnig, M. Pogoda, S. Polak, J. Szczerkowska, R. Tusch, and P. Wójcik. Without them, the initiation and then the advances in the PARMED project would not have been possible.

The work has been supported by Scientific and Technological Cooperation Joint Project between Poland and Austria: KBN-OEADD grant No. A:8N/1999 and by Polish Committee for Scientific Research (KBN) Grant 8T11C 006 15.

References

1. Böszörmenyi, L., Kosch, H., Słota, R., "PARMED - Information System for Long-Distance Collaboration in Medicine", in: Conf. Proc. of Software for Communication Technologies'99 (Third International Austrian-Israeli Technion Symposium with Industrial Forum), Hagenberg-Linz, Austria, April 26-27, 1999, Austrian Technion Society, pp. 157–164.
2. Future Generation Computer Systems, spacial double issue: The Telemedical Information Society - guest editor: A. Marsh, Vol. 14 (1998), Numbers 1–2, June 1998.
3. Kosch, H., Słota, R., Böszörmenyi, L., Kitowski, J., Otfinowski, J., Wójcik, P., " A Distributed Medical Informatuion System for for Multimedia Data - The First Year's Experience of The PARMED Project", in: Conf. Proc. of Conference on High Performance Computing and Networking, Amsterdam, May 8 - 10, 2000, Lecture Notes in Computer Science 1823, pp. 543–546, Springer, 2000.
4. Słota, R., Kosch, H., Nikolow, D., Pogoda, M., Breidler, K., "MMSRS - Multimedia Storage and Retrieval System for a Distributed Medical Information System", in: Conf. Proc. of Conference on High Performance Computing and Networking, Amsterdam, May 8 - 10, 2000, Lecture Notes in Computer Science 1823, pp. 517–524, Springer, 2000.
5. Kosch, H., Breidler, H., Böszörményi, L., "The Parallel Video Server SESAME-KB", in : Peter Kacsuk and Gabriele Kotsis: Distributed and Parallel Systems : From Concepts to Architectures, Kluwer Press, to appear in September 2000.
6. Moustefaoui A., Kosch H, Brunie L.: Semantic Based Prefetching in News-on-Demand Video Servers. Accepted for Publication in Multimedia Tools and Applications, Kluwer Journal.
7. Chiueh, T., Venkatramani, C., Vernick, M : Design and Implementation of the Stony Brook Video Server. SP&E 27(2), pp. 139-154, 1997.
8. Tusch, R., Kosch H., Böszörményi, L. : VIDEX: "An Integrated Generic Video Indexing Approach", in: Proc. of the ACM Multimedia Conference, Los Angelos, California, USA, October-November 2000, ACM Press (long version appeared as Technical Report TR/ITEC/00/2.02 of the Institute of Information Technology, University Klagenfurt, May 2000)
9. Farance, F., Tonkel, J., "Learning Technology Systems Architecture (LTSA) Specification", http://www.edutool.com/ltsa/ltsa-400.html
10. Cetnarowicz, K., Marcjan, R., Nawarecki, E., Zygmunt, M., "Inteligent Tutorial and Diagnostic System.", Expersys'92 - Proceedings, pp. 103–108, Huston - Paris, October 1992.
11. Goldsmith, D.M., "The Impact of Computer Based Training on Technical Training in Industry", Australian Journal of Educational Technology, 1988, 4(2), pp. 103-110; http://cleo.murdoch.edu.au/gen/aset/ajet/ajet4/su88p103.html
12. Esteban, E.P., Agosto, F.J., "Using Web Technology in Undergraduate Research", Computers in Physics, pp. 530-534, Nov/Dec 1998.
13. McGreal, R., Elliott, M., "Learning on the Web", http://teleeducation.nb.ca/lotw/c4.html.

Hydra - Decentralized and Adaptative Approach to Distributed Computing

Marian Bubak[1,2] and Paweł Płaszczak[1]

[1] Institute of Computer Science, AGH, al. Mickiewicza 30, 30-059 Kraków, Poland
[2] Academic Computer Centre – CYFRONET, Nawojki 11, 30-950 Kraków, Poland
email: bubak@uci.agh.edu.pl, p.p@interia.pl
phone: (+48 12) 617 39 64, *fax:* (+48 12) 633 80 54

Abstract. This paper presents Hydra – a system designed for distributed computing on a set of heterogeneous uncommitted machines. It consists of autonomous nodes placed on participating host computers. The nodes attempt to make maximal use of temporarily available resources with a monitoring mechanism. Hydra nodes need no administration. Prototype was implemented in Java. Results of a sample simulation show scalability of the system.

Keywords: distributed computing, decentralization, Java, desktop computing, volunteer computing

1 Introduction

Recently, a new trend is seen in the field of distributed computing. It is an attempt to utilize the power available on the global network, represented by small computers (like PCs) and machines whose resources are normally committed to other work, but can be used during their inactivity. Such an environment for computing is more capricious than professional machines committed to this sort of work. The most nasty features are high granularity, hardware heterogeneity, coordination problems, low resource predictability, high failure factor, and low confidence.

Obviously, a computing system working in that type of environment will be less efficient than one based on a committed supercomputer or a cluster. Heterogeneous and changeable resources demand more time spent on system maintenance. A desirable feature is the continuous monitoring and analysis of available hosts and resources. Changing resource state influences computing strategy. Failure resistance and error checking are other necessary capabilities. Finally, the computing process cannot interfere with other applications on a host.

Feasibility of building such a complicated system is proved by the fact that it can make use of resources that already exist, are free and, so far, useless. Hydra attempts to serve as such a changeable resource environment.

T. Sørevik et al. (Eds.): PARA 2000, LNCS 1947, pp. 242–249, 2001.

2   Inspirations

Today's distributed computing systems are mostly based on ideas that come from PVM [1], MPI [2] or CORBA [3]. Such interfaces can be a base for writing more advanced systems. An advanced approach to the problem is metacomputing that involves special sort of machines and resources and requires serious administration efforts [4].

Organizations like SETI [5] and distributed.net [6] have developed widely-known volunteer computing systems. They are strongly centralized and normally committed to a special sort of computation. Many new systems that originated in the last few years have a common feature which is Java implementation. This is true for ATLAS [7], Javelin [8], Charlotte [9], IceT [10], Ninflet [11] or Jessica, to name only a few.

Agent-based Java systems like Aglets [12], although not intended directly for parallel computing, are closely related to Hydra in the domain of autonomy and intelligence o participating entities.

IceT blends Java code with other languages, attempting to built systems based on heterogeneous resources. Unfortunately it lacks flexibility, requiring administration efforts and having no object-oriented parallel programming mechanisms. Jessica achieves distributed Java code execution by re-implementing Java Virtual Machine and changing the Thread class. Such an approach, although very mature, requires additional effort i.e. installation of JVM.

Ninflet and Javelin represent a three-tier architecture, dividing the participants into Clients, Hosts, and Brokers (or, as in Ninflet, Dispatchers) mediating between them. In Javelin, scalability is in fact restricted by primitive applet-based communication. Ninflet, although implementing advanced techniques like checkpointing, in fact also repeats the Master-Slave communication pattern.

Projects like Javelin, Charlotte or Bayanihan [13,14] make use of browser-based Web Computing. Such systems are server-centered. Users need to log into the server web page every time they want to join. Additionally, applet security model suffers from restricted communication.

In general, the Java-based systems offer ease-of-use even for non-professional users eager to donate their resources, but they require serious administration work on the server side. Usually they lack scalability. The typical star topology means a catastrophe in case of the central server failure.

3   Hydra at a Glance

3.1   Overview

The computing system Hydra attempts to be an environment that requires only a little installation effort (download, save, run) and no administration at all. The only software and hardware requirement is the presence of appropriate JVM. Hydra is decentralized in its design, so that all the nodes are capable of taking part in any system functionality, including mapping tasks to other nodes. The system state is dynamic, at runtime it can load and perform arbitrary tasks, or even change its structure incorporating new nodes. It can perform any sort of parallel program, if only formulated using Hydra task model.

Hydra bears a name of a mythological monster said to have nine heads that grew again when cut off. This picture corresponds to the system design. Hydra can be seen as a set of computing nodes living on separate hosts. A node's capabilities can grow or shrink in time, or it can even completely disappear - it all depends on the state and resources of the hosting machine. Once dead, a node revives and dynamically joins the system when its machine is on again. What's more, new nodes from previously unknown machines can also join Hydra at its runtime and immediately take part in the computing process.

3.2   Decentralization

One of main Hydra paradigms is its decentralization. There system has no main server. Every node is independent and autonomous in its decisions and no hierarchy of them is established (see Fig. 1).

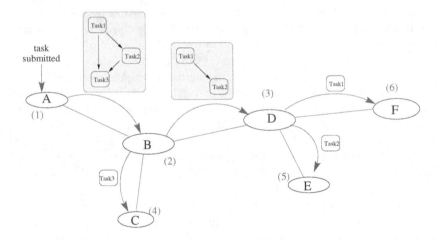

Fig. 1. A decentralized scenario.

A problem is submitted to node A (1). The node decides to send the whole job to its more powerful neighbour B. Node B analyses the problem and splits it in two parts (2). The two subproblems are sent to nodes C and D. While C decides to perform the job itself (4), D prefers to make it parallel. It orders the nodes E and F to perform the subtasks (5,6). It is important that any functionality of the system can be performed by any of the nodes and there is no central service at all. Any node can be charged with a new job, map the tasks among its neighbours or, optionally, perform all the job itself.

3.3 Recursive Task Model

A computing job submitted to Hydra must be defined as a simple task (an inseparable execution unit) or a composite one (whose execution may be divided into subtasks) (Fig. 2). A composite task has a form of a graph of subtasks that defines relations of their input and output data. Subtasks themselves can be simple or composite. Defining the problem as a graph of tasks, a user can suggest which pieces of the algorithm can be executed on separate nodes. The system will decide at runtime if and when to obey these suggestions.

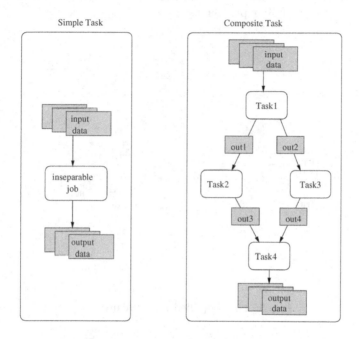

Fig. 2. The recursive task model.

3.4 Monitoring and Information System

Every node continuously collects information about their neighbours and other remote nodes. System topology knowledge includes geographical information on other nodes that compose Hydra. This information is normally (not always) static and serves as a base for contacting other nodes and for resources analysis.

System resource database keeps track of the available resources and current capabilities of home and other nodes. This sort of information is updated quite often and access to this data is necessary when planning the task execution strategy. Knowing the present and past system resources, the node is able to predict the state of the system in the nearest future.

3.5   Node Architecture

The simplified node architecture is shown in the Figure 3. The central object within a node is the Work Manager. The effective computing work is performed by Workers, started by Work Manager. Guardian is the module responsible for communication with other nodes. When a new task is received by Guardian, the Secretary informs Work Manager to start a new Worker. The Worker is given the task and starts performing it. Resource Info Center and Topology Info Center are collecting information about the system. They are necessary for job mapping decisions. Finally, the Initiator is responsible for successful node restart in case of the host reboot or similar problems.

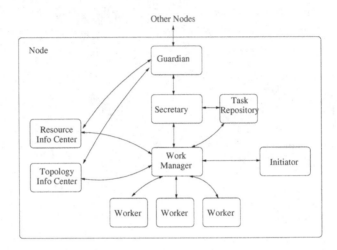

Fig. 3. The simplified node architecture

3.6   Dynamic Structure

The nodes organize themselves logically in a tree-like structure where each one knows its neighbors. New nodes that enter the system have to find their place in the tree. Also, when a node disappears for some reason, it causes changes in the tree structure.

4   Implementation

Hydra in its prototype version has been implemented in Java and RMI. The choice of the language was made mainly because of the byte-code platform independence, ease-of-use and the short implementation time. The obvious execution slowdown caused by Java is balanced by larger number of hosts that can participate in computation thanks to several platforms support.

5   Performance Tests

The Prototype version of Hydra has been tested on a set of PCs with Intel Celeron 366 MHz/64 MB RAM and Celeron 333 MHz/32 MB RAM, running Linux RedHat 5.1 and SUN JDK 1.1.7B. Tests with up to 18 nodes involved only 366 MHz machines, and afterwards, additional 8 Celeron 333 MHz PCs were added. As all the machines have common filesystem, the dynamic code loading option was disabled in the prototype. Four series of simulations were done:

1. a grid of 128 tasks in 4 sequential rows, 32 parallel tasks each,
2. a grid of 64 tasks in 4 rows, 16 tasks each,
3. a single row of 8 parallel tasks,
4. a single row of 16 parallel tasks.

For each simulation series different task sizes were used, but the tasks were always equally sized within a single simulation.

The execution time has been measured between the task submission and arrival of results (Fig. 4). Thus, it includes:

1. time of main task reception and analysis,
2. subtask scheduling time,
3. subtask execution times,
4. synchronization and communication times (between rows of tasks),
5. time of data summary in the final task.

The prototype showed moderate speedup, roughly equal to 1/3 to 1/2 of the ideal one. The simulations (1) and (2) require relatively large communication and synchronization as the tasks are grouped in four stages (Fig. 4, left). In the simulations (3) and (4) there is only one row of tasks so that inter-task communication is reduced to exchanging initial parameters and summary data. These two examples tend to speed up faster (Fig. 4, right).

It is worth to notice that the simulation (3) which consists of 8 tasks should not speed up above 8 hosts. However, in Hydra there is always at least one additional node for scheduling and task transporting to their neighbors. With 16 nodes this functionality is distributed more efficiently what results in an additional speedup.

These preliminary performance results suffer from the simplified prototype implementation. The largest overhead is introduced by:

- inefficient task migration process (currently, two neighbor tasks must be migrated sequentially),
- restricted communication pattern (currently, a node knows only its closest neighbors which lengthens task delegation scenario).

These issues are not limited by the system architecture and we hope to improve them soon.

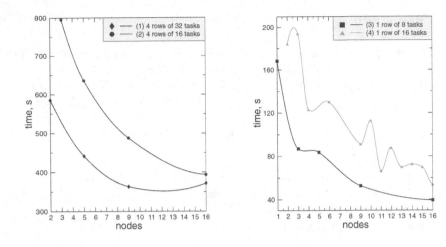

Fig. 4. Performance vs system size

6 Conclusions

The system in its current implementation is experimental. Its aim is to investigate the possible directions of development of distributed computing systems that are more flexible, intelligent and adaptable to the changing environment and therefore require little attention of the hosts administrators. The unique features of Hydra are: decentralization, adaptability, scalability, and fault-tolerance.

The preliminary prototype proves decentralized, scalable and reasonably efficient behavior. It is able to dynamically load and perform arbitrary tasks. The fault-tolerance and efficient adaptability are supported by the system design and will be implemented in the future versions. The future development will focus on security, failure tolerance, advanced dynamic topology, task communication performance, and user interfaces. We believe that Hydra system paradigm opens new perspectives for the world-wide distributed computing.

Acknowledgements. This work has been supported in part by KBN under grant 8 T11C 006 15.

References

1. Geist, A., Beguelin, A., Dongarra, J., Jiang, W., Manchek, R., Sunderam, V.: PVM: Parallel Virtual Machine: A User's Guide and Tutorial for Networked Parallelism. MIT Press. (1994) http://www.netlib.org/pvm3/book/pvm-boook.html
2. Gropp, W., Lusk, E., Skjellum, A.: Using MPI. MIT Press. (1994) http://www.epm.ornl.gov/walker/mpi/index.html
3. Siegel, J. CORBA - Fundamentals and Programming, Wiley, 1997

4. Kesselman, C. and Foster, I. (eds.): *The Grid: Blueprint for a NewComputing Infrastructure*. Morgan Kaufmann Publishers, 1998.
5. SETI@home home page: http://www.bigscience.com/setiathome.html
6. Beberg, A. L., Lawson, J., McNett, D.: distributed.net. http://www.distributed.net
7. Baldeschwieler, J. E., Blumofe, R. D., Brewer, E. A. ATLAS: An Infrastructure for Global Computing. In Proc. of the 7th ACM SIGOPS European Workshop: Systems support for Worldwide Applications, Connemara, Ireland, September 1996
8. Capello P., Christiansen B., Ionescu M. F., Neary M. O., Schauser K. O., Wu, D.: Javelin: Internet-based Parallel Computing Using Java. In Concurrency: Practice and Experience, November 1997.
9. Baratloo A., Karaul M., Kedem Z., Wyckoff, P.: Charlotte: Metacomputing on the Web. In Proc. of the 9th International Conferrence on Parallel and Distributed Computing Systems, Dijon, France, September 1996.
10. Gray, P. A., Sunderam V., S.: Native-Language-Based Distributed Computing Across Network and Filesystem Boundaries, in ACM 1998 Workshop on Java for High-Performance Network Computing
11. Takagi, H., Satoshi, M., Hidemoto, N., Satoshi, S., Satoh, M., Nagashima, U.: Ninflet: A Migrattable Parallel Objects Framework using Java, in ACM 1998 Workshop on Java for High-Performance Network Computing
12. Programming Mobile Agents in Java, http://www.trl.ibm.co.jp/aglets/
13. Project Bayanihan home page: http://www.cag.lcs.mit.edu/bayanihan
14. Sarmenta, L. F. : Bayanihan: Web-Based Vollunteer Computing Using Java Proc. of the 2nd International Conference on World-Wide Computing and its Applications (WWCA'98), Tsukuba, Japan, March 3-4, 1998

Object-Oriented Approach to Finite Element Modeling on Clusters

Roman Wyrzykowski[1], Tomasz Olas[1], and Norbert Sczygiol[2]

[1] Technical University of Czestochowa, Institute of Math. and Computer Science,
Dabrowskiego 73, 42-200 Czestochowa, Poland
{roman, olas}@k2.pcz.czest.pl
http://www.k2.pcz.czest.pl/roman
[2] Technical University of Czestochowa, Institute of Mechanics and Machine Building,
Dabrowskiego 73, 42-200 Czestochowa, Poland
sczygiol@imipkm.pcz.czest.pl

Abstract. This paper presents a concept and implementation of an object-oriented software environment for the parallel finite element modeling on networks (clusters) of workstations. This environment is an extension of the sequential *NuscaS* software, previously developed at the Technical University of Czestochowa. The domain decomposition technique and iterative methods of solving large sparse linear systems are used to develop the parallel kernel of the environment. The application of the object-oriented approach allows mechanisms supporting parallelism to be transparent for developers of new application modules. The performance results obtained for the simulation of solidification of castings are promising, and indicate that a significant reduction in runtime can be achieved for sufficiently large meshes.

1 Introduction

Finite element method (FEM) is a powerful tool for study different phenomena in various areas. However, many applications of this method have too large computational or memory costs for a sequential implementation. Parallel computing allows [11] this bottleneck to be overpassed.

The object-oriented technique [8] is a modern method for software development. It provides better clarity of computer software while assuring higher data security. As a result, object-oriented codes are easy to extend for new and/or more complicated application areas. In development of parallel numerical software, this technique allows combining the speed of parallel/distributed processing and the ease of adapting to needs of end-users.

In the previous work [13], a parallel application for the FEM modeling of solidification of castings has been developed. This application is based on the Aztec parallel library [12] and MPI [5]. The code has been successfully tested on IBM SP2 and HPS2000 parallel computers. However, when running on networks of workstations quite a poor performance has been achieved. Also, it is very time consuming and error prone to adapt the code to modifications in the numerical

T. Sørevik et al. (Eds.): PARA 2000, LNCS 1947, pp. 250–257, 2001.

model chosen for the solidification simulation, not even speaking about re-using this code for another application problem.

This paper presents (Section 2) a concept and implementation of an object-oriented software environment for the parallel finite element modeling on networks (clusters) of workstations. Such clusters built of COTS hardware components and with free or commonly used software are becoming more and more popular alternative to parallel supercomputers [2]. They offer the possibility of easy access to relatively cheap computer power. The rest of this paper is organized as follows. In Section 3, we present performance results obtained for the parallel simulation of solidification of castings. Related works are discussed in Section 4, while conclusions are given in Section 5.

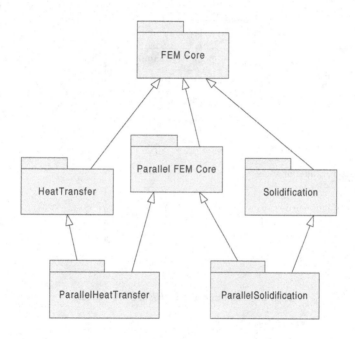

Fig. 1. Processing module of parallel version of *NuscaS*

2   Object-Oriented Implementation on Clusters

The starting point in development of this environment is the *NuscaS* software, designed for sequential computers at the Technical University of Czestochowa. *Nuscas* is dedicated to numerical simulation of such thermo-mechanic phenomena as heat transfer, solidification, stresses, damages of materials, etc. The main part of *NuscaS* is the processing module. It follows the principles of the object-oriented technique, and allows the existing classes to be extend for solving new application problems (see Fig.1). The *FEMCore* kernel of this module (see Fig.2) consists of fundamental classes which are responsible for implementing the finite element method, irrespective of the type of a problem being solved. The mechanism of inheritance is used to develop new application classes for a particular problem.

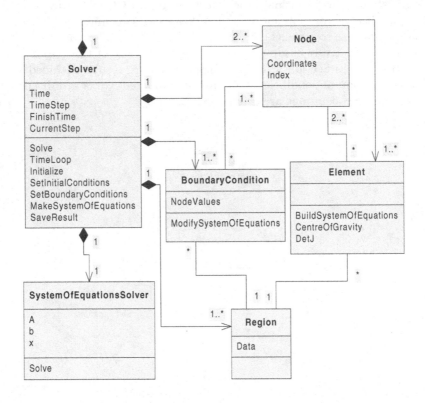

Fig. 2. Contents of the *FEMCore* package

When incorporating cluster computing into the *NuscaS* software, the following basic assumptions have been adopted:

1. developing an application FEM module, a user writing code for the sequential version should have possibility to obtain from this code also its parallel version, without modifying the code;
2. parallelism should not cause additional overheads in the sequential version, e.g. less efficiency or more memory;
3. mechanisms supporting parallelism (communication between processors and allocation of mesh nodes to processors) should be transparent for developers of new application modules;
4. the number of processors in the target message-passing architecture is fixed at run-time.

To reach these aims, mechanisms supporting parallelism are embedded into the core of the environment - *ParallelFEMCore* package (see Fig.3). Their use is accomplished through such object-oriented mechanisms as inheritance and class templates. Another important design solution is to use iterative methods for solving large linear systems with sparse matrices, which arise as a result of FEM discretization. These methods allow [3,9] the sparsity of matrices to be exploited more efficiently than direct methods.

The domain decomposition approach [9,13] is used to parallelize both the creation of stiffness matrix at each time step of modeling, and solution of corresponding linear systems. Consequently, during the parallel realization of FEM modeling two kinds of processes exist:

- *computational process* which has all the information about the corresponding sub-domain, and builds local systems of equations for this sub-domain and solves them;
- *process manager* which manages and synchronizes domain solvers, and performs I/O operations.

For a certain application problem, each computational process is associated with an object created on the base of *DomainSolver* templates from the descendant of the *Solver* class (see *SolidificationDomainSolver* in Fig.3). The same idea is used for creating the *SolidificationProcessManager* object.

The parallel version of *NuscaS* is designed using PVM [4] as a communication layer. At present, the Conjugate Gradient (CG) algorithm with Jacobi preconditioner is used for solving linear systems. A version of the CG algorithm with a single synchronization point [7] is implemented to reduce the idle time of processors. For partitioning FEM meshes generated by *NuscaS*, the Chaco library [6] is currently used. Since the computational kernel of iterative methods is the sparse matrix-vector multiplication, this operation is carefully implemented. In particular, the overlapping of communication with computation is exploited to reduce the parallel runtime.

3   Parallel Modeling of Solidification: Performance Results

The FEM modeling of solidification of castings is chosen for testing the parallel version of *NuscaS*. This physical phenomenon is governed by a quasi-linear heat conduction equation

$$\nabla \cdot (\lambda \nabla T) + \dot{q} = c\rho \frac{\partial T}{\partial t} . \tag{1}$$

The following two systems of linear equations should be solved for the apparent heat capacity formulation of this problem [13]:

$$(\mathbf{M}^n + \Delta t \mathbf{K}^n) \mathbf{T}^{n+1} = \mathbf{M}^n \mathbf{T}^n + \Delta t \mathbf{b}^{n+1} , \tag{2}$$

$$\left(\mathbf{M}^0 + \frac{3}{4}\Delta t \mathbf{K}^0\right) \mathbf{T}^{n+2} = \mathbf{M}^0 \mathbf{T}^{n+1} - \frac{1}{4}\Delta t \mathbf{K}^0 \mathbf{T}^n + \frac{1}{4}\Delta t \left(3\mathbf{b}^{n+2} + \mathbf{b}^n\right) . \tag{3}$$

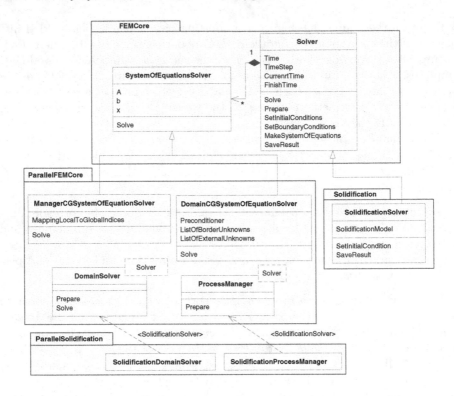

Fig. 3. Modifications in packages for parallel implementation

Here $\dot{q}$ is the term of heat source, which describes the rate of latent heat evolution, T is the temperature, t is time, λ, ρ and c are the thermal conductivity coefficient, density, and specific heat, respectively, while $\mathbf{M}$ and $\mathbf{K}$ are the mass and conductivity matrices. Expression (2) is a result of using the Euler-backward scheme for time integration, which is applied only at the first step of simulation ($n = 1$), while formula (3) corresponds to the Dupont II scheme, used at the remaining steps $n = 2, 3, \ldots$. The superscript (^{0}) indicates that in calculation we use material properties evaluated for the extrapolated temperature $T = (3/2)\, T^{n+1} - (1/2)\, T^n$.

In our experiments, the cast geometry was meshed with 5101, 10225, 20201, 40613, 80401 and 159613 nodes. Every mesh was decomposed using the Chaco library into 2, 3 and 4 sub-domains. For each mesh, both the sequential and parallel versions were tested. The experiments were performed on a network of four Sun Ultra 10 workstations running under the Solaris 2.6 operating systems; these machines were connected by the Fast Ethernet network. The numerical accuracy of simulation was verified by comparison of the computed temperatures with those obtained by the sequential *NuscaS* software.

The performance results of these experiments (see Figs.4, 5) are promising, and indicate that a significant reduction in runtime can be achieved for suffi-

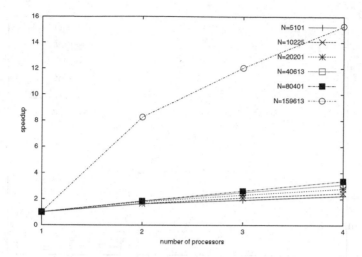

Fig. 4. Speedup for the model problem as a function of number p of processors and number N of nodes in mesh

ciently large meshes. For example, for the mesh containing $N = 40613$ nodes, our parallel code allows speedup $S_p = 3.11$ for $p = 4$ processors, while for $N = 80401$ we have $S_4 = 3.37$. However, for small meshes, communication overheads become more significant, and make the reduction decrease. For example, for the mesh with 5101 nodes, the parallel runtime on four processors is only 2.23 times smaller than the sequential runtime. Note that for our largest mesh with $N = 159613$ nodes, we have a superlinear speedup with $S_p > p$, since in this case $S_2 = 8.27$, $S_3 = 12.04$, $S_4 = 15.26$. This anomalous phenomenon is due to characteristics of hierarchical memory in workstations.

4   Related Works

Another approach to the parallelization of finite element modeling is to use parallel libraries dedicated to solving large sparse linear systems. The P-SPARSLIB library [10] from the University of Minnesota, the Aztec package [12] from the Sandia National Laboratories, and the PETSc library [1] from the Aragonne National Laboratories are representatives of these parallel software packages, built on the top of the MPI communication layer [5]. However, since none of these libraries offers a direct support for finite element modeling applications, their use for this aim is still an extremely difficult task for non-specialists in parallel computing. On the contrary, while building our parallel environment, we employ the "top-down" approach, going from the well-defined application domain to its parallel implementation. Together with using the object-oriented technique, this approach allows the open software system to be built. It integrates the finite element analysis and cluster computing in a way which is in fact transparent for developers of concrete finite element modeling applications.

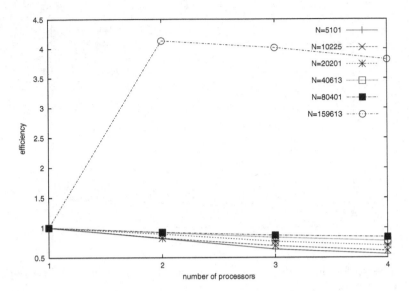

Fig. 5. Efficiency for the model problem versus number of processors, for different numbers of nodes in mesh

5   Conclusions

This paper presents the object-oriented approach to development of the parallel software environment for finite element modeling on networks of workstations. This environment is an extension of the sequential *NuscaS* software, developed in the Technical University of Czestochowa. The domain decomposition technique and iterative methods of solving large sparse linear systems are used to develop the parallel kernel of the environment. The application of the object-oriented approach allows mechanisms supporting parallelism to be transparent for developers of new application modules. The performance results obtained for modeling of solidification of castings are promising, and indicate that a significant reduction in runtime can be achieved for sufficiently large meshes.

References

1. Balay, S., Gropp, W., McInnes, L.C., Smith, B.: PETSc 2.0 User Manual. Tech. Report ANL-95/11 - Revision 2.0.22, Aragonne National Lab., 1998 (available at http:// www.mcs.anl.gov/petsc/petsc.html)
2. Bubak, M.: Cluster Computing. In: Wyrzykowski, R., Mochnacki, B., Piech, H. (eds.): Proc. Third Int. Conf. on Parallel Processing and Applied Math. - PPAM'99, Kazimierz Dolny, Poland, (1999) 92-99
3. Dongarra, J., Duff, J.S., Sorensen D.C., van der Vorst, H.A.: Numerical Linear Algebra for High-Performance Computers. SIAM, Philadelphia (1998)
4. Al Geist et al.: PVM: Parallel Virtual Machine. A User's Guide and Tutorial for Networked Parallel Computing. MIT Press, Cambridge MA (1994)

5. Gropp, W., Lusk, E., Skjellum, A.: Using MPI: Portable Parallel Programming with the Message-Passing Interface. MIT Press, Cambridge MA (1995)
6. Hendrickson, B., Leland, R.: The Chaco User's Guide. Tech. Report SAND95-2344, Sandia National Lab., 1995
7. Meisel, M., Meyer, A.: Hierarchically Preconditioned Parallel CG-Solvers with and without Coarse-Mesh Solvers Inside. SFB399-Preprint 97-20, Technische Universitat Chemnitz, 1997
8. Parallel Programming Using C++. G.V. Vilson, P. Lu, editors. MIT Press, Cambridge MA (1996)
9. Saad, Y.: Iterative Methods for Sparse Linear Systems. PWS Publishing, New York (1995)
10. Saad, Y., Malevsky, A.V.: P-SPARSLIB: A Portable Library of Distributed Memory Sparse Linear Solvers. Tech. Report UMSI 95-180, Univ. Minnesota, 1995.
11. Topping, B.H., Khan, A.I.: Parallel Finite Element Computations. Saxe-Coburg Publications, Edinburgh (1996)
12. Tuminaro, R.S., Heroux, M., Hutchinson, S.A., Shadid, J.N. : Official Aztec User's Guide: Version 2.1. Sandia National Lab., 1999 (available at http://www.cs.sandia.gov/CRF/Aztec_pubs.html)
13. Wyrzykowski, R., Sczygiol, N., Olas, T., Kaniewski, J.: Parallel Finite Element Modeling of Solidification Processes. Lect. Notes in Comp. Sci. **1557** (1999) 183-195

An Object Oriented Framework for Parallel Multiple Optimizations

Beidi S. Hamma

Resource Optimization Consulting
ILOG S.A., 9 Rue de Verdun 94253 Gentilly, France,
and
IAI, Institut Africain d'Informatique, B.P 2263 Libreville, Gabon

bhamma@ilog.fr , http://www.ilog.com

Abstract. This report presents a general Objected-Oriented Framework for Parallel and Multiple Optimizations. We designed a general Objected-Oriented Class Library for Parallel Numerical Optimization. All optimization problems are considered: Unconstrained, Constrained, availability (or not) of Derivatives , availability (or not) of Second Derivatives. We also consider all type of methods for solving those problems: non-derivative methods, gradients methods, global optimisation methods. The report describes the C++ classes dealing with (almost) all possibilities of problems types and methods. This enables to have a very simple and efficient interface which will allow any user to easily define his problem, and to add any new optimization solver to his general framework. Examples using the framework in real-life for solving practical problems such as MBS (MultiBody Systems), are shown.

Keywords. Object-Oriented Programming, C++, Numerical Optimization, HPCN, Parallel Virtual Machine.

1 Introduction

The work presented in this paper was first motivated by the needs of Object Oriented Approach for optimization algorithms within the project ODESIM, a High Performance Computing and Networking (HPCN) project funded by the European Commission within ESPRIT programme. The project acronym stands for Optimum DESIgn of Multi-Body systems(MBS). The starting point for the implementation of the ODESIM system was the CAD and MBS simulation packages available within the consortium. They then had to be improved with new modules for HPCN and optimization. For an efficient and reliable implementation, it has been adopted to use standards like UNIX, C++, X-Windows and other internal tools, based on object oriented programming techniques which will guarantee the openness, portability and scalability of the system.

T. Sørevik et al. (Eds.): PARA 2000, LNCS 1947, pp. 258–267, 2001.

Given that, we designed a general object oriented framework which enables:

1. to encapsulate a parallel simulator, wich simulates the parallel virtual ma-
chime and, at the same time, supervises all the tasks required by the whole
optimization process.
2. to encapsulate any optimization solver.

By doing so, we can easily parallelize the most time consuming parts for any
local optimization solver and also parallelize some global optimization features,
as done in this work.

The paper is organized as follows. We will present:

- In section 2, the Optimization Problems and their classes
- Then, in section 3, the Optimization Solvers and their classes
- In section 4, the HPCN Object-Oriented Classes enabling the Parallel Simu-
lator to handle with parallel optimization as well as multiple optimizations.
- how we encapsulated PVM
- Finally, in section 5, some applications are presented.

2 Classification of Optimization Problems

2.1 Object-Oriented Programming

We do not seek to introduce in this paper, all the underlying concepts of Object-
Oriented Programming (OOP). Let us just recall that the main advantages of
OOP come from the following basic concepts : *abstraction, classes-objects, inheritance, and polymorphism.* For better explanations see [8] , [2] and the (lot of)
books/articles written on the subject.

Those basic concepts enable us for instance to separate the optimization
problems from the solvers dedicated to solve them. It also enable us (thanks
to the polymorphism) to define a abstract *Optimizer* from which are derived
Differentiable Optimizer and *DirectOptimizer* etc ... By doing so, for any problem
we just need to call the method myOptimizer.optimize() (assuming myOptimizer
has been choosen and created according to the probelm to solve); optimize() in
turn, is a overloaded method of the virtual optimize() available from the abstract
class Optimizer. Recall that this abstract class can never be instantiated (i.e. an
object of this class can never be created).

2.2 Optimization Methods

The general form of the problem that we are addressing is:

$$\text{minimize } f(x) \text{ where } x \in \mathbb{R}^n$$

subject to the general:
(possibly non linear) inequality constraints : $C_j(x) \leq 0 \; j = 1, ..., q$
(possibly non linear) equality constraints : $C_j(x) = 0 \; j = q + 1, ...m$
and simple bounds : $l_i \leq x_i \leq u_i$
where l_i and u_i could take the values $l_i = -\infty$ or/and $u_i = +\infty$.

Thus, if we assume that there exits a subset $\mathbf{C}$ of $\mathbb{R}^n$ for which all the above constraints are fullfiled, solving an optimization problem means trying to find, for a given objective (real valued) function f, over a set of decision variables x $\in \mathbf{C}$ (a subset of $\mathbb{R}^n$), find the lowest function value f_* and the corresponding (x_*). Usualy, one just has to find a value f_{local} given by x_{local} such that f_{local} is the the lowest value "just" in a neighborhood of x_{local}.

This is called a *local optimization*. Many methods have been developed for solving local optimization problems. Most of them use the properties of the objective function (gradients , hessians) or/and properties of the set of the decision variables. Good routines and packages are available from NAG, HARWELL, MINOS, LANCELOT, MODULOPT, ... However, many practical engineering applications can be formulated as optimization problems in wich the user could find many local minimizers with different function values. Then, how to decide? How being sure that the best of all possible minimums is reached ? This is the aim of *global optimization*: finding the best of all possible minimums, not determining just a local minimum.

2.3   Problems Classification

We chose to consider that the general case is to have an unconstrained problem. Hence we define the

class OptiProblem {
　　protected
　　　　ObjectiveFunction* myObjective;
　　public
　　　　double Evaluate(vector& X) ; // objective value
}

That is, we can see a constraint as some additionnal information on the problem to be solved. By this, we also leave the possibility to Constraints Programming researchers to use our framework. In fact, they usually may not have an objective to be minimized. They rather try to find a feasible solution satisfying the constraints (or a part of them). This leads us to make the ConstrainedProblem inherit from OptiProblem as can be seen in figure 3.

class ConstrainedProblem public OptiProblem{
　　protected
　　　　ListOfConstraints* myConstraints;
　　public
　　　　doubleVect& EvaluateConstraints(vector& X) ;
　　　　bool EvaluateConstraint(int j, vector& X) ;
　　　　bool areConstraintsSatisfied(vector& X) ; // check all constraints
　　　　bool isConstraintSatisfied(int j, vector& X) ; // check constraint j
}

Remarks. Gradients, Hessian are dealt by the appropriate optimizer which, when needed, could call objective.derivatives() or compute it by its own. Same for the gradients of the constraints. Figures 1 & 2 show the inheritance of variables and functions that are used in the framework. For detailed discussions see [4].

Fig. 1. Optimization Variables Classes and their inheritance. *Inclusion of a given box B into another box A, means B inherits from A*

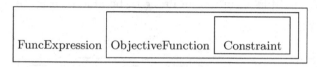

Fig. 2. Functions (objective and constraints) Classes and their inheritance. *Inclusion of a given box B into another box A, means B inherits from A*

Fig. 3. Optimization Problems Classes and their Inheritance. *Inclusion of box B into a box A, means B inherits from A*

3 Classification of Optimization Solvers

The starting point for our classification was the answer to the following question: does a solver/method involve/needs gradients to perform? Hence we first defined two classes : DirectOptimizer and DifferentiableOptimizer which both inherit from an abstract class Optimizer. Afterwards as can be seen in figure 4, we added at the same level a GlobalOptimizer class and derived an HessianOptimizer from DifferentiableOptimizer. The class GlobalOptimizer is designed as a different and separate class because it many often contains some *Strategy* which helps the algorithm to make a kind of global exploration before fiding "good points" from which local searches/explorations are then made. The local searches/explorations may even need a DirectOptimizer or a DifferentiableOptimizer.

class Optimizer {
   protected
      OptiProblem* myProblem;
   public
      virtual bool Optimize( ) ; // to be overloaded in subclasses
}

Remarks. Examples of DirectOptimizer are solvers based on Simulated Annealing, Nelder and Mead, Simplex Method for Linear Optimimization etc .... And, examples of DifferentiableOptimizer are solvers based on Quasi-Newton methods, sequential quadratic programming, etc... Examples of GlobalOptimizer are solvers based on clustering methods, interval analysis etc ...

In figure 4 the class Optimizer is virtual (it has the method *virtual bool optimize()*) so that for a given problem one has to choose a suitable subclass of optimizer like for e.g. a DifferentiableOptimizer in order to be able to use the appropriate optimize() method.

Fig. 4. Solvers Classes and their Inheritance. *Inclusion of a given box B into another box A, means B inherits from A*

3.1   Parallelization of Optimization Algorithms

In the context of ODESIM project we aimed to parallelize the optimization part in order to speed-up the whole package.

Motivations: Speed up the CPU-intensive applications; Use the less possible CPU-time during the optimization process; Exploit the HPCN features of the numerical methods.

What to Parallelize
Local optimization algorithms are iterative $((x_k)_{k=1,2,...})$:
Local Optimization Procedure
   Initialization: x_0, parameters ...
   Repeat   *(current iteration c)*
     STEP 1: Find d_c *(current search/descent direction)*
     STEP 2: Find λ_c *(current step)*
     STEP 3: $x_c = x_{c-1} + \lambda_c d_c$ *(current approximation of the solution)*
     STEP 4: Update.   *(Hessian, Gradients, all datas required for stop,...)*
   Until Test-for-stop

Parallelization: How, When
The cost in nonlinear optimization comes from the following: Function-constraints-derivates can be expensive to evaluate; Number of variables and con-

straints can be large and hence each evaluation has high cost; Many evaluations may be required; Many iterations may be required.

Parallelization possibilities can be: Parallelize individual evaluation of $f(x)$ and/ or $\nabla f(x)$ in steps 2, 4; Parallelize the linear algebraic calculations in steps 1, 4; Perform multiple evaluations ($f(x)$ and/or $\nabla f(x)$) in step 1,4 concurrently, either within the algorithmic framework above, or by designing new algorithms $\Longrightarrow$ (**New Research Area**).

Possible Solutions :

1. Parallelize each evaluation of the objective function, constraints and/or derivates
2. Parallelize the linear algebra involved in each iteration (depending on the algorithm selected)
3. Parallelize the optimization process at a high level, either to perform multiple function, constraints and/or derivate evaluations on multiple processors concurrently and/or to reduce the total number of iterations required.

Limitations

The first possibility above is outside of the domain of optimization research. It is to the user to decide. For instance if the objective function involves the solution of a system of differential equations any tentative parallelization will be the domain of parallel diffential equation research...

In spite of such difficulties, derivates evaluations can be parallelized if gradients are calculated either by finite differences or by automatic differentiation.

$\Longrightarrow$ do not involve change in the optimization algorithm.

The second possibility, parallelizing the linear algebra, is one of the most promizing. But it depends on the algorithm and how the steps 1, 4 are achieved in the algorithm. For example for some quasi-newton algorithms d_c is computed as solution of $A_c.dc = -\nabla f(x_c)$ and in step 4 the update can be $A_+ = A_C +$ rank-two matrix. In this case there may be many possiblities of parallelization. The last possibility is in the domain of optimization research. This possibility deals with the introduction of new algorithms directly devoted to parallel computation and which are not necessary based on known sequential algorithms.

Our Choices. We made the choice of implementing :

Parallel Evaluations of the gradients (for e.g. by finite differences); Multiple Optimizations (e.g. from different selected initial points); Any parallelization possibility that is specific to the method we may select (e.g. for Simulated Annealing we exploited the parallelization possibilities described in [6])

4   The Parallel Simulator for Parallel and Multiple Optimizations

4.1   PVM and SPMD Paradigm Encapsulation

The current version of the Parallel Simulator was implemented according to the SPMD paradigm (Single Programme-Multiple Data). It ensures a unique interface between the parallel code and the end-user. However, by nature the parallel code is MIMD (Multiple Instruction-Multiple Data) implementing a master/slave scheme: the master runs the parallel management tasks while the slaves run the computational simulation tasks. In fact, for sake of maintainability, the parallel MIMD code is implemented with the SPMD paradigm. PVM message passing library [1] has been used in the current implementation of the parallel Simulator. Investigations on other possible parallel implementations were attempted during the project development. The parallel computation of the gradients by finite differentiation only exploits one level of parallelism, spawning tasks on the virtual machine when required. Another way to exploit the HPCN environment in the solution of a constrained optimisation problem may consist in launching several local optimisation procedures from different starting points. Each optimisation process can be considered as a master process that generates several slaves for gradient computation. The parallel Simulator may monitor all the optimisation processes, killing those that have a low convergence rate and maintaining those that are converging to a feasible solution. Following this approach, the parallel simulator manages the "masters" and the "slaves" (i.e., it controls their execution, checks and recovers breakdowns, etc.) until all the results are obtained. For both levels, the parallel simulator must keep all its capabilities regarding dynamic load balancing, fault detection and breakdown recovering.

The parallel simulator has been implemented based on the following :

- HPCN on Clusters of Workstations. We encapsulated all the features in C++ in order to keep an easy maintainability and portability.
- Master/Slaves Processes. We used a master/slave implementation in the following way. A master process starts the parallel simulation by first instantiating, for the virtual network, the following parameters:
  1. the number of physical machines involved in the virtual network (minimum 1)
  2. the host id of each machine
  3. the maximum number of processors available on this machine. This number can be less than the real number of processors (one could decide to allow less processors on some machine due to any other usage, for instance)
  4. the capability of each machine to access or not via *nfs* the data on the network. In case this capability is not available on a given machine, we will eventually have to copy temporaly all the needed data on this machine. That is, we need to perform an encapsulation of some unix system commands like for e.g. mkdir, cd, cp, mv, del, system, ...

Once the virtual network is instantiated, the master will regulary estimate, the tasks that are requested and put them in a queue from which they will be spawned according to a suitable load balacing rule that we implemented [4]. As soon as a machine receives a certain number of tasks to be performed (tasks can be function evaluation, or gradient evaluation, or MBS simulation or a complete optimization process from a given initial point) it will attempt to achieve them by spawning UNIX processes on its available processors. Here again, the master by using the returned values of these UNIX processes will be able to accept, or not, the results of a given task. In case of rejection, the master will reset the task performed by this process and put it again in the queue so that it would be performed later (on any other machine -including even the previous one-).

- UNIX commands encapsulation. As said in the previous paragraph, we need during the parallel simulation to perform some UNIX commands like for e.g.:

  1. copying locally the global data on any machine that can not (or that happened to become unable to) use the *nfs* facilities to read the global data
  2. sending unix processes from anywhere, for any kind of processes (C++ function call, call to another executable, ...)
  3. receiving UNIX returned values saying if a process succeded or not, global information from UNIX system ...

- PVM Encapsulation for Parallel Optimization. The need of PVM encapsulation is rather obvious, since we need PVM in order to make the virtual network consistent: send parallel processes, checking current information on the network... We made the encapsulation of the PVM3 C-library.

5   Applications Examples

5.1   Parallel Optimization within ODESIM

As said in the introduction this work was initiated by the needs of Object Oriented Approach for optimization algorithms within the project ODESIM. Hence one of the first applications we present hereafter concerns local and global optimization of MultiBody Systems (MBS). By these two cases we show how the framework can be used either for (first level) parallel optimization and/or for multiple full optimizations. In the following examples we used an encapsulation of VF13 of HARAWELL as DifferentiableOptimizer. It is used both for local optimization (gradients are computed in parallel) and for the multiple optimizations. In the later case it is used from several starting points, to perform full local optimizations, in parallel.

Local Optimization. For local optimization in ODESIM the only parallelization strategy that we found usefull and that we selected and implemented is the computation of the gradients in parallel (assuming that we use numerical differentiation). This choice is justified by the fact that gradients are not explicitly available and each function evaluation is expensive (*corresponds to a sequence of (long) machanical CAD simulations*). Thus the computation in parallel may speed up the optimization process.

Gradients Computation in parallel. We use the following approximations:

$$f'(x) \approx \frac{f(x+h) - f(x)}{h} \text{ for } h \text{ small enough}$$

Thus, in a N dimensional problem we can evaluate separately each component of the gradient i.e.:

For i=1 to N do {

$$f'_i(x) \approx \frac{f(x+h.e_i)-f(x)}{h} \qquad\qquad // \text{ where } e_i \text{ is the } i^{th} \text{ unit vector}$$

}

This loop is also used for computing the gradients of the constraints functions. In practice, we evaluate the gradients by using twice the formula

$$f'(x) \approx \frac{1}{2}\frac{f(x+h)-f(x)}{h} + \frac{f(x)-f(x-h)}{h} \text{ for } h \text{ small enough}$$

5.2   Parallel Multiple Optimizations within ODESIM

Global Optimization. In a first stage of this work we dealt with some MBS problems given by CEIT ([4]). Those problems by nature are black-boxes problems in point of vue of the optimization process. That is, the way the optimization solver communicates with the problem is the following: for a given set of design variable values, the solver requests from a MBS object the corresponding function and constraints values. This object, in order to get the function and constraints values, calls, in fact, another complete executable which is a software tool for simulation of MBS and for evaluating MBS user-defined parameters.

Then, for the optimization solver point of vue, there is no explicit knowledge of the objective function and its constraints. For the gradients computation we use finite differentiations (i.e. the solver asks for the required function evaluations, and then perfoms its own numerical differentiation).

We used the following very simple and naive strategy: given an initial starting point we draw $2n$ additional starting points. Then we make Multiple Runs from those points. That is from each starting point we run a complete local optimization process. At the end, the best solution found is accepted as the "global".

The Multiple Runs Strategy. Assuming we have:

1. Lower and upper bounds on each design variable i.e. we have

$$a_1 \leq x_1 \leq b_1 , \ a_2 \leq x_2 \leq b_2 , \ ......, \ a_n \leq x_n \leq b_n$$

2. An initial guess, as starting point for the optimization problem,

$$X start^0 = (x_1^0, x_2^0, ...x_n^0)$$

We then **generate** $2n$ additional initial starting points in the following manner. For the the first n points $k = 1, 2, ..., n$

$$k^{th} \text{ starting point } X start^k = (x_1^0, ..., x_{k-1}^0, a_k, x_k^0, ..., x_n^0)$$

that is k^{th} component of the initial guess is replaced by k^{th} lower bound. For the the last n points $j = 1, 2, ..., n$

$$(n+j)^{th} \text{starting point} X start^{(n+j)} = (x_1^0, ..., x_{j-1}^0, b_j, x_j^0, ..., x_n^0)$$

that is j^{th} component of the initial guess is replaced by j^{th} upper bound.

Summary. Now, we have $2n+1$ starting points using the bounds on design variables:

$$X start^0, X start^1, ..., X start^n, ..., X start^{2n}$$

from which we can perfom a complete optimization process for each.

Motivations. By using the bounds constraints we attempt to find minimum located in "corners".

Remarks. Each (full) optimization will, in turn, use gradients evaluations in parallel. Hence we will have to face two levels of parallel tasks (since an optimization task needs the gradients to be computed several times, before terminating itself as a task...). The way this is managed by the parallel simulator will be discussed in a future paper (in preparation).

6   Conclusion

In this paper we presented an Object Oriented Framework for Parallel Multiple Optimizations that gives the following capabilities:

1. better optimization interface in the case where objective functions (and gradients) are evaluated by the outpout of a simulation that is, as a black-box for the optimizer
2. rapid encapsulation of different optimization routines already written in C, Fortran etc. This encapsulation reduces the user's dependency on the way original routines are implemented and work.
3. rapid evaluation of several optimization routines on a same problem.
4. rapid and efficient combination of several optimization routines in the purpose of global optimization.
5. better parallelization using other object-oriented heteregeneous parallel simulators.
6. better code maintenance and re-usability.

References

1. Beguelin, A., Dongarra, J., Geist, A., Jiang, W., Manchek, R., and Sunderam, V., *PVM 3 user's guide and Reference Manual* Tech. Rep. ORNL/TM-12187, Oak Ridge Nat. Lab., Tennessee 37831,1994
2. Budd, T., *An Introduction to Object-Oriented Programming*, Addison-Wesley, Reading, Massachusetts, 1991
3. Byrd, H., Schnabel,R. B., and G. Shultz, *Parallel Quasi-Newton Methods for Unconstrained Optimization*, Math. Prog. 42, pp. 273-306, 1988.
4. Giraud, L., Hamma, B. S., and Jimenez, J. M., *ODESIM: An Object-Oriented Library for the Optimization of the MultiBody Systems (MBS) Design in High Performance Computing and Networking (HPCN)*, Working Paper CERFACS, 1998.
5. Hamma, B., *Global and Local Behaviour of Moving Polytopes*, in "From Local to Global optimization", Pardalos et al. eds, Kluver 1997.
6. Hamma, B., Viitanen, S., and Torn, A., *Parallel Simulated Annealing for Continuous Global Optimization*, Optimization Methods and Software, OMS, Vol. 13, pp. 95-116, 2000
7. Moré, J. J., and Wright, S. J., *Optimization Software Guide*, SIAM Frontiers in Applied Mathematics, vol 14, 1993
8. Stroustrup B., *The C++ Programming Language*, Addison-Wesley, Reading, Massachusetts, 1987

Experiments in Separating Computational Algorithm from Program Distribution and Communication

Shortened version - Full version available at http://shekel.jct.ac.il/~rafi

R.B. Yehezkael[1] (Formerly Haskell), Y. Wiseman[1,2],
H.G. Mendelbaum[1,3], and I.L. Gordin[1]

[1] Jerusalem College of Technology, Computer Eng. Dept.,
POB 16031, Jerusalem 91160, Israel
E-mail: rafi@mail.jct.ac.il, Fax: 009722-6751-200
[2] University Bar-Ilan, Math. and Computer Sc. Dept., Ramat-Gan 52900, Israel
[3] Univ. Paris V, Institut Universitaire de Technologie,
143-av. Versailles, Paris 75016, France

Abstract. Our proposal has the following key features:

1) The separation of a distributed program into a pure algorithm (PurAl) and a distribution/communication declaration (DUAL). This yields flexible programs capable of handling different kinds of data/program distribution with no change to the pure algorithm.

2) Implicit or automatic handling of communication via externally mapped variables and generalizations of assignment and reference to these variables. This provides unified device independent view and processing of internal data and external distributed data at the user programming language level.

3) Programs need only know of the direct binds with distributed correspondents (mailbox driver, file manager, remote task, window manager etc.). This avoids the need for a central description of all the interconnections.

The main short-range benefits of this proposal are to facilitate parallel computations. Parallel programming is a fundamental challenge in computer science, nowadays. Improving these techniques will lead to simplify the programming, eliminate the communication statements, and unify the various communication by using an implicit method for the transfer of data which is becoming essential with the proliferation of distributed networked environment. We present 2 experiments of separation between PurAl and DUAL, using a preprocessor or an object-type library. This new approach might be of interest to both academic and industrial researchers.

1 Introduction

In many cases the same algorithm can take various forms depending on the location of the data and the distribution of the code. The programmer is obliged to take into ac count the configuration of the distributed system and modify the algorithm in consequence in order to introduce explicit communication requests. In a previous paper[21],

T. Sørevik et al. (Eds.): PARA 2000, LNCS 1947, pp. 268–278, 2001.
© Springer-Verlag Berlin Heidelberg 2001

we proposed to develop implicit communication and program distribution on the network, here we present two experiments in this field.

Aim. We want to use any "pure" algorithm written as if its data were in a local virtual memory, and run it either with local or remote data without changing the source code.

1) This will simplify the programming stage, abstracting the concept of data location and access mode.

2) This will eliminate the communication statements since the algorithm is written as if the data were in local virtual memory.

3) This will unify the various kinds of communication by using a single implicit method for transfer of data in various execution contexts: concurrently executing programs in the same computer, or programs on several network nodes, or between a program and remote/local file manager, etc.

For example, the programmer would write something like: x := y + 1;

where y can be "read" from a local true memory (simple variable), or from a file, or from an edit-field of a Man-Machine-Interface (for example, WINDOWS/DIALOG BOX), or through communication with another parallel task, or through a network communication with another memory in another computer. In the same way, x can be "assigned" a value (simple local variable) or "recorded" in a file, or "written" in a window, or "sent" to another computer, or "pipe-lined" to another process in the same computer.

2 Proposal

We propose to separate a program into a pure algorithm and a declarative description of the links with the external data.

1) the "pure algorithm" (PurAl) would be written in any procedural language without using I/O statements.

2) A "Distribution, Use, Access, and Linking" declarative description (DUAL declaration) is used to describe the way the external data are linked to the variables of the pure algorithm.   (i.e. some of the variables of the pure algorithm are externally mapped.) The DUAL declaration is also used to describe the distribution of the programs.

To summarize in the spirit of Wirth [20] :

Program = Pure Algorithm + DUAL declarations of external data distribution/access.

2.1 The Producer Consumer Example and Variations

Pure Algorithm 1:
```
#include "external_types.h"
void   c ( )
{vec seq1;                // vec is defined in external_types.h
 for (int i=1;  ; i++)
         {try {    data_processing (seq1 [i]); }             // A
```

```
        catch (out_of_range_error  ce) {break;};
        }// when up-bound exception on seq1,  i.e. end of data
}   //end c
```

Informal example of Separate DUAL declarations for c : declarations of Distribution, Use, Access, and Linking, (see two different concrete implementation in section 3)
c'site is on computer1;   c.seq1'site is on mailbox2;
c.seq1'access is of type IN with sequential_increasing_subscript;
c.seq1'locking policy is gradual;

Pure Algorithm 2:
```
#include "external_types.h"
void   p ( )
{vec seq2;                // vec is defined in external_types.h
 for (int j=1;  ; j++)
        { if (exit_condition) break;
          seq2 [j]: = <expression>; }                              // B
} // end p
```

Informal example of Separate DUAL declarations for p : declarations of Distribution, Use, Access, and Linking, (see two different concrete implementation in section 5)
p'site is on computer2;   p.seq2'site is on mailbox2;
p.seq2'access is of type OUT with sequential_increasing_subscript;
p.seq2'locking policy is gradual;

Some Comments on the above Programs
The first two algorithms are written at the user programmer level and we have hidden the details of the communication in the package "external_types.h". Local declarations are used inside a loop to define which elements of a vector, are being processed. The variables seq1 and seq2 are externally mapped by the DUAL declarations. The type vec is for a vector of characters of undefined length: this means that externally the vector is unconstrained, but internally the bounds, which are given by the local current value of "i or j", define which elements are being processed. The exception mechanism is used to detect the "end of data" condition ("out_of_range_error").
In the line marked "// A " of the function c, seq1[i] is referenced and this causes a value to be fetched from outside, according to the DUAL declaration where c.seq1'site indicates the location of the vector, a mailbox. Also c.seq1'access indicates the way of accessing the data, "IN sequential_increasing_subscript" means that seq1 is read from outside in sequential manner using a sequentially increasing subscript. Similarly in the line marked "// B " of the function p, the vector seq2[j] is assigned and this causes a value to be sent outside the program according to the DUAL declaration where p.seq2'site indicates the location of the vector, a mailbox.   Also regarding p.seq2'access, "OUT sequential_increasing_subscript" means that seq2 is written outside in sequential manner using a sequentially increasing subscript.

With the previous DUAL declarations, for seq1 and seq2, the programs will behave as a producer/consumer when run concurrently.

Other distributed use of the same algorithms are possible by only modifying the DUAL declarations.

3 Examples of Implementation Prototypes

The DUAL syntax need not to be exactly as it was shown in the section 2.1, it depends on the system where it is realized. To implement the idea of separating the algorithm description and the distribution/communication declarations, we tried   two ways : building a C- preprocessor for Unix, or building a C⁺⁺ object-type library.

3.1 C-Unix-Preprocessor Experiment

In this case the DUAL declarations will be sentences put at the beginning of the program or at the beginning of a bloc. For this method we wrote a preprocessor that reads the DUAL distribution/communication declarations and adds the synchronization, distribution and I/O statements to the pure algorithm (PurAl) according to these DUAL declarations. It fits to static declaration of data and program distribution before execution.

Let's take the classical merge_sort algorithm, the PurAl is written as if the sources and the target were in a virtual local memory without I/O and without explicit distribution calls. Using DUAL declarations, we define separately the two sources and the target as external_types. Furthermore, using a DUAL declaration, we can also define the parallelism of some functions separately from the algorithm, so that the algorithm is written purely as if it was sequential.

Please see below this specific DUAL syntax in this case :

Pure Algorithm 3 : Merge_sort with its DUAL declarations:

```
DUAL parallel    merging  in  one_level; // declaration for the parallelization of the
#define N  10     // algorithmic function 'merging' located in the function 'one_level'
void  main ( )
{ DUAL devin    source_data "./data"; // declare : the 'source_data' variable is external
                            // and is linked to the input sequential file "./data"
   DUAL devout  target "/dev/tty"; // declaration : the 'target' variable is external
                            // and is linked to the output screen "/dev/tty "
         int i;   char x[N ];        // local memory vector  x  to perform the sort
         strncpy ( x , source_data , N ); // 'source_data' is copied to local memory x
         msort ( x , N );   // performing the sort on the local memory vector x
         strncpy ( target , x , N );    // the sorted vector x is copied to the target screen
} //end main------------------------------------------------------------
void  msort (char *x, int n )
         // the merge sort algorithm will sort x using split subvectors aux
```

```
{ DUAL share    aux;
        // declaration : all the split subvectors aux will share the same memory
    int    i, size=1;          // initial size of the subvectors
        while ( size < n )
        {           one_level ( x, aux, n , size );
                        // prepares the subvectors merges at this level of size
            strncpy ( x , aux , n );
                        // copy back the sorted subvectors aux to vector x
            size*=2;          // growing the size  of the subvectors
        }
}//end msort----------------------------------------------------------------
void one_level ( char *x, char *aux, int n , int size )
 //prepares the subvectors to merge
{// 1st vector :    lb1, ub1=low and up bounds;
 // 2nd vector :    lb2, ub2=low and up bounds;
 // lb3 = low bound of the 3rd (merged) vector
    int    lb1=0, ub1, lb2, ub2, lb3=0, i;
        while ( lb1+size < n )
        {           lb2=lb1+size;    ub1=lb2-1;
            if ( ub1+size >= n )        ub2=n -1;
            else                        ub2=ub1+size;
            merging ( x, aux, lb1, lb2, lb3, ub1, ub2 );
                        // can be done sequentially or in parallel
            lb1=ub2+1;        lb3+=size*2;
        }
        i=lb1;    while ( lb3 < n ) aux[lb3++]=x[i++]; //copy the rest of the vector
}//end one_level-----------------------------------------------------------
void merging ( char *x, char *aux, int lb1, int lb2, int lb3, int ub1, int ub2 )
                                            //performs the real merge
{int I=lb1, j=lb2, k=lb3;                   // of one subvector
        while (( i <= ub1 ) && ( j <= ub2 ))
                if ( x[i] < x[j] )    aux[k++]=x[i++];
                else                  aux[k++]=x[j++];
        while ( i <= ub1 )    aux[k++]=x[i++];
        while ( j <= ub2 )    aux[k++]=x[j++];
}// end merging
```

Explanations. The above text of the Pure Algorithm 3 is preprocessed before C-compiling. The preprocessor detects the DUAL declarations and inserts, in the text of the algorithm, the explicit I/O, parallelization and synchronization statements, corresponding to a Unix environment.

a) DUAL for implicit communication (implicit references to I/O variables) :
For instance, when the preprocessor finds : DUAL devin source_data "./data";
it will replace it with the following UNIX compatible text

char * source_data=(_devinp[_devin_i]=fopen ("./data" , "r"), fgets ((char *) malloc (BUFSIZ) , BUFSIZ , _devinp[_devin_i++]));

In the same manner, when the preprocessor finds DUAL devout target "/dev/tty"; it will replace it with the following UNIX compatible text

char *s=(_devoutp[_devout_i++]=fopen ("/dev/tty","w"), (char *) malloc(BUFSIZ));

And when source_data or target are used in the algorithm, the preprocessor will insert the necessary I/O statements, for instance at the end of the main, it will add the UNIX compatible text :   fputs (target , _devoutp[--_devout_i]); free (target); fclose (_devoutp[_devout_i]);

free (source_data); fclose (_devinp[--_devin_i]);

b) DUAL for implicit use of shared local memory :

For instance, when the preprocessor finds :   DUAL share   aux;

it will replace it with the following UNIX compatible text

char * aux=(char *)shmat (_shmid[_share_i++]=shmget(IPC_PRIVATE , BUFSIZ ,0600) , 0, 0600) ;

And at the end of the same bloc, the preprocessor will add

shmdt (aux);   shmctl (_shmid[--_share_i] , IPC_RMID , 0);

c) DUAL for implicit use of parallelism and synchronization :

For instance, when the preprocessor finds the text   DUAL parallel merging   in one_level;

it will add at the beginning of the one_level function, the following text

int parallels=parallel[++parallelI]=0; //prepares the necessary set of parallel branches

Then, the preprocessor will add the parallelization UNIX statement 'fork' before calling the merging functions

_parallel[_parallel_i]++; if (fork( ) == 0) { merging(x, aux, lb1, lb2, lb3, ub1, ub2); exit (0); }

Finally, at the end of the one_level function, the preprocessor will add the synchronization UNIX statement wait according to the number of parallel branches

for (;_parallel[_parallel_i]>0;_parallel[_parallel_i]--) wait(&_parallel_s); _parallel_i-- ;

See appendix I, giving the beginning of the C-program generated after preprocessing this Pure Algorithm 3 .

3.2 C⁺⁺ Object-Type Library Experiment

In this case the DUAL declarations will be variables declarations (using predefined "external_types") put at the beginning of the program or at the beginning of a block. For this method, we wrote a C⁺⁺ Object-type library which handles the distributed variables, implements and masks all the devices declarations and I/O statements, using the C⁺⁺ possibility of overloading the operators. It fits well to dynamic external variable linking, but not to program distribution. Ravid [22] has also made such an experiment in her thesis. Please see below this specific DUAL syntax in this case :

Pure algorithm 4 of sorting elements by the increment index method.

```
#include "External_types.h"         // C++ Object-type library (see below appendix II)
//DUAL declarations to handle the variables linked to implicit communications :
DUALext_float a(in_out,file,"aa",ran);      DUALext_float b(in_out,file,"bb",ran);
DUALext_float t(in_out,file,"tt",ran);      DUALext_float keyboard(in,console);
DUALext_float screenFloat(out,console);    DUALext_char screenChar(out,console);
void main( )
        { int n , i, j;                        // local variables
        screenChar =" Enter number of elements:\n ";// sends message to the 'screen'
        n = keyboard;    // reads the number of elements to sort from the 'keyboard'
                         // and puts it in the local variable 'n '
        screenChar =" Enter your elements:\n "; // sends message to the 'screen'
        for( i=0;i<n ;i++) a[i]=keyboard[i]; // reads, from the 'keyboard',
                         //the elements to sort and stores them in the file a
        for( i=0;i<n ;i++)        // algorithm
           for( j =0;j < n ;j++)     // sorts  vector a[i] using the indexes vector t[i]
                   if (a[i]>a[j]) t[i]=t[i]+1;
        for (i=0;i<n ;i++) b[t[i]]=a[i];      // stores the sorted elements in the file 'b'
        for (i=0;i<n ;i++) {screenChar=\n '; screenFloat[i]=b[i];}
                         //prints sorted file 'b'

        }
```

Explanations. We have written a package " External_types.h " which contains a set of class-types "DUALext_float" "DUALext_int" "DUALext_char" "DUALext_double" etc... which permits to declare variables that can be linked to various devices with various access modes : when you declare an external variable, in a Pure Algorithm, you invoke the constructor of the corresponding class and indicate through the parameters the type of link you desire, the device, the access mode etc... In the above text of the Pure Algorithm 4, there are some DUAL declarations. For instance,

DUALext_float a (in_out,file,"aa",ran);   declares a variable 'a' of type float, which is linked to a random (direct-access) file named "aa" which can be read or written.

DUALext_float keyboard (in,console);   declares a variable ' keyboard ' of type float, which is linked to the input device of the console.

DUALext_float screen (out,console);   declares a variable ' screen ' of type float, which is linked to the output device of the console.

Object-type library

Here, in the Appendix II , we give part of the class DUALext_float , one can see the enum declarations and the constructor functions which permits to declare the device-Name if the variable has to be linked with a console, a file, or a network port (through com1 or com2). One can declare if he wants an input link (using the parameter in), an output link (using the parameter out), or both (using in_out). If the link is with a file, one has to declare the access mode using the parameter ran for random direct access files, or seq for sequential access.

When the external variable is used in the Pure Algorithm, it invokes automatically one of the overloaded operators '=' to output automatically values to the desired device, or the overloaded '[ ] ' for indexing I/O values, or the casting operator to provoke automatic input of values from the desired device.

So in this type of implementation, there is no special compilation, the C^{++} compiler translates as usual. The declarations and the implicit communications are done at run-time.

The other classes DUALext_int, DUALext_char, DUALext_long etc.. are built on the same principle.

4 Related Works and Discussion

Transparent Communication. Kramer et al. proposed to introduce interface languages (CONIC [14], REX [15], DARWIN [16], Magee and Dulay [17]) which allows the user to describe centrally and in a declarative form the distribution of the processes and data links on a network. But the algorithm of each process contain explicit communication primitives.

Hayes et al [18] working on MLP (Mixed Language Programming) proposes using remote procedure calls (RPC's) by export/import of procedure names.

Purtilo [19] proposed a software bus system (Polylith) also allowing independence between configuration (which he calls "Application structure") and algorithms (which he calls "individual components"). The specification of how components or modules communicate is claimed to be independent of the component writing, but the program uses explicit calls to functions that can be remote (RPC) or local.

The Darwin, MLP, and Polylith are oriented towards a centralized description of an application distributed on a dedicated network. So all the binds and instances of programs are defined initially at configuration time.

Our approach in DUAL is aimed towards a non dedicated network in which each program knows only of the direct binds with its direct correspondents. Our claim is that our approach is better suited to interconnected programs in a non dedicated network.

Handling the Man Machine Interface Transparently. Separating the man machine interface from the programming language has been extensively discussed over the years[2] (Hurley and Sibert 1989)[11]. Some researchers considered the application part as the controlling component and the user interface functions as the slave. Others do the opposite: the user interface is viewed as the master and calls the application when needed by the I/O process. Some works (Parnas 1969)[8] described the user interface by means of state diagrams. Edmonds (1992)[9] reports that some researchers describe the user interface by means of a grammar. Some others presented an extension of existing languages, Lafuente and Gries (1989)[10].

5 Conclusion

In our approach, the user interface and the application are defined separately and the link between them is explicitly but separately described. This approach is more general in that the user interface is seen as one part of a unified mechanism in which external data are accessed, the other parts of this unified mechanism being file handling, I/O, and network communication. This separation of the DUAL distribution declaration makes the Pure Algorithm (PurAl) clearer, independent of a network configuration, and versatile in the sense that it can be run in various contexts. The 2 experiments we conduct, show that this idea is feasible and can help the easier development of distributed applications.

References (available on the web at http://shekel.jct.ac.il/~rafi)

Appendix I (Beginning of the executable merge_sort for Unix)

C-program generated after preprocessing the Pure Algorithm 3
(in italic are the additions of the preprocessor to translate the DUALs)
#define MAXEXVAR 100
int _share_i , _shmid[MAXEXVAR]; int _filed_i , _fd[MAXEXVAR];
struct stat _buf[1]; int _send_i , _smsqid[MAXEXVAR]; struct msgbuf _msgp[1];
char *_msgtext; int _rcv_i , _rmsqid[MAXEXVAR]; int _devout_i;
FILE *_devoutp[MAXEXVAR]; int _devin_i; FILE *_devinp[MAXEXVAR];
int _parallel_i=-1 , _parallel[MAXEXVAR];

void main( )   // translation of DUAL source_data and target
{char * source_data =(_devinp[_devin_i]=fopen ("./data" , "r") , fgets ((char *)
malloc (BUFSIZ) , BUFSIZ , _devinp[_devin_i++]));
 char * target =(_devoutp[_devout_i++]=fopen ("/dev/tty" , "w") , (char *)
malloc (BUFSIZ)) ;
   int   i; char x[10];
        strncpy (x , source_data , 10);   msort (x , 10);
        strncpy (target , x , 10);
fputs (target , _devoutp[--_devout_i]);
free (target); fclose (_devoutp[_devout_i]); free(source_data);
fclose (_devinp[--_devin_i]);
}// end main---
void msort(char *x,int n)   //translation of DUAL share
{ char * aux=(char *)shmat (_shmid[_share_i++]=shmget (IPC_PRIVATE, BUFSIZ, 0600), 0 , 0600) ;
        int   i,size=1;

```
        while ( size < n )
        {           one_level ( x, aux, n , size );        strncpy ( x , aux , n );
                    size*=2;}
shmdt ( aux );  shmctl ( _shmid[--_share_i] , IPC_RMID , 0 );
                            //end translation DUAL share
}// end msort-------------------------------------------------------------------------
```

Appendix II (Part of the external_types.h package for PC)

```
enum inout{in,out,in_out}; enum deviceName {console,file,com1,com2 };
enum AccessMode{seq,ran};        int        init_com1 = 0 ,
class DUALext_float        //--------EXTERNAL  F L O A T  TYPE----------
{inout IO; deviceName  DEVICE; char* FileN; AccessMode AccessM; float Value;
long index;
  public: DUALext_float(inout, deviceName , char*, AccessMode);    //constructors
        DUALext_float(inout,deviceName );
        ~DUALext_float( );                           // destructor
        void operator=(float);// overloading of operators for implicit communication
        DUALext_float  operator[ ](long);
        operator float( );
}; //////////////////////////////////////////////////////////////   //CONSTRUCTOR  for files
DUALext_float::DUALext_float(inout D, deviceName  A,
                            char* FileName, AccessMode C)
{IO = D; DEVICE = A; FileN = FileName; AccessM = C; Value = 0;index=0; }
//-----------------------------------------------------------------
DUALext_float:: DUALext_float(inout D, deviceName A)
{IO = D; DEVICE  = A;                //CONSTRUCTOR  for console,com1,com2
        if (DEVICE  == com1)
        { if (!init_com1) open_net( ); init_com1++;}
}//-----------------------------------------------------------------
DUALext _float::~ DUALext _float( )        // DESTRUCTOR
{        if (DEVICE  == com1){ init_com1--; if (!init_com1) close_net( ); }
}//----------------------------------------//OVERLOADING OF operator '[ ]'; for INDEXING
DUALext_float  DUALext ext_float::operator[ ](long count ){
{ if (DEVICE  == file) count *=14;
   if ( AccessM == seq && count < index){cerr <<"Error in sequential \n "; exit(1); }
            index = count ; return(*this);
}//--------------------------//OVERLOADING OF operator '=' ; OUTPUT OF VALUE
void DUALext   DUALext _float::operator=(float value)
{ if (IO != in){   if ( DEVICE == console) { cout << value;}
        if ( DEVICE  == file){    ofstream to; to.open(Ch,ios::ate); to.seekp(index);
                                to.width(13); to << value <<' '; to.close( ); }
        if ( DEVICE  == com1) send_net(value);}
   return;
```

R.B. Yehezkael et al.

```
}//--------------------------//OVERLOADING OF CASTING : INPUT OF VALUE
DUALext_float::operator float( )
{if (DEVICE  == console) { cin >> Value; }
  if (DEVICE  == file ) {ifstream from;  from.open(Ch,ios::ate); from.seekg(index);
                         from.width(13); from >> Value; from.close(); }
  if (DEVICE  == com1 && IO == in)
        { if ( something_in_net( ))Value = receive_net( ); else Value = 0;}
        return(Value);
}//--------------------------------------------------------------------
```

Performance Tuning on Parallel Systems: All Problems Solved?

Holger Brunst, Wolfgang E. Nagel, and Stephan Seidl

Center for High Performance Computing
Dresden University of Technology
01062 Dresden, Germany

{brunst,nagel,seidl}@zhr.tu-dresden.de

Abstract. Performance tuning of parallel programs, considering the current status and future developments in parallel programming paradigms and parallel system architectures, remains an important topic even if the single CPU performance is doubling every 18 months. Based on a brief summary of state of the art parallel programming techniques, new performance tuning aspects will be identified. The main part of the paper concentrates on how to deal with these aspects by means of new performance analysis and tuning concepts. First tool developments are presented where part of these concepts are already implemented. Finally, an existing scientific parallel application will be presented with respect to its performance tuning stages which were carried out at our center.

Keywords: performance visualization, application tuning, parallel programming, MPI-2, OpenMP.

1 Introduction

Todays microprocessor technology provides powerful basic components for both standard workstations and high performance computers. Theoretically, they provide an enormous peak performance.

Looking at the supercomputer development over the past 35 years as depicted in figure 1, we observe an enormous performance increase during the past five years. This dramatic performance increase gives the impression to most people that performance optimization is becoming less and less important. The same graph on a logarithmic scale (figure 2) however, shows that the current situation is not exceptional at all. The available peak performance has doubled every

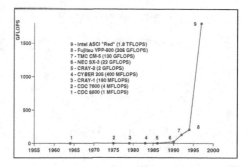

Fig. 1. Supercomputer's Peak Performance over the past 35 Years

18 months for the past 40 years. This is why the assumption that recent and future super computer developments will solve most of our performance problems is simply wrong.

T. Sørevik et al. (Eds.): PARA 2000, LNCS 1947, pp. 279–287, 2001.

Rather, the opposite is true. The huge parallel computers which are currently being developed provide multi-level communication and cash hierarchies. The complexity of these hierarchies makes it more and more complicated to develop applications which perform well.

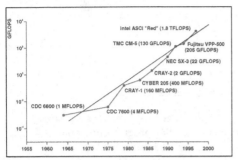

Fig. 2. Supercomputer's Peak Performance (Logarithmic Scale)

In most cases, the performance of applications does not exceed 10 percent of a system's peak performance. Only a few highly optimized applications that were adapted to a certain hardware platform usually transform peak performance gain into real performance [11,18]. A large proportion of the reasons for this disadvantageous trend arise from the complexity of modern microprocessors. Only in rare cases, the system software provided by computer manufacturers will fully support the new parallel execution units and complex memory hierarchies efficiently. Consequently, the full performance potential of a new hardware platform will not be accessible to the majority of applications.

The measurements in figure 3 are meant to be an example for the situation explained above. In figure 3a, the floating point performance of a standard matrix multiplication programmed in Fortran77 with different matrix sizes on an SGI Origin 2000 node is presented. Almost all ijk-loop versions result in a performance (350 MFLOPS) that is quite close to the theoretical peak performance of the MIPS R10000 (390 MFLOPS). Figure 3b shows the measurement for the

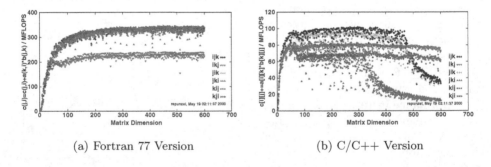

|(a) Fortran 77 Version|(b) C/C++ Version|

Fig. 3. Floating Point Performance of a Standard Matrix Multiplication on an SGI Origin 2000 Node

same matrix multiplication but this time written in C/C++. The performance of the naive algorithm without any manual loop unrolling or blocking never exceeds 100 MFLOPS. We have to state here that we do NOT intend to compare the Fortran and the C/C++ Programming language! This example was chosen to show that using standard programming paradigms like C++ on modern computer platforms, even for simple well known problems, does not automatically guarantee good performance.

2   Evolving Programming Paradigms and Their Tuning Aspects

The scientific community uses parallel programming techniques for a variety of applications. Decreasing prices for parallel computer system hardware makes parallel computing attractive to new users even from the industrial community. Therefore, there is a strong need for simplified, standardized and portable programming paradigms. First steps in this direction include developments like OpenMP [9,13,17] and MPI-2 [7,14,15] which offer abstract parallel constructs to non-expert parallel application developers. As a consequence, most of the applications designed and implemented today make use of MPI or OpenMP. In the future, they may even integrate both concepts into one program. Despite the fact that in principle OpenMP and MPI-2 are relatively simple to use, application performance typically remains far below the theoretical limit. This is where our work has started. From the experience we have had with a large number of parallel applications that were optimized at our center, we identified the following collection of performance tuning aspects:

- OpenMP/Threads
  - efficient data distribution
  - cache optimization
  - process/thread scheduling
  - synchronization
- MPI-2
  - selection of appropriate collective operations
  - parallel file I/O tuning
  - efficient usage of one-sided communication
- MPI combined with OpenMP (Clusters of SMPs)
  - local vs. external computation
  - inter node data distribution
  - inter node synchronization

This paper concentrates on new performance analysis and tuning concepts that allow the adequate handling of the aspects mentioned above.

3   New Analysis/Tuning Concepts

In order to let non-experts benefit from the know-how developed at our center, we designed new analysis and tuning concepts that are suitable to cope with the problems which have been identified above. In the following, these concepts are described in the form of features we would expect from tools that are supposed to help application developers to solve potential performance bottlenecks in future systems.

3.1   Clustering

SMP systems with 8 to 16 processing nodes will become quite common in the parallel computing community due to their good price/performance ratio. The performance of these systems is still limited. One intuitive approach to increase performance would be to link multiple SMPs to a cluster of SMPs. An application running on such a system would most likely be implemented as a mixture of message passing and shared memory programming in order to achieve best performance. The combination of these paradigms and the additional communication layer (inter node communication) requires tool support in the form of navigation and abstraction mechanisms.

3.2   Performance Registers

Microprocessors usually offer counters for a wide range of events including floating point operations, cash misses, etc. These counters are typically used to identify certain event averages like MFLOPS rates. The usability of these rates for performance tuning is limited as they do not provide any location information - that is, the application developer typically does not know which parts of his application to optimize if bad performance has been observed. To increase the usefulness of event counters, cyclic sampling needs to be done with a graphical preparation of the collected data. Combining this with a subroutine/code block tracing mechanism, the user gets an exact insight into which subroutines/code blocks need further code optimization.

3.3   Parallel File I/O

A major novelty in parallel computing is a standardized interface for parallel file I/O like it has been defined and implemented for MPI-2. Therefore, file I/O is no longer a strictly platform dependent issue. Unfortunately, the progress made with respect to code abstraction and code portability does not solve the major problem: File I/O remains an extremely expensive operation with respect to the time it consumes. Applications which regularly need to access files therefore pose the following question when being tuned: *Who* accesses *which* files *when* and for *how long* in order to modify *how much* data with *what* sort of operations.

3.4   Heterogeneous Systems

Several GRID oriented projects [10] like UNICORE [1] propose parallel computing in a seamless fashion. The end users no longer need to know where their applications are actually being executed. Considering a parallel code which is expected to run efficiently on a variety of different platforms, performance tuning gets a new quality. Right now there are no real answers for many problems in this direction. What we do already know is that the generation and collection of performance data poses substantial problems. In the first step, the infrastructure needed to collect, translate, synchronize and edit performance data transparently has to be established.

3.5   Scalability

Relatively cheap 'off the shelf' hardware components allow for quite large parallel systems. This tendency is underlined by initiatives like the ASCI [2,3,4,5, 6] project funded by the US government. Parallel applications running on such large systems generate an enormous amount of profiling data during performance evaluation sessions. The collection, organization, selection and preparation of millions of trace events is one of the major challenges.

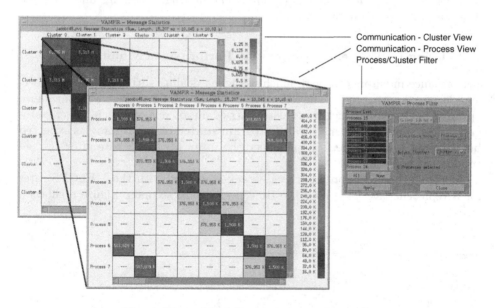

Fig. 4. Hierarchical Communication Statistics

4   A Tool Prototype Implementing the New Concepts

The previous section introduced new concepts for a performance analysis and tuning tool which we will need to optimize next generation applications. Therefore we have already put quite some effort into developing a prototype which implements many of the concepts in the first stage of evolution. The prototype is based on VAMPIR 2.5 [8,16] which is a commercial tool for performance analysis and visualization which is already accepted in the field. The following briefly presents the new concepts in a more practical way by means of our new prototype performing certain visualization tasks of real applications.

4.1   Navigation on Clustered Systems

As clustering is becoming a standard technique in parallel computing, our tool prototype has been equipped with a multi layer visualization hierarchy. This allows us to analyze the performance data on the cluster, node, process, and thread level. As an example, figure 4 depicts how this abstraction works on statistic views for accumulated data. The background window shows the message transfer rates between clusters while the foreground window shows the same metric for inter-process communication. In reality the two windows are one and the same but have different levels of abstraction. The degree of abstraction can be selected in a context menu. In addition, filtering, scrolling, and zooming functions on selected data are available.

4.2   Using CPU Event Counters for Performance Monitoring

Performance monitors can be of much use when dealing with performance bottlenecks of unknown origin. The following example was taken from a performance optimization session, which we recently carried out on one of our customer's programs. For some reason, his parallel program started off performing as expected but suffered a serious performance decrease in its MFLOPS rate after two seconds.

Figure 5 shows the program's call stack combined with the MFLOPS rates for a representative CPU over the time period of 6 seconds. Figure 6 shows a

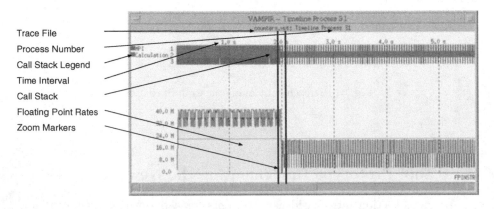

Fig. 5. Performance Monitor combined with Call Stack

close-up of the time interval when the program behavior changes, as marked in figure 5. We see two similar program iterations separated by a communication step. The first one is twice as fast as the second one. We can also see that the amount of work carried out in both iterations is identical, as their integral surfaces are the same (effect 1). A third aspect can be found in the finalizing part (function DIFF2D) of each iteration. Obviously the major problem resides here, as the second iteration is almost 10 times slower than the first one (effect 2).

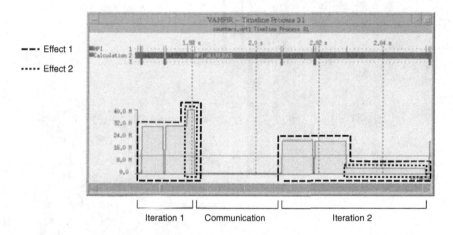

Fig. 6. Performance Monitor based Diagnosis

4.3   Parallel I/O Support

Many scientific parallel applications have moderate I/O requirements. Often, only a small set of input parameters is needed for the time consuming main calculation which results in a small number of output parameters only. There is, however, an increasing number of applications (parallel databases, molecular biology, etc.) which operate on very large data sets. Choosing the right way to access the data is crucial for the performance of this type of application.

Figure 7 shows how I/O performance can be visualized with respect to file access patterns and the resulting overall performance. The example shows the optimized version of a program that repeatedly accesses a number of files to perform its calculation [12]. Originally developed as a sequential program running on a workstation, its I/O access scheme did not quite scale on our parallel machines (Cray T3E and Origin 2000) and performed badly (200 times slower than shown here). As the re-design of the code was out of scope, we had to think of something else. Applying a relatively simple cache strategy, finally improved the overall application performance significantly.

We do not want to stress the tuning process of this particular application at this point. Our focus is on showing that the visualization of parallel I/O behavior is already an important aspect of performance tuning today and will continue to increase in significance.

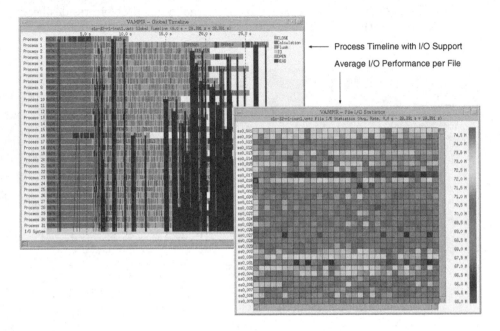

Fig. 7. Parallel File I/O Visualization

5   Conclusion

This paper has presented performance aspects of current and future parallel computer architecture for both software and hardware. New performance analysis and tuning concepts were presented. The key issues are clustering, performance monitoring, parallel file I/O, heterogeneous systems, and scalability. The existing implementation of parts of these within the Vampir framework has been discussed. Further extensions to Vampir that are currently being worked on will serve as a proof of concept, demonstrating the benefits of event based performance analysis to real-world users with large applications on the upcoming clustered SMP systems.

References

[1] J. Almond and D. Snelling. *UNICORE: Secure and Uniform Access to Distributed Resources via the World Wide Web*, 1998.
http://www.fz-juelich.de/unicore/index.html.

[2] Accelerated Strategic Computing Initiative (ASCI).
http://www.llnl.gov/asci.

[3] ASCI Academic Strategic Alliances Program: Centers of Excellence.
http://www.llnl.gov/asci-alliances/centers.html.

[4] Accelerated Strategic Computing Initiative (ASCI) PathForward Project.
http://www.llnl.gov/asci-pathforward.

[5] Blue Mountain ASCI Machine.
http://w10.lanl.gov/asci/bluemtn/bluemtn.html.

[6] Blue Pacific ASCI Machine.
http://www.llnl.gov/asci/platforms/bluepac.

[7] S. Bova, C. Breshears, H. Gabb, R. Eigenmann, G. Gaertner, B. Kuhn, B. Magro, and S. Salvini. Parallel programming with message passing and directives. *SIAM News*, 11 1999.

[8] S. Browne, J. Dongarra, and K. London. Review of performance analysis tools for mpi parallel programs.
http://www.cs.utk.edu/~browne/perftools-review.

[9] D. Dent, G. Mozdzynski, D. Salmond, and B. Carruthers. Implementation and performance of OpenMP in ECWMF's IFS code. In *Proc. of the 5th SGI/CRAY MPP-Workshop*, Bologna, 1999.
http://www.cineca.it/mpp-workshop/abstract/bcarruthers.htm.

[10] I. Foster and C. Kesselman. *The Grid*. Morgan Kaufman, 1998.

[11] F. Hoßfeld and W. E. Nagel. Per aspera ad astra: On the way to parallel processing. In H.-W. Meuer, editor, *Anwendungen, Architekturen, Trends, FOKUS Praxis Informationen und Kommunikation*, volume 13, pages 246–259, Munich, 1995. K.G. Saur.

[12] L. N. Labzowsky, I. A. Goidenko, and P. Pyykko. Estimates of the bound-state QED contributions of the g-factor of valence ns elecrons in alkali metal atoms. In *Processdings of ICAP 2000*, 6 2000.
http://www.df.unipi.it/~fuso/icap2000/Session_A/Labzowsky_Leonti.pdf.

[13] Lund Institute of Technology. *Proceedings of EWOMP'99, 1st European Workshop on OpenMP*, 1999.

[14] Message Passing Interface Forum. *MPI: A Message-Passing Interface Standard*, 1.1 edition, March 1995.
http://www.mpi-forum.org/index.html.

[15] Message Passing Interface Forum. *MPI-2: Extensions to the Message-Passing Interface*, August 1997.
http://www.mpi-forum.org/index.html.

[16] W. E. Nagel, A. Arnold, M. Weber, H.-C. Hoppe, and K. Solchenbach. VAMPIR: Visualization and Analysis of MPI Resources. *Supercomputer 63*, XII(1):69–80, January 1996.
http://www.pallas.de/pages/vampir.htm.

[17] *Tutorial on OpenMP Parallel Programming*, 1998.
http://www.openmp.org.

[18] L. Smarr. Special issue on computational infrastructure: Toward the 21st century. *Comm. ACM*, 40(11):28–94, 11 1997.

Performance Measurement Support for MPI Applications with PATOP

Marian Bubak[1,2], Włodzimierz Funika[1],
Bartosz Baliś[1], and Roland Wismüller[3]

[1] Institute of Computer Science, AGH, al. Mickiewicza 30, 30-059 Kraków, Poland
[2] Academic Computer Centre – CYFRONET, Nawojki 11, 30-950 Kraków, Poland
[3] LRR-TUM – Technische Universität München, D-80290 München, Germany
{bubak,funika}@uci.agh.edu.pl, balis@icsr.agh.edu.pl, wismuell@in.tum.de
phone: (+48 12) 617 39 64, fax: (+48 12) 633 80 54

Abstract. This paper presents a motivation and issues concerning the adaptation of PATOP, an OCM-based on-line performance analysis tool, to the MPI parallel programming library. It covers the general structure of an OCM-based monitoring environment, instrumentation of the MPI library and other enhancements needed both at low-level parts of the monitoring environment and at the user interface level in order to achieve full support of performance analysis of MPI applications.

Keywords: monitoring, performance analysis, instrumentation, OMIS.

1 Introduction

Performance analysis is an important part of parallel program development. A performance measurement tool enables to analyze an application's execution, especially focusing on such problems as utilization of system resources (CPU, memory) or inter-process communication. Thus, being aimed at finding potential bottlenecks, performance analysis may help to improve the overall performance of the parallel algorithm.

There are two kinds of performance measurement tools: *on-line* and *off-line*. In case of an on-line tool, the parallel application runs concurrently with the tool - data analysis and visualization are performed immediately. In off-line environments, performance data is gathered during the application's execution and stored as a file on disk. This file usually contains trace data and therefore is called *trace file*. Subsequent analysis and visualization, which are performed after the application's execution has ended, are based on the data gathered in the trace file. Both kinds of tools, off-line and on-line ones have their advantages and disadvantages when applied, depending on specific analysis goals.

An advantage of the on-line approach is that the user can decide during runtime, which data are to be monitored, based on an evaluation of the already available data. Thus, data acquisition is much more focused than with off-line tools, which allows fine grained monitoring with reasonable perturbation.

T. Sørevik et al. (Eds.): PARA 2000, LNCS 1947, pp. 288–295, 2001.

During recent years, a large number of performance measurement tools has emerged [1]. The bulk of them follow the off-line approach, the representative examples being Vampir, Pablo, and ParaGraph. The on-line tools are much fewer and even fewer are the tools which feature some well defined interface to a monitoring facility, which could provide prerequisites for interoperability. Among the initiatives to define monitoring interfaces are DAMS[2], PARMON[3], and DPCL[4]. However, DAMS does currently not address performance monitoring, while PARMON and DPCL do not offer direct support for applications using de-facto standard message passing environments like PVM [5] and MPI [6].

PATOP is an on-line performance analysis tool, being originally developed for PARIX parallel environment [7] and now constituting part of THE TOOL-SET. It was recently enhanced to work on top of the OCM[1] universal monitoring system. Its capabilities were widely extended and it was adapted to the PVM environment [8]. Performance analysis offered by PATOP provides several types of measurement operations which can be activated via a GUI. Among these are utilization of computational resources, time spent in a send/receive call and the volume of messages exchanged between processes. PATOP works on the system, node, process and even source code level. It offers a wide range of performance displays like scatter plots, pie charts, bar graphs, multicurves and others.

Each monitoring-bound tool needs a *monitoring system* via which it can observe and possibly influence the application state. Since virtually all of the existing tools are supplied with their own proprietary monitoring systems, they are not capable of cooperating with each other. Owing to the OMIS[2] [9] specification and the OCM, PATOP is enabled to interoperate with other on-line monitoring-bound, OMIS-compliant tools like debuggers, visualizers, etc. The OCM, originally developed to to monitor PVM applications, has recently been enhanced to support MPI applications [10]. The next natural step is to adapt PATOP in the same way. In this paper we present some issues concerning this task and methods with which they were resolved.

2 Architecture of the OCM-Based Monitoring Environment

The OCM provides a layer of abstraction between a parallel application and tools, being composed of a high level module, called NDU, interfaced to a tool and a number low level modules, called local monitores, interfaced to an application. Fig. 1 shows the general architecture of the OCM-based tool environment with MPI-specific parts and performance analysis modules being exposed. The OCM gathers information on application processes and, on request, passes it to PATOP. In fact, PATOP communicates with the OCM indirectly using a library called ULIBS [8], which translates high level measurement requests to lower level monioring requests and accordingly transforms the resulting replies.

---

[1] the OCM stands for OMIS Compliant Monitoring system
[2] OMIS stands for On-line Monitoring Interface Specification

290   M. Bubak et al.

The instrumented MPI library (for details of instrumentation, see section 3.1) is linked to a parallel application. The PAEXT extension to the OCM provides efficient mechanisms for storing the performance data. Another OCM extension, PATOPEXT, provides some additional higher level services for PATOP. These extensions are based on a well defined interface of the OCM and do not depend on the OCM's internal implementation. They are linked to the monitoring system mainly for efficiency reasons to avoid excessive communication between the OCM and ULIBS. PERF is the part of ULIBS responsible for performance analysis services. It contains, among others, definitions of performance analysis measurements, which are expressed in terms of OMIS requests.

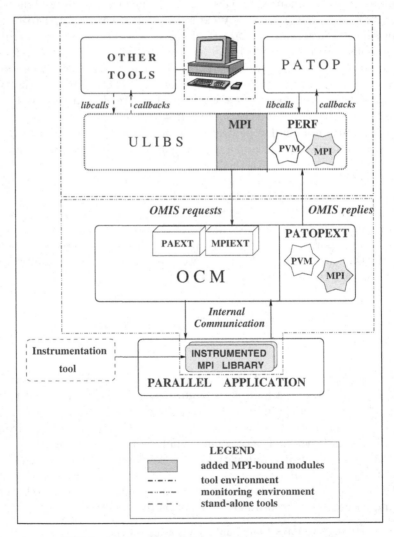

Fig. 1. Architecture of the monitoring environment

All these modules were developed or changed while porting PATOP to the OCM, but until recently only PVM applications could be monitored with the OCM-based tools. The succeeding step in extending the PATOP functionality is porting the latter to MPI [8,10].

3   Adaptation of PATOP to the MPI Environment

Although the OCM was ported to MPI, this port was incomplete from PATOP's viewpoint since it was primarily aimed at enabling an OCM-based debugger, DE-TOP, to support MPI applications, the lacking part of it being instrumentation of the MPI library. The other modules of the tool environment also needed to be extended by MPI-specific parts (for MPI-related objects, see Fig. 1) to enable the full MPI support.

3.1   MPI-Related Extensions within the OCM Infrastructure

PATOP's measurements that deal with inter-process communication may involve tracing the parallel library communication calls[3]. The OCM gathers information about the parallel library calls via instrumentation of the parallel library. The instrumentation in the OCM consists in building a wrapper for each library function. The wrapper performs registration actions prior to and after invocation of the original library function. In *mpich* [11], an MPI implementation being supported by the OCM, a profiling interface as defined by the MPI standard is provided, i.e. each function can be called in the normal mode (MPI_XXX entry points) or by means of the profiling interface (PMPI_XXX entry points). Thus, *mpich* can be instrumented by replacing the 'standard' *mpich* library (with MPI_ entry points) with the library consisting of wrapper functions and linking the *mpich* profiling library (with PMPI_ entry points) to the application.

Since the event registration code is quite complex and *mpich* consists of over a hundred functions, a kit of universal tools for automatic generation of the individual wrappers was developed. Owing to these tools, the only information needed to perform the library instrumentation, were the *mpich* function headers. Another minor change has been done to the PATOPEXT extension to the OCM, where MPI function calls have to be registered.

3.2   Extensions to ULIBS

ULIBS contains the PERF module designed especially for services intended for performance analysis. This part contains information relevant to the performance measurements definitions. These definitions are, in fact, requests to the monitoring system for tracing parallel library calls. Depending on the type of the measurement, proper data is requested from the OCM. In its turn, the PERF

---

[3] In this context, the term 'tracing' does not (necessarily) mean that we create an event trace. Depending on the request, the OCM may just increment a counter.

module needed to be extended by the measurement definitions which depend on the MPI communication subroutines semantics.

Another important extension to ULIBS is a new MPI-specific module which maintains information on MPI communicators created in an application. This issue is described in Section 3.5.

3.3 MPI-Specific Enhancements in PATOP

MPI provides more sophisticated mechanisms for point-to-point and collective communication, while PVM provides the multicast operation (pvm_mcast) and broadcast operation (pvm_bcast). MPI provides an abstract entity called *communicator*, which is a collection of processes together with additional attributes[4]. Moreover, MPI even enables groups of processes to communicate with each other (via so called inter-communicators). Since all these sorts of communication in MPI are to be tracked, PATOP should be provided with some new functionality. Although it is possible in PATOP to specify a subset of processes to apply a measurement to, it does not support such an abstraction as a group of processes. PATOP must be supplied with the capability to recognize MPI communicators and enable the user to access them at the user interface level on a basis similar to those in case of processes or nodes.

3.4 New Measurements in PATOP

Currently, the communication measurements offered by PATOP always pertain to a particular participant of a call, i.e. sender or receiver. However, in case of MPI collective communication, in practice it is not always possible to separate the sender from the receiver. Let us consider an example of the MPI_Bcast routine.

```
int MPI_Bcast ( buffer, count, datatype, root, comm )
void            *buffer;
int             count;
MPI_Datatype    datatype;
int             root;
MPI_Comm        comm;
```

This routine is a collective call, since it is invoked in multiple processes, i.e. all members of the communicator comm. The process with rank root sends a message to every other process within comm, while other processes receive the message. Thus, the same routine can be used for a send or a receive operation. In general, it is hardly possible to distinguish the actual case. However, the global communication is interesting by itself and it is arguable whether such a distinction is necessary. Therefore, two new measurements have been added to PATOP, described as follows:

---

[4] e.g. *virtual topology*

1. *Integrate between the start and the end of collective call*, i.e. measure the delay of a global communication call.
2. *Count size at the end of a collective call*, i.e. measure the volume of data transferred within a collective call.

These measurements are expressed in terms of the OMIS requests similarly as those related to point-to-point communication, but different functions are traced. For instance, the message volume transferred within a global `MPI_Bcast` call can be measured by means of the following request:

```
thread_has_ended_lib_call([], "MPI_Bcast"):
         pa_counter_local_increment(pa_c_1, $len)
```

The semantics of this request is as follows: *whenever the MPI_Bcast function call has ended, increment a value of the associated counter pa_c_1 by the value of the length of the message transferred.* The *counters* are special objects for effective storing of performance data [8].

3.5 New Scope of System Objects - Process Groups

Collective communication always involves a communicator, therefore it would be convenient to have a possibility to apply a global measurement at once to a process group involved, instead of selecting the processes one-by-one. Therefore, the tool environment was enhanced to provide information on communicators created in an application. The information on communicators is gathered by means of instrumentation of the MPI communicator-related subroutines.

Fig. 2 shows the PATOP *New Measurement* dialog extended with new features.

In the *Type of Measurement* column, two new items were added: *Delayed in Global* and *Global Volume*. In the column, where the scope of a measurement is to be chosen, beside the *Node Level* and *Process Level* lists, a new *Group Level* list containing identifiers of existing process groups (communicators) was added. If a process group is selected, the *Process Level* list is automatically activated and all member processes of the selected group are shown. A measurement will apply to processes selected from the *Process Level* list, or - if nothing is selected - to groups selected from the *Group Level* list.

4 Concluding Remarks

MPI is a powerful communication library which is widely used in parallel applications. Therefore, MPI support is a highly desirable capability of parallel monitoring-bound tools. PATOP's development is aimed at creation of a universal environment for parallel programming support. Such an environment should support quite a wide range of inter-process communication patterns and provide a number of tools, like performance analyzers, debuggers, visualizers, etc.

294   M. Bubak et al.

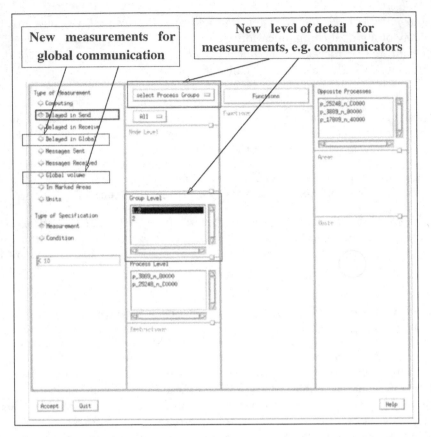

Fig. 2. Changes in the *New Measurement* dialog in PATOP

Porting PATOP to MPI makes it a universal on-line performance measurement tool, capable of tracing both PVM and MPI applications. Owing to the OCM's versatility and flexibility, PATOP is enabled to interoperate at some level with other tools, e.g. with DETOP, i.e. run concurrently on the same instance of parallel application, which is an indispensable feature for providing extensibility and power of performance analysis.

Acknowledgements. This work has been carried out within the Polish-German collaboration and supported, in part, by KBN under grant 8 T11C 006 15.

References

1. Browne, S.: Cross-Platform Parallel Debbuging and Performance Tools. In: Alexandrov, V., Dongarra, J., (eds.): Recent Advances in Parallel Virtual Machine and Message Passing Interface, Proc. 5th European PVM/MPI Users' Group Meeting, Liverpool, UK, September 7-9, 1998, Lecture Notes in Computer Science **1497**, Springer, 1998, pp. 257-264.

2. Cunha, J., Lourenço, Vieira, J., Moscão, B., and Pereira, D.: A Framework to Support Parallel and Distributed Debbuging. In: Sloot, P., Bubak, M., Hertzberger, B., (eds.): *Proc. Int. Conf. High Performance Computing and Networking*, Amsterdam, April 21-23, 1998, 708-717, Lecture Notes in Computer Science **1401**, Springer, 1998.
3. Rajkumar, B., Krishna Mohan, K.M., Gopal, B.: PARMON: A Comprehensive Cluster Monitoring System. In: Proceedings of High Performance Computing on Hewlett-Packard Systems, Zurich, Switzerland, 298 – 311 ETH Zürich, 1998; HPCC System Software PARMON Group: PARMON User Manual, C-DAC,1998 `http://www.dgs.monash.edu.au/~rajkumar/parmon/ParmonManual.pdf`
4. Pase, D. Dynamic Probe Class Library: tutorial and reference guide, Version 0.1. Technical Report, IBM Corp., Poughkeepsie, NY, June 1998. `http://www.ptools.org/projects/dpcl/tutref.ps`
5. Geist, A., et al.: PVM: Parallel Virtual Machine. A Users' Guide and Tutorial for Networked Parallel Computing. MIT Press, Cambridge, Massachusetts (1994)
6. MPI: A Message Passing Interface Standard. In: Int. Journal of Supercomputer Applications, **8** (1994); Message Passing Interface Forum: MPI-2: Extensions to the Message Passing Interface, July 12, (1997) `http://www.mpi-forum.org/docs/`
7. Wismüller, R., Oberhuber, M., Krammer, J. and Hansen, O.: Interactive Debugging and Performance Analysis of Massively Parallel Applications. *Parallel Computing*, **22**(3), (1996), 415-442
8. Bubak, M., Funika, W., Iskra, K., Maruszewski, R., and Wismüller, R.: Enhancing the Functionality of Performance Measurement Tools for Message Passing Applications. In: Dongarra, J., Luque, E., Margalef, T., (Eds.), Recent Advances in Parallel Virtual Machine and Message Passing Interface. Proceedings of 6th European PVM/MPI Users' Group Meeting, Barcelona, Spain, September 1999, Lecture Notes in Computer Science **1697**, Springer, 1999. pp. 67-74.
9. Ludwig, T., Wismüller, R., Sunderam, V., and Bode, A.: OMIS – On-line Monitoring Interface Specification (Version 2.0). Shaker Verlag, Aachen, vol. 9, LRR-TUM Research Report Series, (1997) `http://wwwbode.in.tum.de/~omis/OMIS/Version-2.0/version-2.0.ps.gz`
10. Bubak, M., Funika, W., Gembarowski, R., and Wismüller, R.: OMIS–Compliant Monitoring System for MPI Applications. In: R. Wyrzykowski, B. Mochnacki, H. Piech, J. Szopa (Eds.), PPAM'99 - The 3th International Conference on Parallel Processing and Applied Mathematics, Kazimierz Dolny, Poland, 14 - 17 September 1999, pp. 378-386, IMiI Czestochowa (1999).
11. Gropp, W., Lusk, E.: User's Guide for `mpich`, a Portable Implemention of MPI. ANL/MCS-TM-ANL-96/6, 1996.

A Parallel Volume Visualization Using Extended Space Leaping Method

Sung-Up Jo and Chang-Sung Jeong*

Department of Electronics Engineering, Korea University
1-5Ka, Anam-dong, Sungbuk-ku, 136-701, Korea
luco@snoopy.korea.ac.kr
csjeong@charlie.korea.ac.kr

Abstract. In this paper we present a new extended space leaping method which allows the drastic speed up and the efficient load balancing by performing skipping processes not only in data space domain but also in screen space domain, and based on the method, present a fast and well balanced parallel ray casting algorithm. We propose a novel forward projection technique for computing information on skipping processes in both domains very fast by combining run-length encoding and line drawing algorithm, and shall show that it can be implemented in parallel with ease and efficiency by the proper distribution of the encoded data while the information produced from the technique being exploited to provide the load balancing. Also, we implemented our algorithms on PVM(Parallel Virtual Machine), and show our experimental result.

Keywords: parallel ray casting, volume rendering, space leaping, run-length encoding, forward projection, line drawing algorithm

1   Introduction

Volume rendering is a technique that extracts 3D shape of objects from volume data. It provides a powerful tool to scientists and engineers, allowing them to examine a complex three dimensional volume from a variety of orientations and to investigate its structure and complexity. However, volume rendering requires high computational costs, which motivated the development of parallel implementation. Among the techniques developed for volume rendering, ray casting is considered the most simple and well suited for parallel processing. In ray casting, a ray is passed through each pixel of the screen. While following a ray, the voxels it passes through are sampled and the resulting colors and opacities are accumulated to yield the ray's final color. Rendering speeds in ray casting can be improved by a variety of acceleration methods such as adaptive termination[1, 3], adaptive sampling rate along the ray, adaptive refinment, spatial coherency between rays, and space leaping method[4]. In this paper, we are concerned with the design of the parallel ray casting algorithm based on space leaping method.

---

* This work has been supported by KISTEP and Brain Korea 21 project

T. Sørevik et al. (Eds.): PARA 2000, LNCS 1947, pp. 296–305, 2001.

So far, many parallel ray casting algorithms have been reported [6,5]. They can be classified as either image space partitioning or data space partitioning depending on how these computational elements are combined as tasks in parallel. However, the previous parallel algorithms have some difficulties and limits in load balancing when exploiting image or data space partitioning, since the computation time taken for each ray traversal is different according to whether it intersects non-empty volume space or not and when it intersects non-empty volume data if it does. In this paper we present a fast and well balanced parallel ray casting algorithm based on a new extended space leaping technique. We shall show that it allows the drastic speed up and the efficient load balancing of the whole parallel algorithm by performing skipping processes not only in data space domain but also in screen space domain.

The outline of our paper is as follows: In section 2, we examine some existing space leaping methods briefly. In section 3, we describe the basic idea of our new extended space leaping method and in section 4, the details of algorithm for our new method. Then in section 5, we give a parallel algorithm using extended space leaping method and in section 6, we'll explain the experimental result which shows the signficant speed up of our method.

2 Existing Space Leaping Method

One of the most widely used and effective acceleration techniques for ray casting is to efficiently traverse or altogether skip the empty data space and implement resampling process only in non-empty data space. This method is called space leaping. In this section we describe some existing space leaping methods such as *octree*, *flat pyramid* and *vicinity flag*.

The well known method for space leaping is to reconstruct original volume data into hierarchical data structure such as octree and pyramid[2,8]. Octree method decomposes the original volume data into eight sub-volumes recursively until all voxels contained in sub-volume satisfy uniform condition. When a ray propagates into volume data, adjusted ray traversal algorithm skips uniform empty space by maneuvering through the hierarchical data structure. When using simple octree, we must perform neighbor search in hierarchical data structures to get the information about empty space whenever ray meets empty data space and this takes a great time consuming over entire octree-based algorithm.

Instead of traversing hierarchical data structure directly, the uniformity information obtained by octree can be stored in additional 3D volume grid as in figure 1.

In this type of volume grid, called flat pyramid, each empty voxel is assigned a pointer that indicates information on empty sub-volume to which it belongs. When ray encounters a non-empty voxel, it is handled by usual ray casting algorithm but when encountering a voxel with a pointer to sub-volume, ray performs a leap forward that will bring it to the first voxel beyond the current empty sub-volume. In flat pyramid, we can drive the information on empty sub-

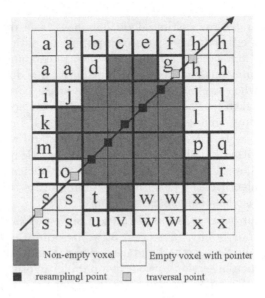

Fig. 1. Flat pyramid. Each empty voxel has a pointer to empty sub-volume to which it belongs.

volume directly from the pointer without neighbor search that is the most time consuming operation when using hierarchical data structures.

It is obvious that the empty data space does not have to be resampled to accelerate ray casting. Thus, ray can skip empty data space as fast as possible without doing anything. Vicinity flag[3] method is based on this idea. Vicinity flag algorithm surrounds the non-empty voxels with a one-voxel-deep cloud of adjacent empty voxels. That is, all empty voxels neighboring non-empty voxels are assigned a special "vicinity flags" to represent the boundary of non-empty voxels. Each ray through screen pixel rapidly traverses the empty space until it encounters a voxel with "vicinity flag". When encountering the first vicinity voxel, we switch to the more accurate traversal algorithm until it encounters the empty voxel that indicates the end of non-empty voxel. Then it rapidly traverses the volume space again until it encounters another vicinity voxel.

3   Basic Idea of Extended Space Leaping Method

When a ray is cast through a pixel in the screen, it may or may not intersect non-empty voxels in data space. If the ray intersects any non-empty voxel during the traversal through data space, the pixel through which it passes contributes to the final image, and is called *active pixel*; otherwse *nonactive pixel*. It is obvious that nonactive pixels do not contribute to the final image and we don't have to cast rays for those nonactive pixels. That is, we can skip those nonactive pixels not to waste time on ray traversal through the empty data space. In our algorithm, we speed up the algorithm by skipping nonactive pixels in the screen domain and casting rays only for active pixels while skipping empty space for each ray as in the previous space leaping method.

Our method stores, in each pixel, the coordinates of the first and last non-empty voxels encountered by the ray emitted at that pixel. The first and last non-empty voxels are called *neareast and farthest active voxels* and their coordinates *nearest and farthest active depths* respectively. Then we can start the ray traversal for each pixel directly from the nearest active depth and stop it at the farthest active depth instead of traversing the entire propagation path. Our method makes use of active pixels and active depths to perform skipping processes in screen and data space domain respectively.

The basic idea of our ray casting algorithm is to achieve the speed up by finding active pixels as well as active depths at once and performing skipping processes in both domains of screen and data space based on the information on active pixels and active depths. Our algorithm can be described in two phases. In the first phase, find active pixels and active depths for each active pixel. In the second phase, calculate, for each active pixel on the screen, its value as follows: Generate a ray into the data space. Starting at the nearest active depth where the ray intersects non-empty voxel, follow the ray while sampling the volume at constant interval by applying vicinity flag. Accumulate the color and opacities of these sampled values. Stop following the ray when it is known that it cannot significantly change its value, or when it intersects the farthest active depth. Therefore, the main problem lies in how to find active pixels and active depths as fast as possible in the first phase to accerlate our ray casting algorithm. We can determine whether a pixel is active or not by forward projection which generates a ray propagating from the eye through the pixel toward volume data and then checks if it intersects any non-empty voxel. Conversely, we can find all the active pixels by forward projection which generates a ray propagating from every non-empty voxel toward eye and determines the pixel it intersects on the screen. In the subsequent sections we shall show that active pixels as well as active depths can be computed at once very fast by exploiting run-length encoding and line drawing algorithm in the forward projection.

4   Extended Space-Leaping Method

In this section, we shall first briefly describe our extended space leaping method using forward projection, and then show how to improve its performance using run-length encoding and line drawing algorithm.

4.1   Forward Projection

Our acceration technique is based on forward projection which maps each voxel in volume data onto the screen in order to find active pixels and active depths. It can be described as follows: Suppose an arbitrary viewing direction is given. Whenever encountering non-empty voxel during the traversal of volume data, we rotate the voxel to allign it in the viewing direction, and project it onto

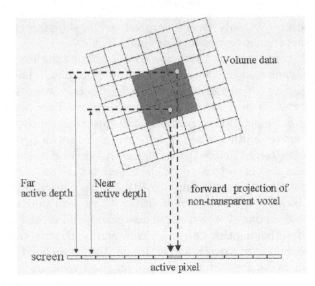

Fig. 2. Forward Projection - project nontransparent voxels onto image screen to find active pixel and active depth

the image screen. After the projection, the pixel intersected by the projection is determined as an active pixel, and the coordinate of the projected voxel is used to update the nearest and farthest depths of the pixel so far. After finishing the traversal of volume data, we can determine, for each pixel, whether it is active or not and its final active depths if it is an active pixel. Our forward projection method is similar to that of splatting algorithm, but differs from it in that it does not require decomposing volume data into slices that is most parallel to the image screen and traversing in special order such as front to back or back to front. Also, our forward projection method makes use of run-length encoded volume data and line drawing algorithm to improve its speed, which shall be explained in detail in the subsequent sections.

4.2   Run-Length Encoding

In forward projection, we need to traverse the volume data in order to find non-empty voxels to be projected onto the screen. However, traversing all the voxels one by one is not efficient even though it is easy and simple, since it may waste a great amount of time on traversing empty voxels. In our algorithm, we transform the original volume data into run-length encoded one, and make use of it to traverse only through the non-empty voxels while skipping over the empty voxels during forward projection.

Run-length encoding is a data compression technique which groups a maximal set of concatenated empty voxels (respectively non-empty voxels) as one *empty voxel run* (respectively *non-empty voxel run*) each identified with its flag set to 1(respectively 0) and length. As in figure 3, volume data consists of two

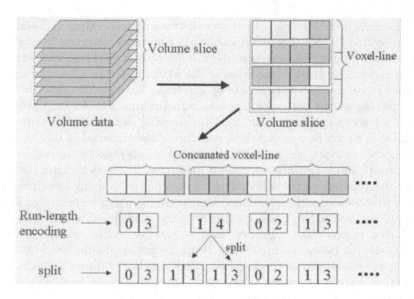

Fig. 3. Data structure: run-length encoded volume data, voxels that represented by gray-color are nontransparent ones

dimensional volume slices, each of which is in trun composed of voxel lines. By traversing line by line and then slice by slice along volume data, we can generate a run-length encoded data which is a series of empty or non-empty voxel runs. Thus, run-length encoded data allows us to skip over concatenated empty voxels at a time and hence reduce the time taken to find non-empty voxels.

In our run-length encoding scheme, each non-empty voxel run covering more than one voxel line is splitted into disjoint pieces so that each piece consists of voxels in the same voxel line. This can be done by slightly modifying the run-length encoding scheme, that is, by generating a new non-empty voxel run whenever the currently traversing non-empty voxel run extends to the next voxel-line. This modification in run-lengh encoding is necessary to accelerate forward projection using line drawing algorithm, which shall be described in the subsequent section.

4.3   Line Drawing Algorithm

By using run-length encoded volume data, we can accelerate the forward projection algorithm by skipping all the empty voxel runs at once. However, it still takes some time to process non-empty voxel run, since we need to traverse each voxel in the non-empty voxel run one by one to project it to the screen. In order to accelerate the process of non-empty voxel run further, we exploit line drawing algorithm. It is a technique which connects two separated screen pixels as one straight line by a sequence of pixels. It produces the sequence by selecting the pixels which are closest to the straight line.

In the run-length encoded volume data, each voxel run can be thought of as a simple line in 3D space, since it consists of consecutive non-empty voxels aligned along the same direction. We splited general run-length encoding as described in previous section for this. Therefore, each voxel run leaves a straight line footprint on the screen when all the voxels in the run are projected. Thus, if we know the starting and ending pixels of the straight line footprint on the screen, we can find the other active pixels between them by executing line drawing algorithm without projecting all the corresponding voxels in the run. The forward projection method using run-length encoding and line drawing algortihm can be described as follows: For each non-empty voxel run during the traversal of the run-length encoding volume data, project its first and last voxels onto the screen, and find their corresponding two active pixels respectively. Then, determine the active pixels corresponding to the other voxels of the run by applying line drawing algorithm to those two active pixels as start and ending pixels respectively. Finally, calculate, for each active pixel, its depth to the corresponding voxel by linear interpoation between the depths of the first and last active pixels, and update its active depths by comparing it with them. In our algorithm, we do not store all (x, y, z) coordinates of the voxel as active depth, but just z value as depth information, since if we know z value, other 2 coordinates can be calculated from the line equation from the eye passing through the pixel.

5   Parallel Algorithm

The basic idea of our parallel ray casting algorithm is to achieve the speed up and the load balancing by finding active pixels as well as active depths, distributing active pixels among the computing nodes, and then performing skipping processes in both domains of screen and data space based on the information on active pixels and active depths. Our parallel algorithm consists of three phases. (See figure 4.) In the first phase we find active pixels and active depths by using forward projection as follows: Initially, the volume data is partitioned among various processors. Each processor finds non-empty voxel runs for its partitioned volume slice, and then executes forward projection using line drawing algorithm for each voxel run. In the second phase, the active pixels obtained in the first phase are distributed among processors. Here we assign the same amount of active pixels to each processor to achieve efficient load balancing. Then the value of each active pixel on the screen is calculated as follows: Generate a ray through the active pixel into the data space. Starting at the nearest active depth where the ray intersects non-empty voxel, follow the ray while sampling the volume at constant interval. Accumulate the color and opacities of these sampled values. Stop following the ray when it is known that it cannot significantly change its value, or when it intersects the farthest active depth. In the third phase, the resulting partial images obtained from the second phase are merged to yield the final image. Since each computing node has equal amount of active pixels and the computation time of each active pixel has little difference by exploiting the active depth, the overall load of the parallel algorithm becomes well balanced.

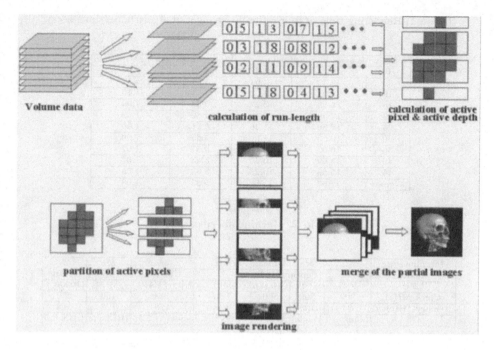

Fig. 4. Illustration of Parallel Ray Casting

6   Experimental Result

We implemented the sequential ray casting algorithm to see acceleration improvement of our extended space leaping method on SGI O2 workstation. Test data set is a 256 x 256 x 225 human head, and image screen measures 256 x 256 pixels. Table 1 shows the rendering time for volume images generated by incrementally rotating a data by 5 degrees of for different algorithms. Time is measured in seconds and the maximum opacity value that non-empty voxel can have is set to 1.0 (perfect opaque surface).

Our algorithm shows significant speed-acceleration improvement over the other methods: 80% over flat pyramid, 90% over vicinity flag method, and 94% over general methods respectively.

We implemented the parallel ray casting algorithm on PVM which consists of 16 heterogeneous machines, two Ultrasparc1, two SGI O2, a SGI Octane, three Pentium II PCs running Windows NT/98 and 8 Pentium II PCs running Linux connected by 100 Mbps Ethernet. Our test data set is a 256 x 256 x 225 human head, and image screen measures 1024 x 1024 pixels. The details of hardware and software information for each machine are shown in table 2.

Since each machine has different computing power, we have measured the relative performance with M_1 as reference machine by comparing the execution time of the identical sequential ray casting program on each machine. Then, the expected speed up is computed as a sum of each relative performance of

Table 1. Rendering time of four different sequential algorithms.(time: secs)

direction	general method	vicinity flag	flat pyramid	our new method
0	19.34	12.62	4.72	1.03
5	19.42	13.01	4.54	1.10
10	19.07	12.68	5.21	1.10
15	18.55	12.27	5.33	1.09
20	17.93	11.46	4.82	1.09
25	17.30	11.46	5.11	1.09
30	16.63	11.04	5.65	1.09
35	16.07	10.73	6.57	1.07
40	15.69	10.43	5.45	1.07
45	15.36	10.28	5.82	1.05
average	17.54	11.60	5.32	1.08

Table 2. Machine specifications

Machine type	M_1	M_2	M_3	M_4	M_5	M_6	M_7
Model	Pentium II PC				USparc1	O_2	$Octane$
CPU	P II	P II	Celeron	P II	UltraSPARC	MIPS R10000	
Clock(MHz)	300	300	366	300	143	150	250
Memory(MBytes)	64	128	128	128	128	128	512
OS	Linux 2.0	Win98	Win98	WinNT 4.0	Solaris 2.5	IRIX 6.3	IRIX 6.5

participating machines. The relative performance of the machines obtained by executing the identical sequential ray casting program is shown in table 3. Table 4 shows the execution time, speed up and efficiency of ray casting according to the number of machines. The efficiency represents the ratio of speedup with respect to expected speedup. As the number of machines increases, the parallel algorithm shows relatively good speed up with efficiency around 80% without degrading its performance due to the communication overhead. This results from the efficient load balancing from the computation of active pixels and depths and their proper data distribution.

7   Conclusion

In this paper we have proposed a new extended space leaping technique which allows the drastic speed up and the efficient load balancing by performing skipping processes not only in data space domain but also in screen space domain, and based on the technique, have presented the fast and well balanced parallel ray casting algorithm. We have presented a novel forward projection method for computing the active pixels and active depths very fast by using run-length encoding and line drawing algorithm, and have shown that the active pixels and depths can be exploited to perform the skipping processes in space and data domains to speed up the whole algorithm, and that the load balance can be efficiently achieved through the proper distribution of active pixels together with information on active depths. Also, we have shown that the forward projection

Table 3. Measurement of relative performance with respect to M_1 for ray casting

machine i	M_1	M_2	M_3	M_4	M_5	M_6	M_7
OS	Linux	Win98	Win98	WinNT	Solaris2.5	IRIX6.3	IRIX6.5
(spec.)	(PII-300)	(PII-300)	(PII-366)	(PII-300)	(USparc1)	(O_2)	(Octane)
running time	195.759	144.879	122.341	133.041	413.812	205.390	128.690
relative perf.	1.0	1.351	1.600	1.471	0.473	0.953	1.521

Table 4. Performance results of parallel ray casting

number of objects		$1(M_1)$	$2(M_2)$	$4(M_{3,5})$	$8(M_{1,1,4,6})$	$12(M_{1,2,5,7})$	$16(M_{1,1,1,1})$
expected speedup		1.0	2.0	4.073	8.497	12.842	16.842
PVM	time (sec)	195.759	119.292	57.977	29.140	19.330	14.860
	speedup	1.0	1.641	3.376	6.718	10.127	13.174
	efficiency (%)	100.0	82.05	82.90	79.06	78.86	78.22

method can be implemented in parallel with ease and efficiency by the volume data partitioning for the computation of the encoded data and the proper distribution of the encoded data among processors. We have implemented the parallel ray casting algorithm on PVM which consists of various heterogenous machines, and have shown that it has relatively good speedup due to the well organized load balancing.

References

1. Danskin, J. and Hanrahan, P. Fast algorithms for volume ray tracing, 1992 workshop on Volume Visualization, Boston, MA, 1992, pp. 91-98.
2. J. Danskin, R. Bender, and G. T. Herman, Algebraic reconstruction techniques (ART) for three-dimensional electron microscopy and X-ray photography, J. Theoretical Biology, vol.29, pp.471-482, 1970.
3. Yagel, R., Cohen, D., Kaufman, A. And Zhang, Q., Volumetric Ray Tracing, TR 91. 01. 09, Computer Science, SUNY at Stony Brook, January 1991.
4. R. Yagel, and Z. Shi, Accelerating Volume Animation by Space-Leaping, Visualization '93, 1993, pp. 63-69.
5. V. Goel and A. Mukherjee, An Optimal Parallel Algorithm for Volume Ray Casting, Visual Comput, Vol. 12, 1996, pp. 26-39.
6. C. Kose and A. Chalmers, Profiling for efficient parallel volume visualization, Parallel Computing Vol. 23, 1997, pp. 943-952.
7. M. Levoy, Display of Surface from Volume Data, IEEE Computer Graphics & Application Vol.8, No.3, 1990, pp. 29-37.
8. M. Levoy, A hybrid ray tracer for rendering polygon and volume data, IEEE Computer Graphics & Application Vol.10, No.2, 1990, pp. 33-40.

Hands-On Training for Undergraduates in High-Performance Computing Using Java

Christian H. Bischof[1,2], H. Martin Bücker[1], Jörg Henrichs[2], and Bruno Lang[2]

[1] Institute for Scientific Computing, Aachen University of Technology,
D-52056 Aachen, Germany,
{bischof,buecker}@sc.rwth-aachen.de
http://www.sc.rwth-aachen.de
[2] Computing Center, Aachen University of Technology, D-52056 Aachen, Germany,
{bischof,henrichs,lang}@rz.rwth-aachen.de
http://www.rz.rwth-aachen.de

Abstract. In recent years, the object-oriented approach has emerged as a key technology for building highly complex scientific codes, as has the use of parallel computers for the solution of large-scale problems. We believe that the paradigm shift towards parallelism will continue and, therefore, principles and techniques of writing parallel programs should be taught to the students at an early stage of their education rather than as an advanced topic near the end of a curriculum. A certain understanding of the practical aspects of numerical modeling is also a useful facet in computer science education. The reason is that, in addition to their traditional prime rôle in computational science and engineering, numerical techniques are also increasingly employed in seemingly non-numerical settings as large-scale data mining and web searching. This paper describes a practical training course for undergraduates, where carefully selected problems of high-performance computing are solved using the programming language Java.

1 Reasons for Java in High-Performance Computing

With the evident success as a programming language, Java is also emerging as the de facto standard language for platform independent software development in higher education. While scientific and engineering computation is obviously not its primary target area, Java has more to offer for large-scale numerical computing than the promise of platform independence, the availability of an extensive standard library, and the rapid development time in coding. Java can be viewed as a "universal" language enabling a rich set of tasks involved in high-performance computing. It provides a standardized infrastructure for distributed applications in a potentially heterogeneous environment, a well-integrated graphical interface for data visualization, and the option to integrate existing applications written in other programming languages [1].

Early implementations of Java exhibited poor performance, and thus Java has often been blamed for its inefficiency. However, there has been tremendous

T. Sørevik et al. (Eds.): PARA 2000, LNCS 1947, pp. 306–315, 2001.

progress in increasing its performance over the last years. Some key ideas of the Java Grande Forum [2] have found their way into the language and, today, it does make sense to seriously consider Java for large-scale applications [3,4,5]. In a few years from now, Java might even obtain competitive or better performance than C++ or Fortran. This—and not just the fact that Java is appealing to students—is the reason why we believe that the time is ripe for teaching Java in the context of high-performance computing.

2   Reasons for an Undergraduate Course at Aachen

For the following discussion it is helpful to keep in mind the structure of higher education in Germany. Typically, students obtain the so-called "Abitur" after 13 years of schooling and are then entitled to attend, tuition-free, a university of their choice in a field of study of their choice. Restrictions of attendance exist in a few fields, with computer science, mathematics, and most of the engineering subjects currently not being affected. As a result, enrollment can be subject to large fluctuations. For example, first-year enrollment in computer science at the University of Technology in Aachen sky-rocketed from 220 students for the winter semester 1998/1999 to now 379 in 1999/2000.

This increase reflects the outstanding professional opportunities for computer science graduates currently and for the foreseeable future. One aspect of this demand is the need to incorporate computer science techniques in more and more application fields, one of them being computational science and engineering. Successful numerical computing techniques for partial differential equations for example require non-numerical algorithms such as graph partitioning to support efficient data structure management, whereas on the other hand, applications such as data mining, information retrieval and web searching, make increased use of numerical modeling. No matter what the application, though, parallel computing techniques are a necessary ingredient for dealing with large problems. In response to this development, Aachen University of Technology created a professorship in high-performance computing within the computer science department.

In the first two years of study, all computer science students in Aachen attend a core curriculum of computer science and mathematics courses whose successful completion is a precondition for the final degree. In the second part of their studies, students can choose courses pretty much according to their preferences. It is only required that at any rate a certain percentage of "theoretical" and "practical" courses be included in order to avoid too narrow an education. Parallel computing, implementation-oriented numerics courses, and issues of algorithmic scalability in this context are at the moment only offered in the second part of the studies. Thus, we saw the need to provide an avenue for interested students to get acquainted at an earlier stage with principles of parallel computing, numerical software, and high-performance computing. This opportunity arises through the project-oriented "practical course in software" that every student must complete in the first two years of study. Like in the later years, students here have the choice between a variety of offerings.

We developed a practical training course for undergraduates taught at Aachen University of Technology within the curriculum of Computer Science. It assumes neither a knowledge of Java nor an extensive background in numerical mathematics. Moreover, no experience with parallel computing is assumed. Of course, a certain familiarity with serial programming in some programming language is a prerequisite in a broader sense. During the course, the students acquire a basic understanding of object oriented programming and of some numerical methods, but the focus is different. Our course, which we describe in this paper, is an introduction to parallel computing and a journey highlighting different aspects involved in high-performance computing.

3   The Structure of the Course

The course takes off with a one-and-a-half-days Java tutorial that is held together with the organizers of two other Java-related courses. The tutorial introduces the students to the programming language Java and its strengths and peculiarities.

The core of the course consists of three units of one or two exercises each, to be worked out by groups of two or three students. For each unit, handouts are prepared providing the student with detailed background information. In particular, the numerical methods are fully described so that no additional literature is required. (All methods may be found in standard textbooks on numerical analysis, e.g., [6,7].)

After each exercise every group of students must demonstrate and explain its implementation to its supervisor. In addition, one group has to present its design (and in particular the preceding decisions) to the remaining groups in order to strengthen the communication skills.

The overall structure of the course is depicted in Figure 1. The units are described in more detail in the following three sections, starting with Unit 1 that is aimed at introducing the students to the Java language and to object-oriented design. Task parallelism comes into play in Unit 2; see Section 5, and finally, Unit 3 is based on data parallelism; see Section 6. Future directions of the course are outlined in Section 7.

4   Introduction to Java and Object-Oriented Design

The first unit is aimed at introducing the students to the basics of the Java language and to object-oriented design.

Instead of the ubiquitous "hello world" example, our students start by implementing a class `RealVector`. This class provides real vectors of arbitrary, but not dynamically changing length, together with methods for accessing such a vector $\mathbf{x}$:

- determine the dimension d of $\mathbf{x}$,
- read or set a selected entry $\mathbf{x}_i$,
- linear combination $\alpha\mathbf{x} + \beta\mathbf{y}$ with another `RealVector` $\mathbf{y}$ and $\alpha, \beta \in \mathbb{R}$,

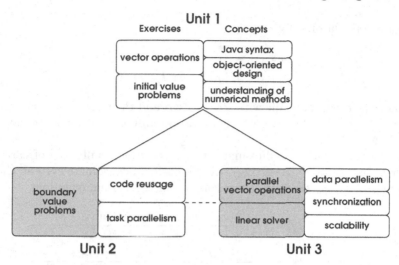

Fig. 1. The structure of our course. For each unit, the contents and the concepts of its exercises are indicated. Exercises involving parallelism are shaded grey.

- inner product $\mathbf{y}^T\mathbf{x} = \sum_{i=1}^{d} \mathbf{y}_i\mathbf{x}_i$ with another `RealVector` $\mathbf{y}$,
- Euclidean norm $\|\mathbf{x}\|_2 = \sqrt{\mathbf{x}^T\mathbf{x}}$,

and appropriate constructors. This straight-forward exercise recalls the basic Java syntax and the exception handling facilities (e.g., attempts to combine vectors of different dimensions must be caught).

The `RealVector` class is reused in the second, substantially more complex, exercise of the first unit, which consists of implementing initial value problems and solving them with two different numerical methods. More precisely, the task is to compute the solution $\mathbf{u}(t) \in \mathbb{R}^d$, for $t \in [u, b]$, of an explicit first-order initial value problem

$$\mathbf{u}'(t) = f(t, \mathbf{u}(t)) , \qquad \mathbf{u}(t_0) = \mathbf{u}_0 , \tag{1}$$

where $f : \mathbb{R} \times \mathbb{R}^d \to \mathbb{R}^d$ is a suitable function (the *right-hand side* of the *differential equation*), $t_0 \in [a, b]$ (the *initial time*), and $\mathbf{u}_0 \in \mathbb{R}^d$ (the *initial values*) are given.

The practical relevance of this exercise is easily motivated with the fact that many time-dependent processes can be modeled as initial value problems, e.g.,

1. Malthus' law of population growth,

$$p'(t) = \alpha p(t) ,$$

which assumes a constant replication rate α,
2. Verhulst's model for population growth, which also incorporates contention proportional to the probability $p^2(t)$ of two individuals meeting each other,

$$p'(t) = \alpha p(t) - \beta p^2(t) ,$$

3. a simple predator–prey model,

$$x'(t) = -\alpha x(t) + \beta x(t)y(t) ,$$
$$y'(t) = \gamma y(t) - \delta x(t)y(t) ,$$

where the predators x die with some rate α and their birth rate is proportional to the probability of predators meeting prey y (vice versa for the prey), and finally

4. the trajectory of a satellite under the gravitational influence of earth and moon. In a suitable (revolving) two-dimensional coordinate system, the trajectory $(x_s(t), y_s(t))$ obeys the second-order differential equations

$$x_s'' = x_s + 2y_s' - \frac{\lambda(x_s + \mu)}{((x_s + \mu)^2 + y_s^2)^{3/2}} - \frac{\mu(x_s - \lambda)}{((x_s - \lambda)^2 + y_s^2)^{3/2}} ,$$

$$y_s'' = y_s - 2x_s' - \frac{\lambda y_s}{((x_s + \mu)^2 + y_s^2)^{3/2}} - \frac{\mu y_s}{((x_s - \lambda)^2 + y_s^2)^{3/2}} ,$$

which in turn are easily transformed into a system of four first-order equations by introducing x_s' and y_s' as additional variables. This initial value problem is often used as a test-case for numerical methods.

The handout briefly explains how the differential equations are derived from the underlying problems and thus gives the students some feeling for the modeling process. These examples also show the diversity of problems that a solver must be able to handle: problem dimensions ranging from 1 in the first two examples to 2 in the third and 4 in the last example, and widely varying shape of the solution(s): from simple exponential growth or decay in the first example to curves with heavily varying curvature in the last example.

For the numerical solution of these initial value problems two different numerical algorithms are to be implemented:

1. the simple method by Euler, which works on an equidistant grid $t_i = t_{i-1}+h$, $i = 1, 2, \ldots$, and updates the solution via $\mathbf{u}(t_i) = \mathbf{u}(t_{i-1})+h \cdot f(t_{i-1}, \mathbf{u}(t_{i-1}))$, and

2. a more sophisticated order-5/4 Runge–Kutta–Fehlberg solver, which relies on an order-5 Runge–Kutta method for updating the solution and on another order-4 Runge–Kutta method for adapting the step size $h_i = t_i - t_{i-1}$ in order to meet some prescribed accuracy criterion.

One possible design allowing arbitrary combinations of initial value problems and solvers is based on the following decomposition (cf. the discussion above):

– An initial value problem IVP *contains* an ODE object describing the ordinary differential equation $\mathbf{u}' = f(t, \mathbf{u})$, a double initial time t_0, and a RealVector $\mathbf{u}_0$ of initial values. IVP also provides a method Solve(IVPSolver s, double $t_{\max}$), which calls the initial value solver s to integrate the current problem (this) from t_0 to the time $t_{\max} \geq t_0$.

- Noting that the only operation with the various differential equations is evaluating their right-hand sides f, ODE is an interface (or an abstract class) declaring the presence of a method EvaluateRHS(double t, RealVector u).
- The differential equations from the above examples give rise to classes ODE_Malthus, ODE_Verhulst, ODE_PredatorPrey, and ODE_Satellite, which *implement* the interface ODE (or *extend* the abstract class), providing appropriate code for the method EvaluateRHS.
- IVPSolver is again an interface (an abstract class) stating that a method Integrate(IVP p, double $t_{\max}$) is available for integrating the initial value problem p up to the time $t_{\max}$.
- Finally, the two numerical methods are realized in classes IVPSolver_Euler and IVPSolver_RungeKuttaFehlberg, which *implement/extend* IVPSolver and provide the respective code for Integrate.

Having the students apply different solvers to a variety of initial value problems is beneficial in two ways. First, the separation of "problem" and "solver" naturally leads to abstractions and enforces a multiple-class design, in contrast to the "everything-in-one-class" code we were confronted with in an earlier similar course. Second, comparing the performance (with respect to accuracy as well as execution time) of the two solvers, the students get some understanding why simple solvers fail on harder problems like the satellite. In fact, the students told us that our hands-on lab taught them more about numerical methods than a very condensed numerical analysis course had done.

5   Task Parallelism

In the second unit the IVP class hierarchy is embedded into a larger framework for solving boundary value problems.

As a particular application, we prescribe the initial position of the satellite (cf. Section 4) and seek an initial velocity that leads to the satellite flying in a closed orbit with a given periodicity T. This requirement results in a nonlinear system of equations,

$$g(\mathbf{v}_0) = \mathbf{0} , \qquad (2)$$

where $g(\mathbf{v}_0) \in \mathbb{R}^2$ is the deviation of the satellite's position at time $t = t_0 + T$, given the initial velocity $\mathbf{v}_0 \in \mathbb{R}^2$ at t_0, from its initial position.

While Newton's method for solving nonlinear systems (and Gaussian elimination for solving the linear systems occurring in each Newton step) should be known to the students, one feature of this system is certainly new to them: the function g is not given by an explicit formula. Instead, an evaluation of g amounts to the numerical solution of an initial value problem. Thus the Jacobi matrix $g'(\mathbf{v}_0)$, which is needed in Newton's method, cannot be determined via symbolic differentiation. Therefore, other techniques for the evaluation of the derivatives of a function are discussed. The basics of numerical differentiation and automatic differentiation [8] and their pros and cons are explained in order to show the students the diversity of available techniques.

Accepting all disadvantages of numerical differentiation, forward finite differences are chosen for the approximation of the 2×2 Jacobian matrix. Thus the resulting algorithm for solving the boundary value problem involves three calls to a solver for an initial value problem of type (1). These three initial value problems dominate the execution time and, furthermore, are independent of each other so that they can be solved concurrently using Java threads. Therefore, the concept of task parallelism is introduced in Unit 3, the three tasks consisting in the time-consuming numerical solution of independent initial value problems.

This kind of parallelism poses no problem for the students because it comes naturally and requires no synchronization besides waiting for the completion of the threads.

6 Data Parallelism

The focus of the third unit is on iterative solvers for linear systems $A\mathbf{u} = \mathbf{b}$. First, it is shown how such large sparse systems naturally emerge during the solution of partial differential equations. Then a comparison with Unit 2 is drawn, where a 2×2 linear system was solved with the well-known Gaussian elimination procedure. It is shown that a method suitable for a small problem size is not necessarily adequate for larger problem sizes. The fill-in problem of direct methods is explained and its connection to theoretical computer science is highlighted. Hence, the need for iterative methods relying only on matrix–vector products is motivated when switching from small dense to large sparse systems.

The students are asked to implement an iterative solver for the solution of a large sparse linear system of equations. For the sake of simplicity, it is assumed that the coefficient matrix is symmetric positive definite such that the conjugate gradients (cg) method is applicable. In the undergraduate course we also do not require to include preconditioning in the implementation. It is, however, explained that preconditioning is crucial in practice because it significantly reduces the number of cg steps and thus the overall solution time.

The cg method allows for parallelism in two places. First, each vector–vector operation (linear combinations $\alpha\mathbf{x} + \beta\mathbf{y}$ and inner products) can be distributed over multiple processors. And second, the computation of each matrix–vector product $\mathbf{z} = A\cdot\mathbf{x}$ can be parallelized in several ways. In both cases, suitable data decompositions are the key to exploiting the available parallelism. As compared with the task parallelism in Unit 2, this data parallelism is typically finer grained, i.e., synchronization is required after a lower number of concurrent operations.

In order to make use of the parallelism at the vector–vector level, one exercise in this unit consists of extending the `RealVector` class (introduced in the very first exercise) from a serial to a concurrent implementation. More precisely, we are concerned with using p threads to manipulate vectors of dimension d where, in general, $d \geq p$ is assumed. To this end, the space $\mathbb{R}^d$ is interpreted as the Cartesian product of p subspaces $\mathbb{R}^{d_i}$ with $\sum_{i=1}^{p} d_i = d$ and $d_i \approx d/p$; that is, each vector $\mathbf{x} \in \mathbb{R}^d$ is decomposed as

$$\mathbf{x} = \left(\widehat{\mathbf{x}}_1^T, \widehat{\mathbf{x}}_2^T, \ldots, \widehat{\mathbf{x}}_p^T\right)^T,$$

where $\widehat{\mathbf{x}}_i \in \mathbb{R}^{d_i}$ for $i = 1, 2, \ldots, p$ are the block components. Using this decomposition, the vector operations are reformulated in terms of their block components. For instance, given scalars α and β, the linear combination

$$\mathbf{z} = \alpha\mathbf{x} + \beta\mathbf{y}$$

of two vectors, $\mathbf{x}$ and $\mathbf{y}$, is computed by p threads such that a thread i, where $i = 1, 2, \ldots, p$, carries out the computation of the ith block component

$$\widehat{\mathbf{z}}_i = \alpha\widehat{\mathbf{x}}_i + \beta\widehat{\mathbf{y}}_i$$

of the result vector $\mathbf{z}$. Another example includes the computation of the inner product of two vectors, $\mathbf{x}$ and $\mathbf{y}$, that can be rearranged as

$$\mathbf{y}^T\mathbf{x} = \sum_{i=1}^{p} \widehat{\mathbf{y}}_i^T\widehat{\mathbf{x}}_i$$

due to the associativity of addition. In the latter example, the addition of the thread-local partial sums $\widehat{\mathbf{y}}_i^T\widehat{\mathbf{x}}_i$ into the final result requires working with either a **synchronized** variable or a **synchronized** method.

Parallelization of the matrix–vector product $\mathbf{z} = A\mathbf{x}$ is achieved by having thread i compute its block component $\widehat{\mathbf{z}}_i \in \mathbb{R}^{d_i}$ of the result vector, which requires accessing d_i consecutive rows of the matrix A as well as several of $\mathbf{x}$'s entries whose number and positions are determined by the sparsity pattern of A. This decomposition is explained with the pentadiagonal matrix that arises from the centered differences discretization of the 2D Laplace equation, together with a row-wise numbering of the variables. Here we also show that the "standard" representation of matrices and vectors is not necessarily the most appropriate one: In this example the vector entries should have double indices $\mathbf{x}_{i,j}$ reflecting the grid position of the respective variable, and the matrix A is not stored at all but occurs only implicitly in the method for multiplying A with a vector.

Besides introducing data parallelism and basic decomposition schemes, this unit also brings two scalability issues into focus: Are methods that are suitable for a small problem also adequate for larger problem sizes (*algorithmic scalability*), and how many processors can be kept busy for a given problem size without severe performance degradation (*parallel scalability*)?

7   Future Directions

Responses from the students revealed that the course is attractive to students and essentially fulfills their expectations. There is, however, potential for improvement.

Given the graphics capabilities built into Java, it would be natural and motivating to have the computations be followed by a visualization of the results. So far we had restrained from including graphics because this would require substantial learning effort not directly related to the main focus of the course. For the future we plan to provide an easy-to-use interface for visualization.

Another avenue for improvement is the increased use of tools for monitoring the temporal behavior and the performance of serial and multithreaded applications, like TAU [9] and VAMPIR [10].

The students would also like to experiment with simulations that are not based on modeling the system with differential equations. We are planning to include a finite automaton approach to multi-population simulation.

Finally, we consider having the groups work on different projects, which are put together at the end of the course.

8 Concluding Remarks

Nowadays, large-scale scientific applications are predominantly implemented using programming languages other than Java. However, Java's efficiency has been significantly improved over the last years eliminating its primary disadvantage with respect to high-performance computing. Java's appeal to students as well as its standardized infrastructure for distributed applications, its support for data visualization, and its ease of combining existing applications written in different programming languages are the reasons why we began to teach Java in the context of high-performance computing.

The practical training course gives an introduction to concurrent programming for undergraduates. In particular, concepts such as task and data parallelism and synchronization are practiced. In addition, the students acquire a basic understanding of object oriented programming and of several numerical methods and are made aware of scalability issues.

Our experience has shown that due to the shared memory paradigm underlying the thread parallelism of Java, students very quickly get a first running parallel program. This success makes them highly motivated and very interested in parallel computing.

References

1. A. J. C. Bik and D. B. Gannon. A note on native level 1 BLAS in Java. *Concurrency: Practice and Experience*, 9(11):1091–1099, 1997.
2. Java Grande Forum. http://www.javagrande.org/.
3. R. F. Boisvert, J. J. Dongarra, R. Pozo, K. A. Remington, and G. W. Stewart. Developing numerical libraries in Java. *Concurrency: Practice and Experience*, 10(11–13):1117–1129, 1998.
4. J. E. Moreira, S. P. Midkiff, M. Gupta, P. V. Artigas, M. Snir, and R. D. Lawrence. Java programming for high-performance numerical computing. *IBM Systems Journal*, 39(1):21–56, 2000.

5. J. E. Moreira, S. P. Midkiff, and M. Gupta. From flop to megaflops: Java for technical computing. To appear in *ACM Transactions on Programming Languages and Systems*.
6. R. L. Burden and J. D. Faires. *Numerical Analysis*. Brooks/Cole Publishing Company, Pacific Grove, CA, 6th edition, 1997.
7. D. Kincaid and W. Cheney. *Numerical Analysis*. Brooks/Cole Publishing Company, Pacific Grove, CA, 2nd edition, 1996.
8. A. Griewank. *Evaluating Derivatives: Principles and Techniques of Algorithmic Differentiation*. SIAM, Philadelphia, PA, 2000.
9. A. Malony and S. Shende. Performance technology for complex parallel and distributed systems. In G. Kotsis and P. Kacsuk, editors, *Distributed and Parallel Systems: From Concepts to Applications—Proc. Third Austrian-Hungarian Workshop on Distributed and Parallel Systems, DAPSYS 2000*, pages 37–46, Norwell, MA, 2000. Kluwer.
10. W. E. Nagel, A. Arnold, M. Weber, H.-C. Hoppe, and K. Solchenbach. VAMPIR: Visualization and analysis of MPI resources. *Supercomputer 63*, XII(1):69–80, 1996.

A Parallel 3-D FFT Algorithm on Clusters of Vector SMPs

Daisuke Takahashi

Department of Information and Computer Sciences, Saitama University
255 Shimo-Okubo, Urawa-shi, Saitama 338-8570, Japan
daisuke@ics.saitama-u.ac.jp

Abstract. In this paper, we propose a high-performance parallel three-dimensional fast Fourier transform (FFT) algorithm on clusters of vector symmetric multiprocessor (SMP) nodes. The three-dimensional FFT algorithm can be altered into a multirow FFT algorithm to expand the innermost loop length. We use the multirow FFT algorithm to implement the parallel three-dimensional FFT algorithm. Performance results of three-dimensional power-of-two FFTs on clusters of (pseudo) vector SMP nodes, Hitachi SR8000, are reported. We succeeded in obtaining performance of about 40 GFLOPS on a 16-node Hitachi SR8000.

1 Introduction

The fast Fourier transform (FFT) [1] is an algorithm widely used today in science and engineering. Three-dimensional FFTs are important in many applications. Parallel three-dimensional FFT algorithms on distributed-memory parallel computers have been well studied [2,3,4,5].

Many vendors support parallel FFTs on clusters of scalar symmetric multiprocessors (SMPs) [6], but few vendors support parallel FFTs on clusters of vector SMP nodes.

In this paper, we propose a high-performance parallel three-dimensional FFT algorithm on clusters of vector SMP nodes.

Our proposed parallel three-dimensional FFT algorithm is based on the multiple FFT algorithm [7,8]. On a vector SMP node, an innermost loop length is particularly important to get high performance. The three-dimensional FFT algorithm can be altered into a multirow FFT algorithm [7] to expand the innermost loop length. We use the multirow FFT algorithm to implement the parallel three-dimensional FFT algorithm.

We have implemented parallel three-dimensional FFT algorithm on clusters of (pseudo) vector SMP nodes, the Hitachi SR8000, and we report the performance in this paper.

Section 2 describes the Hitachi SR8000 architecture and its unique features. Section 3 describes the three-dimensional FFT algorithm. In section 4, we propose a parallel three-dimensional FFT algorithm. Section 5 describes a power-of-two FFT algorithm on a single vector SMP node. Section 6 gives performance results. In section 7, we provide some concluding remarks.

T. Sørevik et al. (Eds.): PARA 2000, LNCS 1947, pp. 316–323, 2001.

2   Hitachi SR8000 Architecture

The Hitachi SR8000 Series is a distributed-memory parallel computer with (pseudo) vector SMP nodes. An SR8000 system consists of from 4 to 128 nodes. Each node contains eight RISC microprocessors which have the "pseudo vector processing" [9] facility. This allows data to be fetched from main memory, in a pipelined manner, without stalling the succeeding instructions. The result is that data is fed from memory into the arithmetic units as effectively as in a vector supercomputer. Peak performance of a node is 8 GFLOPS and maximum memory capacity is 8 GB.

The nodes on the SR8000 are interconnected through a multi-dimensional crossbar network. The communication bandwidth available at a node is 1 GB/s (single direction).

3   Three-Dimensional FFT

The three-dimensional discrete Fourier transform (DFT) is given by

$$y(k_1, k_2, k_3) = \sum_{j_1=0}^{n_1-1} \sum_{j_2=0}^{n_2-1} \sum_{j_3=0}^{n_3-1} x(j_1, j_2, j_3)\omega_{n_3}^{j_3k_3}\omega_{n_2}^{j_2k_2}\omega_{n_1}^{j_1k_1}, \qquad (1)$$

where $\omega_{n_r} = e^{-2\pi i/n_r}$ $(1 \le r \le 3)$ and $i = \sqrt{-1}$.

The three-dimensional FFT based on the multirow FFT algorithm is as follows:

Step 1: n_1n_2 simultaneous n_3 point multirow FFTs

$$x_1(j_1, j_2, k_3) = \sum_{j_3=0}^{n_3-1} x(j_1, j_2, j_3)\omega_{n_3}^{j_3k_3}.$$

Step 2:   Transpose

$$x_2(k_3, j_1, j_2) = x_1(j_1, j_2, k_3).$$

Step 3: n_3n_1 simultanetous n_2 point multirow FFTs

$$x_3(k_3, j_1, k_2) = \sum_{j_2=0}^{n_2-1} x_2(k_3, j_1, j_2)\omega_{n_2}^{j_2k_2}.$$

Step 4:   Transpose

$$x_4(k_2, k_3, j_1) = x_3(k_3, j_1, k_2).$$

Step 5: n_2n_3 simultanetous n_1 point multirow FFTs

$$x_5(k_2, k_3, k_1) = \sum_{j_1=0}^{n_1-1} x_4(k_2, k_3, j_1)\omega_{n_1}^{j_1k_1}.$$

Step 6:   Transpose

$$y(k_1, k_2, k_3) = x_5(k_2, k_3, k_1).$$

The distinctive features of the three-dimensional FFT can be summarized as:

- If n_1, n_2 and n_3 are equal ($n_1 = n_2 = n_3 \equiv n^{1/3}$), the innermost loop length can be fixed to $n^{2/3}$. This feature makes the algorithm suitable for vector processors.
- A matrix transposition takes place three times (steps 2, 4 and 6).

4   Parallel Three-Dimensional FFT Algorithm

The parallel three-dimensional FFT algorithm we implemented is based on the multirow FFT algorithm.

Let $N = N_1 \times N_2 \times N_3$. On a distributed-memory parallel computer which has P nodes, the array $x(N_1, N_2, N_3)$ is distributed along the first dimension N_1. If N_1 is divisible by P, each node has distributed data of size N/P. We introduce the notation $\hat{N}_r \equiv N_r/P$ and we denote the corresponding index as $\hat{J}_r$ which indicates that the data along J_r are distributed across all P nodes. Here, we use the subscript r to indicate that this index belongs to dimension r. The distributed array is represented as $\hat{x}(\hat{N}_1, N_2, N_3)$. At node m, the local index $\hat{J}_r(m)$ corresponds to the global index as the *cyclic* distribution:

$$J_r = \hat{J}_r(m) \times P + m, \quad 0 \leq m \leq P-1, \quad 1 \leq r \leq 3. \tag{2}$$

To illustrate the all-to-all communication it is convenient to decompose N_i into two dimensions $\tilde{N}_i$ and P_i, where $\tilde{N}_i \equiv N_i/P_i$. Although P_i is the same as P, we are using the subscript i to indicate that this index belongs to dimension i.

Starting with the initial data $\hat{x}(\hat{N}_1, N_2, N_3)$, the parallel three-dimensional FFT can be performed according to the following steps:

Step 1: $(N_1/P) \cdot N_2$ simultaneous N_3 point multirow FFTs

$$\hat{x_1}(\hat{J}_1, J_2, K_3) = \sum_{J_3=0}^{N_3-1} \hat{x}(\hat{J}_1, J_2, J_3)\omega_{N_3}^{J_3 K_3}.$$

Step 2:   Transpose

$$\hat{x_2}(K_3, \hat{J}_1, J_2) = \hat{x_1}(\hat{J}_1, J_2, K_3).$$

Step 3: $N_3 \cdot (N_1/P)$ simultaneous N_2 point multirow FFTs

$$\hat{x_3}(K_3, \hat{J}_1, K_2) = \sum_{J_2=0}^{N_2-1} \hat{x_2}(K_3, \hat{J}_1, J_2)\omega_{N_2}^{J_2 K_2}.$$

Step 4:   Transpose

$$\hat{x_4}(\tilde{K}_3, \hat{J}_1, K_2, P_3) = \hat{x_3}(P_3, \tilde{K}_3, \hat{J}_1, K_2)$$
$$\equiv \hat{x_3}(K_3, \hat{J}_1, K_2).$$

Step 5:   All-to-all communication

$$\hat{x_5}(\tilde{K}_3, \tilde{J}_1, K_2, P_1) = \hat{x_4}(\tilde{K}_3, \hat{J}_1, K_2, P_3).$$

Step 6: Rearrangement
$$\hat{x}_6(\hat{K}_3, K_2, J_1) \equiv \hat{x}_6(\hat{K}_3, K_2, P_1, \tilde{J}_1)$$
$$= \hat{x}_5(\hat{K}_3, \tilde{J}_1, K_2, P_1).$$

Step 7: $(N_3/P) \cdot N_2$ simultaneous N_1 point multirow FFTs
$$\hat{y}(\hat{K}_3, K_2, K_1) = \sum_{J_1=0}^{N_1-1} \hat{x}_6(\hat{K}_3, K_2, J_1)\omega_{N_1}^{J_1 K_1}.$$

The distinctive features of the parallel three-dimensional FFT algorithm can be summarized as:

- $N^{2/3}/P$ simultaneous $N^{1/3}$ point multirow FFTs are performed in steps 1, 3 and 7 for the case of $N_1 = N_2 = N_3 = N^{1/3}$. Therefore the innermost loop length can be fixed to $N^{2/3}/P$.
- The all-to-all communication occurs just once.
- The output data $\hat{y}(\hat{K}_3, K_2, K_1)$ is transposed output.

If both of N_1 and N_3 are divisible by P, the workload on each node is also uniform.

5   Power-of-Two FFT Algorithm on a Single Vector SMP Node

Parallel FFT algorithms on vector SMPs have been proposed [10,11,12].

On a single vector SMP node we use FFT algorithms of the Stockham autosort algorithm [13] for radix-2, 4 and 8.

The SR8000's microprocessors have a three-operand multiply-add instruction, which computes $a = \pm a \pm bc$, where a, b and c are floating-point registers. Goedecker [14] reduced the number of instructions necessary for radix-2, 3, 4 and 5 FFT kernels by maximizing the use of fused multiply-add instructions. Although Goedecker's technique works well for a four-operand multiply-add instruction which computes $d = \pm a \pm bc$, where a, b, c and d are floating-point registers, it does not work well for the three-operand multiply-add instruction. Therefore we use the conventional *decimation-in-frequency* (DIF) version of the Stockham autosort algorithm.

Table 1 shows the number of operations required for radix-2, 4 and 8 FFT kernels on a single node of the Hitachi SR8000. The higher radices are more efficient in terms of both memory and floating-point operations. A high ratio of floating-point instructions to memory operations is particularly important on a vector SMP node. In view of a ratio of floating-point instructions to memory operations, the radix-8 FFT is more advantageous than the radix-4 FFT. The power-of-two point FFT (except for 2 point FFT) can be performed by a combination of radix-8 and radix-4 steps containing at most two radix-4 steps. That is, the power-of-two FFTs can be performed as a length $n = 2^p = 4^q 8^r$ ($p \geq 2$, $0 \leq q \leq 2$, $r \geq 0$).

Table 1. Real inner-loop operations for radix-2, 4 and 8 FFT kernels based on the DIF Stockham FFT on a single node of the Hitachi SR8000.

	Radix-2	Radix-4	Radix-8
Loads and stores	8	16	32
Multiplications	4	12	32
Additions	6	22	66
Total floating-point operations ($n \log_2 n$)	5.000	4.250	4.083
Floating-point instructions	8	28	84
Floating-point/memory ratio	1.000	1.750	2.625

5.1   The Radix-4 FFT

Let $n = 4^p$ and $w_q = e^{-2\pi i/q}$. An ns simultaneous n point radix-4 multirow FFT based on the DIF Stockham FFT is as follows:

```
do t = 1, p
  l = 4^{p-t}, m = 4^{t-1}

  complex*16 X_{t-1}(ns, m, 4 * l), X_t(ns, 4 * m, l)
  do j = 1, l
    do k = 1, m
      do row = 1, ns
        c_0 = X_{t-1}(row, k, j)
        c_1 = X_{t-1}(row, k, j + l)
        c_2 = X_{t-1}(row, k, j + 2 * l)
        c_3 = X_{t-1}(row, k, j + 3 * l)
        d_0 = c_0 + c_2
        d_1 = c_0 - c_2
        d_2 = c_1 + c_3
        d_3 = -i(c_1 - c_3)
        X_t(row, k, j)         = d_0 + d_2
        X_t(row, k + m, j)     = w_{4l}^{j-1}(d_1 + d_3)
        X_t(row, k + 2 * m, j) = w_{4l}^{2(j-1)}(d_0 - d_2)
        X_t(row, k + 3 * m, j) = w_{4l}^{3(j-1)}(d_1 - d_3)
      end do
    end do
  end do
end do
```

Here the variables c_0–c_3 and d_0–d_3 are temporary variables.

5.2   The Radix-8 FFT

Let $n = 8^p$ and $w_q = e^{-2\pi i/q}$. An ns simultaneous n point radix-8 multirow FFT based on the DIF Stockham FFT is as follows:

```
do t = 1, p
  l = 8^{p-t}, m = 8^{t-1}

  complex*16 X_{t-1}(ns, m, 8*l), X_t(ns, 8*m, l)
  do j = 1, l
    do k = 1, m
      do row = 1, ns
        c_0 = X_{t-1}(row, k, j)
        c_1 = X_{t-1}(row, k, j+l)
        c_2 = X_{t-1}(row, k, j+2*l)
        c_3 = X_{t-1}(row, k, j+3*l)
        c_4 = X_{t-1}(row, k, j+4*l)
        c_5 = X_{t-1}(row, k, j+5*l)
        c_6 = X_{t-1}(row, k, j+6*l)
        c_7 = X_{t-1}(row, k, j+7*l)
        d_0 = c_0 + c_4
        d_1 = c_0 - c_4
        d_2 = c_2 + c_6
        d_3 = -i(c_2 - c_6)
        d_4 = c_1 + c_5
        d_5 = c_1 - c_5
        d_6 = c_3 + c_7
        d_7 = c_3 - c_7
        e_0 = d_0 + d_2
        e_1 = d_0 - d_2
        e_2 = d_4 + d_6
        e_3 = -i(d_4 - d_6)
```

$$e_4 = \frac{\sqrt{2}}{2}(d_5 - d_7)$$

$$e_5 = -\frac{\sqrt{2}}{2}i(d_5 + d_7)$$

```
        e_6 = d_1 + e_4
        e_7 = d_1 - e_4
        e_8 = d_3 + e_5
        e_9 = d_3 - e_5
        X_t(row, k, j)       = e_0 + e_2
        X_t(row, k+m, j)     = ω_{8l}^{j-1}(e_6 + e_8)
        X_t(row, k+2*m, j)   = ω_{8l}^{2(j-1)}(e_1 + e_3)
        X_t(row, k+3*m, j)   = ω_{8l}^{3(j-1)}(e_7 - e_9)
        X_t(row, k+4*m, j)   = ω_{8l}^{4(j-1)}(e_0 - e_2)
        X_t(row, k+5*m, j)   = ω_{8l}^{5(j-1)}(e_7 + e_9)
        X_t(row, k+6*m, j)   = ω_{8l}^{6(j-1)}(e_1 - e_3)
        X_t(row, k+7*m, j)   = ω_{8l}^{7(j-1)}(e_6 - e_8)
      end do
    end do
  end do
end do
```

Here the variables c_0–c_7, d_0–d_7 and e_0–e_9 are temporary variables.

Table 2. Performance of parallel three-dimensional FFTs on the Hitachi SR8000.

$N_1 \times N_2 \times N_3$	$P = 1$		$P = 2$		$P = 4$		$P = 8$		$P = 16$	
	Time	GFLOPS	Time	GFLOPS	Time	GFLOPS	Time	GFLOPS	Time	GFLOPS
$2^6 \times 2^7 \times 2^7$	0.02876	3.646	0.01845	5.683	0.01177	8.909	0.00663	15.816	0.00448	23.406
$2^7 \times 2^7 \times 2^7$	0.06010	3.664	0.03839	5.736	0.02404	9.160	0.01277	17.244	0.00782	28.159
$2^7 \times 2^7 \times 2^8$	0.12150	3.797	0.07568	6.096	0.04692	9.833	0.02499	18.462	0.01437	32.107
$2^7 \times 2^8 \times 2^8$	0.24938	3.868	0.15415	6.258	0.09734	9.911	0.04981	19.367	0.02782	34.676
$2^8 \times 2^8 \times 2^8$	0.51265	3.927	0.31493	6.393	0.20052	10.040	0.10035	20.062	0.05518	36.485
$2^8 \times 2^8 \times 2^9$	1.02915	4.076	0.63086	6.649	0.40801	10.280	0.20011	20.960	0.10895	38.498
$2^8 \times 2^9 \times 2^9$	2.09089	4.172	1.27866	6.823	0.79872	10.923	0.40348	21.622	0.21736	40.137

5.3   Parallelism on a Single Vector SMP Node

On a single node of the Hitachi SR8000, the outer loop of each FFT kernel is distributed across the node's eight microprocessors. In the radix-4 and 8 FFT kernels, this corresponds to the **do** j loop. The difficulty with distributing the outer loop is that its range ends at one. When the range of j is less than eight, it is necessary to interchange the loop indices of j and k. This permits the compiler to distribute the **do** k loop in a way that keeps all eight microprocessors active.

6   Performance Results

To evaluate our parallel three-dimensional FFT algorithm, m of $N = N_1 \times N_2 \times N_3 = 2^m$ and the number of nodes P were varied. We averaged the elapsed times obtained from 10 executions of complex forward FFTs. The parallel three-dimensional FFTs were performed on double-precision complex data and the table for twiddle factors was prepared in advance.

The Hitachi SR8000 (128 nodes, 1024 GB total main memory size, 1024 GFLOPS peak performance) was used as a distributed-memory parallel computer. In the experiment, we used 1 node $\sim$ 16 nodes on the SR8000.

MPI was used as a communication library on the SR8000. All routines were written in FORTRAN. The compiler used was optimized FORTRAN77 V01-00 of Hitachi Ltd. The optimization option, `-nolimit -rdma -W0,'opt(o(ss))'`, was specified.

Table 2 shows the results of the average execution times of the three-dimensional parallel FFT. The column headed by n shows the number of points of FFTs. The next ten columns contain the average elapsed time in seconds and the average execution performance in GFLOPS.

We note that on the SR8000 with 16 nodes, about 40 GFLOPS was realized with size $N = 2^8 \times 2^9 \times 2^9$ in the parallel three-dimensional FFT algorithm as in Table 2.

7   Conclusion

In this paper, we proposed the parallel three-dimensional FFT algorithm on clusters of vector SMP nodes.

Our parallel three-dimensional FFT algorithm has resulted in high-performance three-dimensional parallel FFT transforms suitable for clusters of vector SMP nodes.

We succeeded in obtaining performances of about 40 GFLOPS on a 16-node Hitachi SR8000. The performance results demonstrate that the proposed algorithm has low communication cost and long vector length.

Implementation of the GPFA (generalized prime factor FFT algorithm) [15] which has a lower operation count than conventional FFT algorithms for any $N = 2^p 3^q 5^r$ on clusters of vector SMP nodes is one of the important problems for the future.

References

1. J. W. Cooley and J. W. Tukey, "An algorithm for the machine calculation of complex Fourier series," *Math. Comput.*, vol. 19, pp. 297–301, 1965.
2. A. Brass and G. S. Pawley, "Two and three dimensional FFTs on highly parallel computers," *Parallel Computing*, vol. 3, pp. 167–184, 1986.
3. R. C. Agarwal, F. G. Gustavson, and M. Zubair, "An efficient parallel algorithm for the 3-D FFT NAS parallel benchmark," in *Proceedings of the Scalable High-Performance Computing Conference, May 23-25, 1994, Knoxville, Tennessee*, pp. 129–133, IEEE Computer Society Press, 1994.
4. M. Hegland, "Real and complex fast Fourier transforms on the Fujitsu VPP 500," *Parallel Computing*, vol. 22, pp. 539–553, 1996.
5. C. Calvin, "Implementation of parallel FFT algorithms on distributed memory machines with a minimum overhead of communication," *Parallel Computing*, vol. 22, pp. 1255–1279, 1996.
6. IBM Corporation, *Parallel Engineering and Scientific Subroutine Library Version 2 Release 1.2 Guide and Reference (SA22-7273)*, 3rd ed., 1999.
7. C. Van Loan, *Computational Frameworks for the Fast Fourier Transform*. SIAM Press, Philadelphia, PA, 1992.
8. M. Hegland, "An implementation of multiple and multi-variate Fourier transforms on vector processors," *SIAM J. Sci. Comput.*, vol. 16, pp. 271–288, 1995.
9. K. Nakazawa, H. Nakamura, T. Boku, I. Nakata, and Y. Yamashita, "CP-PACS: A massively parallel processor at the University of Tsukuba," *Parallel Computing*, vol. 25, pp. 1635–1661, 1999.
10. P. N. Swarztrauber, "Multiprocessor FFTs," *Parallel Computing*, vol. 5, pp. 197–210, 1987.
11. D. H. Bailey, "FFTs in external or hierarchical memory," *The Journal of Supercomputing*, vol. 4, pp. 23–35, 1990.
12. K. R. Wadleigh, G. B. Gostin, and J. Liu, "High-performance FFT algorithms for the Convex C4/XA supercomputer," *The Journal of Supercomputing*, vol. 9, pp. 163–178, 1995.
13. P. N. Swarztrauber, "FFT algorithms for vector computers," *Parallel Computing*, vol. 1, pp. 45–63, 1984.
14. S. Goedecker, "Fast radix 2, 3, 4, and 5 kernels for fast Fourier transformations on computers with overlapping multiply-add instructions," *SIAM J. Sci. Comput.*, vol. 18, pp. 1605–1611, 1997.
15. C. Temperton, "A generalized prime factor FFT algorithm for any $N = 2^p 3^q 5^r$," *SIAM J. Sci. Stat. Comput.*, vol. 13, pp. 676–686, 1992.

High-End Computing on SHV Workstations Connected with High Performance Network

Lars Paul Huse and Håkon Bugge

Scali AS,
Olaf Helsets vei 6,
P.O. Box 70, Bogerud,
N-0621 Oslo, Norway
{lph,hob}@scali.com http://www.scali.com

Abstract. In this paper, we present performance of a set of typically compute intensive benchmarks on a 16 node SCI based PC cluster. The performance is compared with other currently used high-end computers. Dedicated clusters with high-performance interconnect are shown to be competitive in absolute performance to more traditional high-end computers, offering good scalable performance at a low price.

1 Introduction to Clusters

A *cluster* is a machine that consists of a number of workstations (often low-cost PCs) interconnected with one or more networks adapters to act as a single computing resource. This paper compares *dedicated clusters* [2] with other high-end computers. The top 3 entries in the 1999 fall top 500 list of supercomputers [17] are dedicated clusters, even though their network and processor granularity varies. These are the ASCI Red (9632 proc.) by Intel, the ASCI Blue-Pacific SST (5808 proc.) by IBM and the ASCI Blue Mountain (6144 proc.) by SGI. Scali is represented on 351. place on the top 500 list with a 192 processor system (a SCI cluster with 96 dual Intel Pentium II workstations). The machine was made in co-operation with Fujitsu-Siemens, and is marketed as *hpcLine*.

Moore's law (introduced in 1965 [10], revised in 1975 [11] and still valid) states that the capacity of semiconductors (e.g. microprocessor) doubles every 18 months. Clusters based on *SHV* (standard high-volume) hardware benefit both from pricing that comes with the high volumes, and from short time to market for new processor speed upgrades. Recycling cluster workstations as personal workstations, provides a sound lifecycle while maintaining leading performance of a cluster, and reduces the total cost of ownership of high performance clusters to a minimum. Given the small market for *MPP*s (massive parallel processor) and the technological evolution, it is difficult to keep up for specially designed solutions both with respect to price and performance compared to SHV.

For commodity clusters based on off-the-shelf workstations, the internal network is usually the limiting factor for scalable performance. The Beowulf [1] project's approach for improving network performance is to stack multiple Ethernet cards in each PC and add Ethernet switches. This approach scales up to a

T. Sørevik et al. (Eds.): PARA 2000, LNCS 1947, pp. 324–332, 2001.

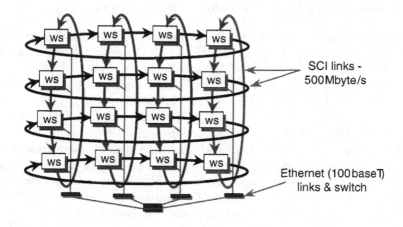

Fig. 1. 16 workstations (WS) interconnected with a 4x4 2D SCI torus

moderate number of nodes (e.g. 16 nodes), but not without considerable cost
to large configurations. This is why Scali clusters use *SCI* (Scalable Coherent
Interface) [16] as interconnect. SCI is a standardized high-speed interconnect
based on shared memory, with the adapters connected in closed rings. SCI's
error-checking mechanisms in hardware enable reliable data communication with
minimal software intervention, and hence enable very low latency communication. The Dolphin PCI-SCI adapters [4] can connect up to three SCI rings to
the same adapter, with traffic routing between the rings done in hardware. This
distributed switch solution eliminates the use for expensive switches that other
high performance networks (e.g. ATM, Myrinet, Gigabit-Ethernet, Servernet,
Memory Channel) need, and gives scalable performance for large configurations [3]. Figure 1 shows a 16 node cluster interconnected with a 4x4 2D SCI
torus.

MPI (Message Passing Interface) [12] is a well-established communication
standard. *ScaMPI* [6] is Scalis thread-hot & -safe high performance implementation of MPI. ScaMPI currently runs over local and SCI shared memory on Linux
and Solaris for x86-, IA-64-, Alpha- and SPARC-based workstations. ScaMPI
over SCI (SSP 2.0.2) has a latency of 6.0 μs and a peak unidirectional bandwidth of 86 MByte/s (bidirectional 90 MByte/s), and 1.7 μs - 320 MByte/s SMP
internal with dual 733 MHz Intel Pentium IIIs on an i840 motherboard.

The major arguments against using clusters for high-end computing has been
that even if processor performance for small SMPs often are better than their
larger SMP/MPP relatives, scalability has been poor. In this paper we will show
that this is not true for SCI based clusters. In the rest of this paper we will
present performance and scalability of a set of compute intensive benchmarks
(section 2) & applications (section 3) on a cluster of PCs and compare them
against other commercial high-end computers.

1.1   Available and Used Technology - Spring 1999

As of June 2000, there are several alternatives for dual processor workstations, e.g. Compaq AP550 (866 MHz Intel Pentium III), Compaq AlphaStation DS20E (500 MHz Dec Alpha 21264 or 667 MHz 21264A) and Sun Ultra 60 (450 MHz Sun UltraSPARC-II). The SMP, SGI Origin 2000 is currently available with ut to 512 processors (300 MHz R12K or 250 MHz R10K MIPS). The MPP, Cray T3E-1200 can be configured with up to 2048 600 MHz Alpha 21164 processors. These are the high-end numbers for all architectures, fetched from the internet.

Since SCI networks have high bandwidth and scalable performance [3], it can supply very high aggregated bandwidth. On a large SMP even though it has an impressive capacity on its backbone/interconnecting bus, the bandwidth and storage capacity of the memory banks will have to be divided between all processors. A Origin 2000-300 has a local copy bandwidth of 186 Mbyte/s/proc.(128 proc.) [9], T3E-1200 474.5 Mbyte/s/proc.(512 proc.) [9] & dual 733 MHz Pentium IIIs on an i840 motherboard 211 Mbyte/s/proc (363 Mbyte/s for single proc), i.e. similar processor bandwidth for small and large SMP. The impressive processor bandwidth of the T3E is due to using only static memory, explaining the 'low' memory capacity of most T3Es.

All tests were run on a cluster with 16 Intel 440BX PCs each with dual Pentium III (Katmai) 450 MHz and 256 MBytes memory running Linux 2.2.14 interconnected with Dolphin D311/D312 32 bit/33 MHz SCI-PCI cards [4] in a 4x4 torus.

2   CFD Examples - NPB 2.3 Kernels

NPB (NAS Parallel Benchmarks) [14] is a collection of benchmarks to test the performance of highly parallel computers. NAS 2.3 constitutes eight *CFD* (computational fluid dynamics) problems, coded in MPI and standard C and Fortran 77/90. The benchmark has to be run as is, i.e. no algorithmic modification of the code is allowed, but parallel directives can be added [14]. Each benchmark comes in 4 sizes, W, A, B & C (in increasing order) with a four fold increase in compute requirement between sizes. Each benchmark is also associated with an operation count, *op*, which scale with the problem size. All test results are given in Mop/s/node, i.e. the accumulated performance of all processes running on a node. *Speedup*, $SU(N)$, is calculated as the aggregated application performance of N nodes divided by the single node performance.

As a references performance numbers from Cray T3E-1200 [14] and SGI Origin 2000 (195 MHz R10K processors) [8] are given. MPPs & SMPs are marketed by their number of processes, while clusters by their number of workstations. Performance is therefore given per processor for the SMP & MPP, and per node for the cluster. The Origin 2000 tests were run sequential on an otherwise unloaded machine. Running with multiple instances of an application concurrently on Origin 2000 might reduce performance, [18] indicate up to 50% performance reduction for communication intensive tests.

Table 1. Performance of the NAS kernels size A [Mop/s/node] ppn=proc/node.

Test	Machine	ppn	1 node	2 nodes	4 nodes	8 nodes	16 nodes	SU(16)
EP	Cray T3E	-	-	3.40	3.40	3.40	3.40	-
EP	Origin 2000	-	4.34	4.33	4.33	4.33	4.30	15.8
EP	SCI cluster	1	1.25	1.25	1.25	1.25	1.23	15.7
EP	SCI cluster	2	2.49	2.46	2.46	2.48	2.46	15.8
MG	Cray T3E	-	-	96.30	88.50	101.40	90.10	-
MG	Origin 2000	-	72.92	62.80	65.33	64.90	60.84	13.3
MG	SCI Default	1	51.79	53.64	48.66	56.07	52.47	16.2
MG	SCI OpenMP	1	51.13	53.11	48.56	54.97	51.68	16.1
MG	SCI Optimized	1	62.00	55.64	54.28	67.74	55.59	14.3
MG	SCI Default	2	83.02	77.62	91.20	82.64	68.58	13.2
MG	SCI OpenMP	2	81.37	82.95	73.15	83.48	74.48	14.6
MG	SCI Optimized	2	94.28	90.09	82.94	99.85	78.31	13.3
CG	Cray T3E	-	-	25.90	24.10	32.80	19.50	-
CG	Origin 2000	-	33.71	26.30	27.46	41.24	35.05	16.6
CG	SCI cluster	1	36.94	35.93	32.76	30.48	23.36	10.2
CG	SCI cluster	2	62.20	57.06	50.16	38.92	30.98	7.9
FT	Cray T3E	-	-	47.50	47.20	46.50	45.50	-
FT	Origin 2000	-	53.50	43.25	43.20	41.68	39.47	11.9
FT	SCI cluster	1	48.11	40.80	40.36	38.41	37.86	12.5
FT	SCI cluster	2	74.78	72.42	65.72	63.56	59.54	12.7
IS	Cray T3E	-	-	3.80	3.60	3.30	2.40	-
IS	Origin 2000	-	2.20	2.00	2.10	2.40	1.90	13.8
IS	SCI cluster	1	5.11	4.30	3.90	3.63	3.16	9.8
IS	SCI cluster	2	7.58	6.46	5.44	4.56	4.24	8.9
LU	Cray T3E	-	-	76.60	77.50	76.30	73.80	-
LU	Origin 2000	-	77.05	81.58	86.71	93.99	100.90	20.9
LU	SCI cluster	1	77.76	78.85	76.96	89.45	89.56	18.4
LU	SCI cluster	2	125.08	131.82	166.72	162.06	155.02	19.8

Table 2. Performance of the NAS application size A [Mop/s/node] ppn=proc/node.

Machine	Class	ppn	1 N	2 N	4 N	4.5 N	8 N	9 N	12.5 N	16 N	SU(16)
SP	Cray T3E	-	-		49.00			46.20		46.10	-
SP	Origin 2000	1	51.89		51.49			61.18		60.11	18.5
SP	SCI cluster	1	55.67		54.48			47.87		50.10	14.4
SP	SCI cluster	2		77.10		74.34	64.52		47.52		
BT	Cray T3E	-	-		66.20			65.70		64.00	-
BT	Origin 2000	-	64.45		54.04			47.88		47.12	11.7
BT	SCI cluster	1	59.88		67.70			66.74		65.23	17.4
BT	SCI cluster	2		101.62			88.58		77.20		

2.1   Embarrassingly Parallel (EP)

EP is an embarrassingly parallel kernel. The EP kernel compute Gaussian deviates by acceptance-rejection method and tally counts in concentric square annuli of random-numbers. The code is embarrassingly parallel, in that no communication is required for the generation of the random numbers. Table 1 shows the performance of the NAS EP kernel. The reduced performance of the cluster compared to the Origin & T3E is because none of our Fortran compilers (GNU, Portland Group or Fujitsu) recognized the 64 bit arithmetic in the random number generator in randi8. We therefore had to use the more resource demanding randdp for random generation. The scaling of EP is as expected very good.

2.2   Multigrid (MG)

MG is a simplified multigrid kernel. MG uses highly structured (both short and long distance) communication. Table 1 shows the performance of the NAS MG kernel. The default entries are the unmodified NPB kernel, hence utilizing more than one processor per node is achieved through multiple MPI processes on each node. Since ScaMPI is thread-hot & -safe, is it possible to mix MPI and OpenMP [15] in the same application. Node internal parallelism is then achieved with OpenMP parallel regions, while MPI is still used for communication between nodes. The OpenMP compiled code performs best for large configurations, possibly due to the improved node internal communication and synchronization. The optimized run is a slight code modification of the OpenMP source (loop-fusion between two sub-calculations). This is an optimization modern compilers should have detected! The single node performance of the cluster and T3E are similar, while the Origin is lower. Scaling is similar for all machines.

2.3   Conjugate Gradient (CG)

In CG a conjugate gradient method is used to compute an approximation to the smallest eigenvalue of a large, sparse, symmetric positive definite matrix. This kernel is typical of unstructured grid computations in that it tests irregular long distance communication, employing unstructured matrix vector multiplication. Table 1 shows that the performance of the NAS CG kernel on all machines show super-linear speedup. The Origin shows good scaling, but falls off to 26.20 Mop/s/process for the 32 process run [8] (speedup[32]=24.8).

2.4   3 Dimensional FFT PDE (FT)

The FT kernel solves a 3-D partial differential equation solution using FFTs. This kernel performs the essence of many spectral codes. It is a rigorous test of all-to-all communication performance. Table 1 shows that all machines have similar single processor performance of the NAS FT kernel, but more surprisingly; the scaling is similar for all machines as well.

2.5   Integer Sort (IS)

IS kernel is a large integer sort. This kernel performs a sorting operation that is important in 'particle method' codes. It tests both integer computation speed and communication performance. Table 1 shows the performance of the NAS IS kernel. Since [8] does not provide numbers for the IS benchmark, the Origin performance numbers are taken from the NPB website [14]. For the IS benchmark the cluster outperforms the 'heavy iron'.

2.6   LU Solver (LU)

The LU kernel makes a triangular factorization of a matrix. It is the only bench-mark in the NPB suite that sends large numbers of very small (40 byte) messages. Table 1 shows the super-linear performance of the NAS LU kernel. Since SCI has very low latency, the LU kernel gets very good scaling on the cluster. The cluster also has low latency node internal communication, resulting in a performance increase between 80-100% when utilizing both processes for LU size B.

2.7   Pentadiagonal Solver (SP)

The SP application benchmark solves three sets of uncoupled systems of equations, first in the x, then in the y, and finally in the z direction. These systems are scalar pentadiagonal. Table 2 shows the performance of the NAS SP application. SP must be run on a square number of processes, explaining the somewhat sparse table. The single process performance of all machines are similar, but the Origin scales best.

2.8   Block Tridiagonal Solver (BT)

The SP and BT algorithms have a similar structure, but while systems are scalar pentadiagonal in the SP code, it is block tridiagonal with 5x5 blocks in the BT code. BT must also be run on a square number of processes. Table 2 shows the performance of the NAS BT application. The cluster & T3E outperforms the Origin for BT, both in terms of scaling and absolute performance.

2.9   Architecture of the Processing Node

To achieve high performance for a complete cluster it is important to have good single node/workstation performance. In addition to the workstations used in rest on the paper, we tested tree workstations. Table 3 shows the performance of the NAS size W compiled with Gnu gcc/g77. The Gnu compiler doesn't always produce optimal code, but was the only compiler available on all platforms. Since EP, FT & MG didn't compile with g77, numbers for these are not presented. Given available technology (section 1.1) comparing the 3 first entries in table 3 should be relevant. Since the Alpha based workstation delivers 50-100% better performance than similar Pentium based workstations, using Alpha based SCI

Table 3. Single processor performance of NAS class W [Mop/s].

Workstation	Processor	MHz	BT	CG	IS	LU	SP
PC (Intel 440BX)	PIII (Katmai)	450	68.86	37.76	5.68	80.57	50.68
Sun Ultra-2	UltraSPARC-II	300	48.18		9.84	53.52	40.81
COMPAQ AlphaServer	Alpha 21264 (EV67)	466	114.56		10.13	163.33	76.00
PC (Intel i840)	PIII (Coppermine)	733	87.79	56.31	8.12	104.16	60.24

clusters for number-crunching would be interesting. Given the general technology update with new workstations a performance increase of additional 20-50% can be added now, and additional 100% each 18 months (section 1).

3   Industrial Applications Example - FEKO

FEKO [5] is a comprehensive implementation of the Method of Moments (MoM) technique and hybridised with asymptotic high-frequency techniques (physical optics - PO and uniform theory of diffraction - UTD) for the analysis of electrically large problems. The application is dominated by double precision (64 bit) linear algebra. Our example is an in-core problem solved in a 13308 x 13308 matrix. To get comparable results, the application was run both with ScaMPI over SCI and *MPICH* [13] over 100 Mbit Ethernet (LiteOn LNE100TX (rev 32)) on the same cluster. MPICH is a freely available, portable implementation of MPI from Argonne National Laboratory.

The ScaMPI version used totally 1157.79 s on 16 nodes with two processes per node, while MPICH used 2340.52 s. The LU decomposition phase takes the dominant part of the execution time, and is timed separately. The ScaMPI version achieved totally 7535.93 MFLOPS (834.17 s) in the LU decomposition, while the MPICH version achieved only 3053.51 MFLOPS (2058.69 s) i.e. only 40% of the ScaMPI performance. The ScaMPI LU performance corresponds to 235.5 MFLOPS per processor, i.e. more than 50% of peak performance.

4   Summing Up

All presented tests perform calculations in double precision floating point numbers (64 bit). 32 bit or lower floating point vector computations performance on PCs can be boosted even further by using extensions like the Intel SIMD instruction set [7] (introduced with Pentium III). As shown in section 2.9, using Alpha based workstations in SCI clusters is a compact alternative to PCs. Also as shown in section 2.9, by upgrading the used cluster to more powerful workstations currently available, performance increase up to 300% can be expected with the same number of processes/workstations.

Comparing performance as seen from a user is simple, if the application run faster, the performance is better. Comparing scalability of single against dual process per node, is on the other hand not a straightforward task. Given that work partition does not introduce overhead, the non-communication part of the application would run twice as fast - increasing (doubling) the communication pressure to the network. Since the data workset is divided between the two processes residing on the same node, the message length usually reduces to half (potentially reducing communication performance) and the totally exchanged datavolume increase. Using OpenMP (or similar paradigms) to achieve SMP internal parallelism is therefore an attractive approach, preserving the communication characteristics of the single process per node, and as shown in section 2.2 the combination of ScaMPI and OpenMP performs very well. All of the used benchmarks and application increase performance 50-100% when increasing number of processes/threads from one to two per node (dual processor workstations).

All in all clusters of powerful and low cost workstations interconnected with a scalable high performance SCI network are for all studied benchmarks and applications an alternative to large MPPs & SMPs in high-performance computing.

References

1. The Beowulf Project home page at http://www.beowulf.org.
2. H. Bugge, P.O. Husøy:, Dedicated Clustering: A Case Study. *Proceedings of 4th International Workshop on SCI-based High-Performance Low-Cost Computing* (1995).
3. H. Bugge, K. Omang: Affordable Scalability using Multicubes. *SCI. Architecture & software for high-performance compute clusters. LNCS 11734 - Springer-Verlag,* pp 167-175(1999)
4. Dolphin ICS: PCI-SCI Bridge Functional Specification. *Version 3.01* (1996).
5. FEKO (FEldberechnung bei Körpern mit beliebiger Oberfläche) from ElectroMagnetic Software & Systems (Pty) Ltd. *http://www.feko.co.za/.*
6. L.P. Huse, K. Omang, H. Bugge, H. Ry, A.T. Haugsdal, E. Rustad: ScaMPI - Design and Implementation. *SCI. Architecture & software for high-performance compute clusters. LNCS 11734 - Springer-Verlag,* pp 249-261 (1999)
7. Intel Cooperation: Intel Architecture Software Developer's Manual (1999)
8. H. Jin, M. Frumkin, J. Yan: The OpenMP Implementation of NAS Parallel Benchmarks and Its Performance. *NAS Technical Report NAS-99-011* (1999)
9. J.D. McCalpin: The STREAM Memory Bandwidth benchmark. *http://www.cs.virginia.edu/stream* (May 2000).
10. G.E. Moore: Cramming More Components Onto Integrated Circuits, *Electronics, Vol. 38, No. 8,* pp. 114-117 (1965).
11. G.E. Moore: Progress in digital integrated electronics. *In Proceedings IEEE Integrated Electron Devices Meeting,* pp. 11-13 (1975).
12. MPI Forum: MPI: A Message-Passing Interface Standard. *Version 1.1* (1995)
13. MPICH *Version 1.2.0.* Available from http://www.mcs.anl.gov/mpi/mpich (1999)
14. NAS Parallel Benchmarks home page at http://www.nas.nasa.gov/Software/NPB.
15. The OpenMP API home page at http://www.openmp.org.(2000)
16. IEEE standard for Scalable Coherent Interface IEEE Std 1596-1992 (1993)
17. The 14th TOP500 Supercomputer Sites list *http://www.top500.org* (Nov. 1999).

18. F.C. Wong, R.P. Martin, R.H. Arpaci-Dusseau, D.E. Culler: Architectural Requirements and Scalability of the NAS Parallel Benchmarks. *In proceedings of Supercomputing '99* (1999)

A special thank to Ulrich Jakobus (EMSS) for advise regarding the FEKO application.

From the Big Bang to Massive Data Flow: Parallel Computing in High Energy Physics Experiments

C. Adler[3], J. Berger[3], D. Flierl[3], H. Helstrup[2], J.S. Lange[3], J. Lien[2],
V. Lindenstruth[4], D. Röhrich[1], D. Schmischke[3], M. Schulz[4], B. Skaali[5],
H.K. Sollveit[1], T. Steinbeck[4], R. Stock[3], C. Struck[3], K. Ullaland[1],
A. Vestbø[1], and A. Wiebalck[4]

[1] Department of Physics, University of Bergen,
Allegaten 55, 5007 Bergen, Norway
[2] Bergen College, Postboks 7030, 5020 Bergen, Norway
[3] IKF, University of Frankfurt,
August-Euler-Strasse 6, D-60486 Frankfurt, Germany
[4] KIP, University of Heidelberg,
Schröderstraße 90, D-69120 Heidelberg, Germany
[5] Department of Physics, University of Oslo,
Postboks 1048 Blindern, 0316 Oslo, Norway

Abstract. Tracking detectors in high-energy physics experiments produce hundreds of megabytes of data at a rate of several hundred Hz. Processing this data at a bandwidth of 10-20 Gbyte/sec requires parallel computing. Reducing the huge data rate to a manageable amount by realtime data compression and pattern recognition techniques is the prime task. Clustered SMP (Symmetric Multi-Processor) nodes, based on off-the-shelf PCs and connected by a high bandwidth, low latency network, provide the necessary computing power. Such a system can easily be interfaced to the front-end electronics of the detectors via the internal PCI-bus. Data compression techniques like vector quantization and data modeling and fast transformations like conformal mapping or the adaptive, generalized Hough-transform for feature extraction are the methods of choice.

1   Probing the Early Universe in the Laboratory: Ultrarelativistic Heavy Ion Collisions

By colliding heavy nuclei at the highest energies available (e.g. ALICE experiment at the Large Hadron Collider (LHC) at CERN and STAR experiment at the Relativistic Heavy Ion Collider (RHIC) at BNL) a new state of matter is created, in which quarks, instead of being bound up into more complex particles such as protons and neutrons, are liberated to roam freely.

Such a state must have existed just a few microseconds after the Big Bang, before the formation of particles of matter as we know them today. By smashing lead nuclei into each other, those conditions can be recreated. The collisions

T. Sørevik et al. (Eds.): PARA 2000, LNCS 1947, pp. 333–341, 2001.

create temperatures over 100 000 times as hot as the centre of the sun and energy densities twenty times that of ordinary nuclear matter.

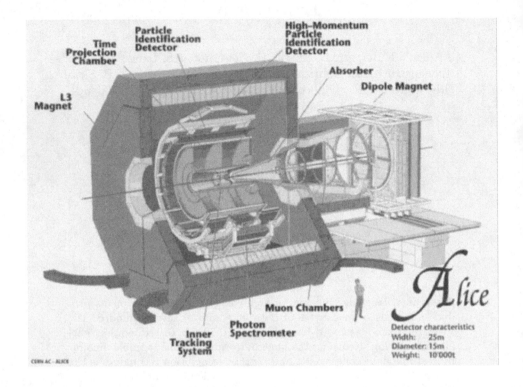

Fig. 1. The ALICE detector at the Large Hadron Collider at CERN.

The ALICE experiment [1] will measure such violent collisions of two heavy nuclei at the LHC at CERN in year 2005. The detector consists of a 10,000 ton magnet and several subdetectors (figure 1). The main detector is a 3-dimensional tracking device, the Time Projection Chamber (TPC). Charged particles traversing the detector initiate clouds of charge - called clusters - along their path through the detector volume. The detector is read-out by 600,000 ADC channels, producing a data rate of about 15 GByte/sec. The data from the various subdetectors (upper row in figure 2) are transferred into the receiver processors via optical fibres and PCI. These processors constitute the first level of a parallel PC-farm [3], which will reduce the data rate by a factor of 10 to 100. The data acquisition system (see figure 2) [2] collects the subevents and assembles them into a single event for permanent storage.

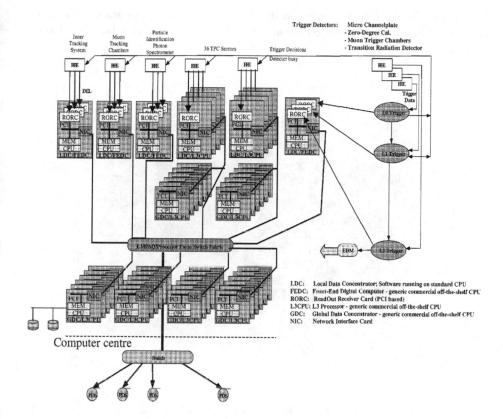

Fig. 2. Sketch of the data acquisition and Level-3 system architecture.

2   Handling the Massive Data Flow: The Level-3 Trigger System

Tracking detectors in high-energy physics experiments produce hundreds of megabytes of data at a rate of several hundred Hz. Collisions of heavy nuclei at the highest energies available for example produce thousands of particles, which are recorded by tracking detectors. To illustrate the amount of data, a typical collision between two gold nuclei at RHIC recorded by the TPC detector of the STAR experiment [4,5] is shown in figure 3. The data rate which is expected for ALICE is far higher than what can be transferred to the permanent storage system. Online processing is necessary in order to select interesting events and subevents or to compress the data efficiently by modeling techniques. Processing detector information at a bandwidth of 10-20 Gbyte/sec or even higher (the peak rate into the front-end electronics is 6 Tbyte/sec) requires a massive parallel computing system (figure 4).

Such a system (Level-3 trigger system) [3] is located in the data flow between the front-electronics of the detector and the event-builder of the data acquisition system.

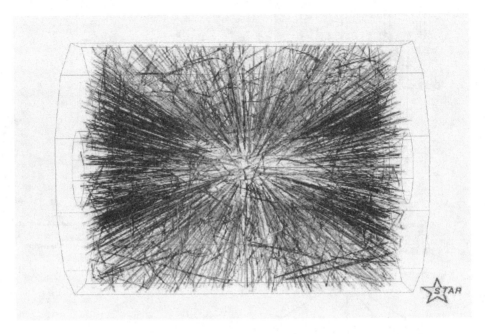

Fig. 3. Particle tracks in the 3D tracking detector TPC. The event shown is a gold-gold collision at RHIC, recorded by the TPC-detector of the STAR experiment.

3   System Architecture: Clustered SMP Farm

Clustered SMP nodes, based on off-the-shelf PCs, and connected by a high bandwidth, low latency network provide the necessary computing power for online pattern recognition and data compression.

The system nodes are interfaced to the front-end electronics of the detector via their internal PCI-bus or whatever will be it's substitute by the time the system will be built, such as PCI-X.

3.1   Network Topology

The intelligent Level-3 layer of the ALICE experiment will make use of the inherent granularity of the tracking detector readout. The detector consists of 36 sectors, each sector being divided into 5 subsectors. The data from each subsector are transferred via an optical fibre from the detector into 180 receiver nodes of the PC farm.

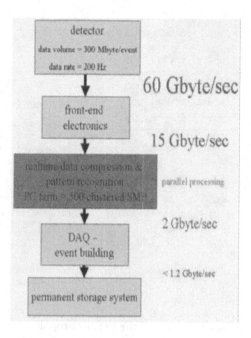

Fig. 4. Data volume and event rate: data flow from detector via the parallel Level-3 system into the data acquisition.

Figures 2 and 5 show a sketch of one possible Level-3 network topology [8]. The receiver processors are all interconnected by a hierarchical network. Each sector is processed in parallel, results are then merged on a higher level, first locally in 144 sector and 72 sextet nodes and finally in 12 event nodes which have access to the full detector data.

3.2 Farm Hierarchy

The hierarchy of the parallel farm is adapted both to the parallelism in the data and to the complexity of the pattern recognition. The first layer of nodes receives the data from the detector and performs the pre-processing task, i.e. cluster and track seed finding on the subsector level. The next two levels of nodes exploit the local neighbourhood: track segement finding on a sector level. Finally all local results are collected from the sectors and combined on a global level: track segment merging and final track fitting.

3.3 Farm Specifications

The farm has to be fault tolerant and scalable. The cluster nodes are commercial off-the-shelf SMP boxes. Possible network technology candidates are amongst others SCI [9] or ATOLL [10]. The operating system will be LINUX, using

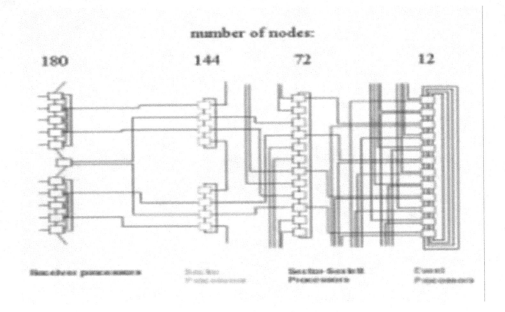

Fig. 5. Layout of the network topology.

a generic message passing API for communication with the exception of data transfers, which use directly remote DMA.

4   Realtime Data Compression: Data Modeling Based on Fast Pattern Recognition

Data compression and realtime pattern recognition of tracking detectors used in high-energy physics experiments are the key tasks in the operation of such experiments. About 20,000 particles per interaction each produce about 100 clusters in the detectors. These signals have to be readout, processed, recognized and grouped into track segments. The event rate sets an upper limit for the time budget of these operations: about 5 milliseconds.

Data rate reduction can be achieved by either reducing the event rate ("Level-3 trigger") or the data volume. Based on the detector information - clusters and tracks - interesting events or subevents can be selected for further processing, thus reducing the event rate. Reduction in the size of the event data can be achieved by compression techniques.

Data compression techniques can be divided into two major families: lossy and lossless. Lossy data compression concedes a certain loss of accuracy in exchange for greatly increased compression. Lossy compression proves effective when applied to graphical images and digitized voice. Lossless compression consists of those techniques guaranteed to generate an exact duplicate of the input data stream after a compress/expand cycle.

In general, data compression is a two step process: modeling and coding. Compression consists of taking a stream of symbols and transforming them into codes. If the compression is effective, the resulting stream of codes will be smaller than the original symbols. The decision to output a certain code for a certain symbol is based on a model. The model is simply a collection of data and rules used to process input symbols and determines which code to output. Lossless data compresssion is generally implemented using either statistical or dictionary-based modeling. Statistical modeling reads in and encodes a single symbol at a time using the probability of that character's appearance [11]. Dictionary-based modeling uses a single code to replace strings of symbols. The Huffman algorithm [12] is an example of a minimum redundancy coding based on probabilities; the vector quantizer [14,15] is a method based on a codebook.

General lossless or slightly lossy methods like entropy coders and vector quantizers can compress tracking detector data only by factors 2-3 [3]. The choice of the data model [13] is of uttermost importance in order to achieve high compression factors (about 15). All correlations in the data have to be incorporated into the model. The precise knowledge of the detector performance, i.e. analog noise of the detector and the quantization noise, is necessary to create a minimum information model of the data. Realtime pattern recognition and feature extraction are the input to the model. The recent increase of inexpensive computing power and memory size as well as the availability of high-bandwidth, low-latency networks makes almost lossless data compression based on sophisticated modeling feasible.

The data modeling scheme is based on the fact that the information content of the TPC are tracks, which can be represented by models of clusters and track parameters. Figure 6 shows a thin slice of the detector, clusters are aligned along the particle trajectories (helices). The best compression method is to find a good model for the raw data and to transform the data into an efficient representation. Information is stored as model parameters and (small) deviations from the model. The results are coded using Huffman and vector quantization algorithms. The relevant information given by a tracking detector are the local track parameters and the parameters of the clusters belonging to this track segment. The local track model is a helix; the knowledge of the track parameters is necessary for the description of the shape of the clusters in a simple model. The pattern recognition reconstructs clusters and associates them with local track segments. Note that pattern recognition at this stage can be redundant, i.e clusters can belong to more than one track and track segments can overlap. Once the pattern recognition is completed, the track can be represented by helix parameters. In a second step, the deviation of the cluster centroid position from the track model (residuals), the deviation from the average charge and deviations from the expected model (based on the track parameters) are calculated for each cluster. These numbers are then quantized by a nonlinear transfer function adapted to the detector noise and detector resolution. The event rate sets an upper limit for the time budget of these operations: about 5 milliseconds.

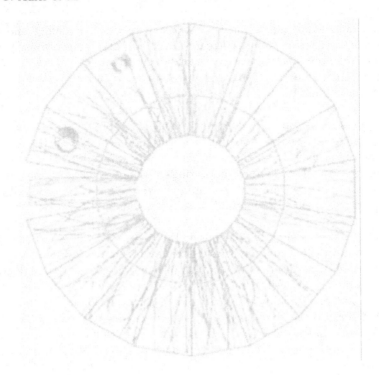

Fig. 6. Simulated collision of two heavy nuclei at LHC. Only a very small fraction of the TPC-detector (thin slice) is shown.

Both Level-3 triggering and data modeling require a fast parallel pattern recognition.

In STAR, this process is done in two steps: first clusters are found and analysed and cluster centroids are then combined to form tracks. The fast track finder employs conformal mapping, i.e. a transformation of a cirlce into a straight line [17], followed by a track follower [16]. The track finding takes between 39 and 88 ms on an ALPHA XP1000 (450 MHZ, 21264 CPU) for about 400 tracks [6,7].

This method has been adapted to ALICE, but due to the higher track density, an alternative method is being developed. The adaptive, generalized Hough-transform is applied to raw data, i.e. before cluster finding. Local track segments are found and based on the track information, cluster shapes are modeled.

5   Outlook

At the moment the architecture of the system is being designed and parallel pattern recognition and data compression algorithms are under development. The system will be installed at the ALICE experiment which will start to take data in 2005. A prototype system (about 5% of the complexity of the final farm) is currently operative in the STAR experiment at RHIC.

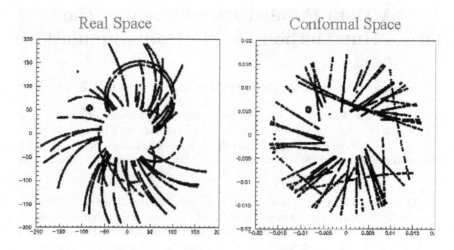

Fig. 7. Conformal mapping of a thin detector slice.

References

1. ALICE Collaboration, Technical Proposal, CERN/LHCC/95-71.
2. M. Arregui *et al.* (ALICE Collaboration), *The ALICE DAQ: current status and future challenges*, Proceedings CHEP2000, (2000), to be published.
3. ALICE collaboration, Technical Design Report of the Time Projection Chamber, CERN/LHCC 200-001 (2000).
4. STAR collaboration, STAR Conceptual Design Report, Lawrence Berkeley Laboratory, University of California, *PUB-5347*, June 1992.
5. http://www.star.bnl.gov.
6. A. Ljubicic, Jr. et al., *Design and Implementation of the STAR Experiment's DAQ*, 10th IEEE Real Time Conference, 9/22-26/97, Beaune, France.
7. C. Adler et al., *The proposed Level-3 Trigger System for STAR*, subm. to IEEE Transactions in Nuclear Science (1999).
8. T. Steinbeck, L3 Developments, ALICE DAQ meeting, CERN, (March 16, 2000).
9. SCI (Scalable Coherent Interface) Standard Compliant, *ANSI/IEEE 1596-1992*.
10. http://www.atoll-net.de.
11. Shannon, C.E., *Mathematical theory of communication*, Bell System Technical Journal, 27 (1948) 379-423.
12. D.A. Huffman, *A method for the construction of minimum-redundancy codes*, Proceedings of the IRE, Vol. 40, No. 9, (1952) 1098-1101.
13. H. Beker and M. Schindler, *Data compression on zero suppressed High Energy Physics Data*, Institute of Computer Graphics, Technical University Vienna (1997).
14. Gray, R. M., *Vector Quantization*, IEEE ASSP Magazine April (1984) 4.
15. Linde, Y., Buzo, A, and Gray, R. M., *An algorithm for vector quantizer design*, IEEE Transactions on Communications **28** (1980) 84.
16. P. Yepes, *A Fast Track Pattern Recognition*, STAR Note 248 (1996); Nucl. Instrum. Meth. **A380** (1996) 582.
17. D. Brinkmann et al., Nucl. Instrum. Meth. **A354** (1995) 419.
18. F. Pühlhofer, D. Röhrich, R. Keidel, Nucl. Instrum. Meth. **A263** (1988) 360.

A Data Parallel Formulation of the Barnes-Hut Method for $N-$Body Simulations[*]

M. Amor[1], F. Argüello[2], J. López[3], O. Plata[3], and E.L. Zapata[3]

[1] Dept. Electronics and Systems, Univ. of La Coruña, Spain
[2] Dept. Electronics and Computation, Univ. of Santiago de Compostela, Spain
[3] Dept. of Computer Architecture, Univ. of Málaga, Spain
margaaml@udc.es, arguello@dec.usc.es, {juan,oscar,ezapata}@ac.uma.es

Abstract. This paper presents a data–parallel formulation for $N-$body simulations using the Barnes-Hut method. The tree-structured problem is first linearized by using *space–filling curves*. This process allows us to use standard data distributions and parallel array operations available in data-parallel languages. A new efficient HPF implementation of the Barnes-Hut method is presented in this paper, characterized by the use of array copy sections to express communications, obtaining efficient point–to–point data transferences. In addition, HPF_LOCAL constructs are used to obtain very compact and efficient local (node) codes.

1 Introduction

The difficulty of the efficient programming of complex applications on parallel computers has motivated much research in recent years towards the goal of (semi–)automatic parallelization. Data-parallel languages like High Performance Fortran (HPF) is one of the most significant cases. However it has not been proven the suitability of HPF to obtain efficient parallel codes for a large class of complex applications, known as irregular applications.

Recent work [3,4] has shown that an irregular problem like the hierarchical $N-$body simulation can be implemented in HPF. All these implementations, however, use non-adaptive and adaptive versions of the Anderson's method for far-field force computations. This paper shows an efficient HPF implementation of a different hierarchical force calculation method, the Barnes–Hut algorithm (BH) [1]. There are significant differences between their implementations and ours. Basically, in our case communications are implemented using array copy sections, obtaining efficient point–to–point data transferences. In addition, HPF_LOCAL constructs are used to obtain very efficient node local codes.

1.1 The Burnes-Hut Method

A number of different methods for force calculation in $N-$body simulations has been proposed in the literature, such as direct integration methods (*particle-particle* methods), mesh methods and hierarchical (tree code) methods. The BH

---

[*] This work was supported by the Ministry of Education and Science (CICYT) of Spain under project TIC96-1125-C03

algorithm is the simplest hierarchical N–body method, widely used in astrophysics. As an astrophysics application simulates a large number of bodies (millions of stars), hierarchical methods employ a divide–and–conquer algorithm to reduce the number of particles in the force sum. A simple physical intuition is used: the force contribution of a far enough galaxy can be completely replaced by a single point mass located at its center of mass. This idea is recursively applied with the help of a tree data structure, an octtree in 3D or a quadtree in 2D. The root of a quadtree (octtree) is a square (cube) comprising the whole body domain. This large box is subdivided into four (eight) equal-sized subboxes. This partitioning process is repeated recursively until at most one body is found in each subbox.

Basically, the BH method follows three stages. First, the quadtree (octtree) for the body domain is built. Second, for each subbox of the tree structure, the center of mass and the total mass for all the bodies inside are computed. And third, for each body, the tree data structure is traversed to compute all the force contributions following a divide–and–conquer strategy. That is, if the distance from the body to a subbox is much larger than its size, then all the bodies contained are approximated by a single point mass. One typical distance test, known as the *Multipole Acceptability Criterion* (MAC), is based on checking if the fraction l/d is less than some user-defined threshold [1], where l is the length of a side of the subbox and d is the distance from the body to the subbox center of mass.

2 Data Parallel Implementation

A data parallel formulation of an irregular problem, like the Barnes-Hut method, requieres a number of phases:

Linearization: The programming model implies the use of linear arrays to represent the data. In our formulation, we linearize the tree data structures (quadtrees or octtrees) through the use of space–filling curves [5]. We use, in particular, a Peano–Hilbert (PH) linear ordering, which has the property that, to a certain degree, two points adjacent on the curve are adjacent in the tree structure. In addition, the PH curve is easy to generate and results in simple indexing. As an example, Fig. 1 shows an 2D ensemble of bodies and its associated quadtree. Below in figure, the corresponding PH curve is depicted together with the corresponding linear ordering of the bodies.

Data Distribution: After linearizing the problem, the linear sequence of bodies is block distributed among processors, to keep the locality exhibited by the PH curve, and the computations of the simulation are organized as array operations.

Communication Organization: It is one of the key steps in order to obtain an efficient parallel implementation of the algorithm. We have used array copy sections to organize communications, instead of array indirections or irregular gather/scatter operations, as in [3,4]. This fact facilitates the compiler work to generate efficient communication operations.

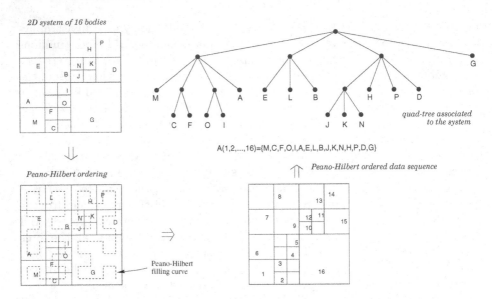

Fig. 1. Example of a set of bodies in a 2D space partitioned to build a quadtree, and the corresponding linearization using a PH space–filling curve

Workload Balancing: Finally, in dynamic computations, a workload balancing scheme should be included in the parallel algorithm, usually at the simulation step level.

As an example, Fig. 2 represents the complete dataflow of one simulation step of the BH method for a 2D system. The upper part of this figure shows the original ensemble of bodies, its corresponding quadtree, the PH linearization of the tree, and the block partitioning of the resulting linear array of bodies.

To simplify the explanation, we have represented the BH method by the following string of operators,

$$\prod_{t=0}^{end}\left[\prod_{i=1}^{U}T_{z,i}\left(\xi\Upsilon_{z,Y}\psi\prod_{i=U+1}^{UU}T_{z,i}\right)FVX\right] \tag{1}$$

The initial product runs over the iterations or steps of the simulation. For each iteration, we have three computing stages.

– In the first stage ($\prod_{i=1}^{U}T_{z,i}$), each processor builds a local quadtree from the assigned block of bodies. Symbol z represents the root of the local tree, which is built incrementally by adding one by one all the local bodies ($U = N/Q$ bodies in each processor). This is a completely local computation that can be easily written into HPF as follows (operations to build a quadtree are not essential in the explanation of the BH method implementation, and thus were encapsulated into a procedure),

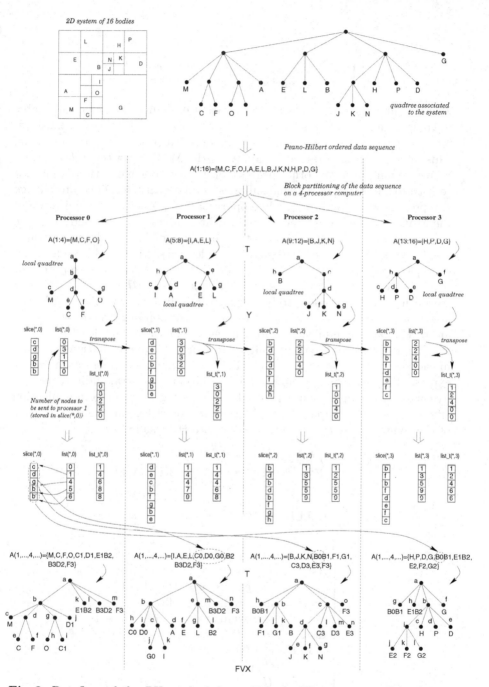

Fig. 2. Dataflow of the BH method for a 16-body 2D system parallelized on a 4-processor computer

$$\prod_{i=1}^{U} T_{z,i} \rightarrow \begin{array}{l} \textbf{do} \ \ i = 1, \ U \\ \quad \textbf{CALL} \ BuidTree(Tree, \ A, \ i) \\ \textbf{end do} \end{array}$$

- The second stage represents an information exchange, in order to be able to compute force contributions on the local bodies. Each processor must determine what information that it owns is needed by the rest of processors. This information is obtained by applying the MAC heuristic to *benchmark* bodies coming from the other processors. In our implementation, each processor broadcasts their extreme subboxes in its local tree. This interchange is represented by operator ξ, while operator Υ represents the application of the MAC test. The results are stored in three tables, slice, list and list_t (see Fig. 2). The $i - th$ entry of table list stores the number of nodes to be transfered to processor i. The corresponding nodes are stored in table slice. As an example, processor 0 in Fig. 2 must send the three nodes c, d, g to processor 1 and the node b to processors 2 and 3. Finally, the $i - th$ entry of table list_t stores the number of nodes to be received from processor i. Afterwards, the above information is communicated to the corresponding processors by means of a global communication interchange, an all–to–all broadcast (operator ψ). The use of those three tables allows to group all communications together in each simulation step and, in addition, such communications can be easily expressed in terms of Fortran90 array operations. No array indirections nor irregular gather operations are needed, obtaining a better communication efficiency. As an example, the explained three operators in this second stage of the simulation step can be translated into HPF as follows,

$$\xi \rightarrow \begin{cases} buffer(1 : 2 * Q : 2) = A(1 : N : U) \\ buffer(2 : 2 * Q : 2) = A(U : N : U) \end{cases}$$

$$\Upsilon \rightarrow \textbf{CALL} \ \ MAC(Tree, buffer, slice, list)$$

$$\psi \rightarrow \begin{cases} list\_t = \text{transpose}(list) \\ list = \text{sum\_prefix}(list, \text{dim} = 1, \\ \qquad\qquad \text{exclusive} = .true.) + 1 \\ list\_t = \text{sum\_prefix}(list\_t, \text{dim} = 1, \\ \qquad\qquad \text{exclusive} = .true.) + U + 1 \\ \textbf{do} \ i = 1, Q \\ \quad \textbf{FORALL}(j = 1 : Q) \\ \qquad A(list\_t(j, i) : list\_t(j + 1, i) - 1, i) = \\ \qquad slice(list(i, j) : list(i + 1, j) - 1, j) \\ \textbf{end do} \end{cases}$$

- In the third stage $(\prod_{i=U+1}^{UU} T_{z,i})$, the information exchanged is added to the corresponding local tree, obtaining this way an extended local tree. UU is the total number of bodies that each processor now has, considering also the owned ones. Finally, each processor is now able to compute forces on the owned bodies, and update their velocities and positions (F, V and X operators). All these operations are also local computations, as in the first stage, and can be written into HPF in a similar way, as next,

$$\prod_{i=U+1}^{UU} T_{z,i} \rightarrow \begin{cases} \textbf{do } k = 1, Q \\ \quad \textbf{do } i = U + 1, UU \\ \quad\quad \textbf{CALL } BuildTree(Tree, A, i) \\ \quad \textbf{end do} \\ \textbf{end do} \end{cases}$$

$$F \rightarrow CALL\ Force(A)$$
$$V \rightarrow CALL\ Velocity(A)$$
$$X \rightarrow CALL\ Position(A)$$

Fig. 3 shows a HPF program that directly implements the above explained process. This code is static, that is, no workload balancing is guaranteed. To overcome this problem we have also implemented a dynamic load balanced version of the BH method, using a diffuse type dynamic scheduling algorithm [2], which preserves some data locality. At the end of each simulation step, the scheduling piece of code tests if the workload imbalance is greater than some threshold. The imbalance is calculated taking the difference between the loads of the most-loaded and the least-loaded processors divided by the average load in the full system. In case of high imbalance, the most-loaded processor sends a fraction of its set of bodies to the least-loaded of its processor neighbors (a linear processor array is considered). The workload is estimated by counting the total number of force contributions regarding its set of bodies.

3   Evaluation

The HPF code for the BH method (Fig. 3) was compiled using the PGI's PGHPF compiler on a Cray T3E. Input bodies were generated randomly according to the Plummer model with monopolar approximation and potential softening, which is widely used to generate spherical galaxies made up of equal mass bodies. The softening parameters avoids singularities when particles get too close to each other. The code for the particle generator was lifted from the SPLASH2 Barnes application [6]), and modified to be used in our simulation.

The granularity of computation is high, particularly for the center of mass calculation ($O(\frac{N \log N}{P})$), force computation ($O(\frac{N \log N}{P})$) and update stages ($O(\frac{N}{P})$). There is no synchronization in these stages. A small number of barriers are used to maintain data dependencies throughout some stages. However, this number is independent of problem size and number of processors.

Fig. 4 (a) shows the performance of the static HPF code. Speedups are calculated with respect to the sequential program running on one processor, and

```
!HPF$ PROCESSORS linear(Q)
    TYPE(BODY), ALLOCATABLE :: A(:)
    TYPE(QUADTREE), ALLOCATABLE :: Tree(:)
    INTEGER, ALLOCATABLE :: PH(:)
    INTEGER :: slice(max,Q), list(Q+1,Q), list_t(Q+1,Q)
!HPF$ ALIGN PH(i) WITH A(i)
!HPF$ ALIGN (*,i) WITH slice(*,i) :: list, list_t
!HPF$ DISTRIBUTE(BLOCK) ONTO linear :: A, Tree
!HPF$ DISTRIBUTE(*,BLOCK) ONTO linear :: slice
    READ (*,*) A, PH
    A(PH) = A   ! Redistribute A using the PH ordering
    do t=0, end   ! Iterate the simulation
! First stage: quadtree building
    do i=1,U
        CALL BuildTree(Tree,A,i)
    end do
! Second stage: Extreme subboxes interchange
    buffer(1:2*Q:2) = A(1:N:U)
    buffer(2:2*Q:2) = A(U:N:U)
! Second stage: MAC heuristic computation
    CALL MAC(Tree,buffer,slice,list)
! Second stage: Global communication
    list_t = TRANSPOSE(list)
    list = SUM_PREFIX(list,DIM=1,EXCLUSIVE=.TRUE.) + 1
    list_t = SUM_PREFIX(list_t,DIM=1,EXCLUSIVE=.TRUE.) + U + 1
    do i=1,Q
        FORALL (j=1:Q) A(list_t(j,i):list_t(j+1,i)-1,i) = slice(list(i,j):list(i+1,j)-1,j)
    end do
! Second stage: Local tree expansion
    do k=1,Q
        do i=U+1,UU
            CALL BuildTree(Tree,A,i)
        end do
    end do
! Third stage: Compute forces, update velocities and positions
    CALL Force(A)
    CALL Velocity(A)
    CALL Position(A)
    end do
```

Fig. 3. HPF program for solving N–body problems using the BH method

100 timesteps were executed. Superlinearity is obtained as the dataflow reorganization introduces high data locality. Beyond 8 processors there is a significant performance reduction (see Fig. 4 (b)) due to the fact that the tree building stage does not speedup quite as well as the other stages. This performance knee depends on the size of the input data set. Concerning the load balancing diffuse algorithm, performance gains is presented in the table at the lower part of Fig. 4 (c) compared to the extreme case of "no balance".

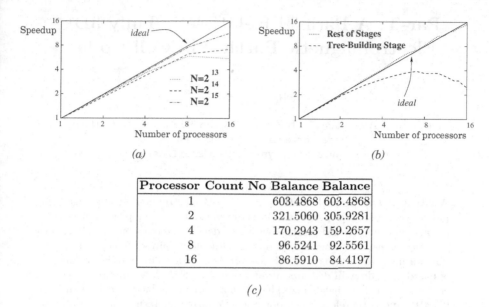

Processor Count	No Balance	Balance
1	603.4868	603.4868
2	321.5060	305.9281
4	170.2943	159.2657
8	96.5241	92.5561
16	86.5910	84.4197

(c)

Fig. 4. (a) Speedup for the BH method (with no dynamic worload balancing), (b) breakdown for the tree-building and the rest of stages and (c) execution times (in sec.) on 16-processor Cray T3E and $N = 2^{14}$, comparing dynamic load balancing and no balancing schemes

References

1. Barnes, J., Hut, P.: A Hierarchical $O(N \log N)$ Force-Calculation Algorithm. Nature **324** (Dec. 1986) 446–449.
2. Fonlupt, C., Marquet, P., Dekeyser, J.-L.: Data–Parallel Load Balancing Strategies. J. Parallel Computing **24** (1998) 1665–1684.
3. Hu, Y.C., Johnsson, S.L., Teng, S.-H.: High Performance Fortran for Highly Irregular Problems. ACM SIGPLAN Symp. Principles and Practice of Parallel Programming (PPoPP'97) (Las Vegas, NV, Jun. 1997) 13–24.
4. McCurdy, C., Mellor–Crummey, J.: An Evaluation of Computing Paradigms for N–body Simulations on Distributed Memory Architectures. ACM SIGPLAN Symp. Principles and Practice of Parallel Programming (PPoPP'99) (Atlanta, GA, May 1999) 25–36.
5. Ou, C., Ranka, S., Fox, G.: Fast and Parallel Mapping Algorithms for Irregular Problems. J. of Supercomputing **10 (2)** (1996) 119–140.
6. Woo, S.C., Ohara, M., Torrie, E., Singh, J.-P., Gupta, A.: The SPLASH2 Programs: Characterization and Methodological Considerations. ACM/IEEE Int'l. Symp. Computer Architecture (ISCA'95) (Jun. 1995) 24–26.

Par-T: A Parallel Relativistic Fully 3D Electromagnetic Particle-in-Cell Code

Peter Messmer

Institute of Astronomy, ETH Zentrum, CH-8092 Zürich, Switzerland
messmer@astro.phys.ethz.ch,
http://www.astro.phys.ethz.ch/staff/messmer

Abstract. The parallel relativistic fully 3D electromagnetic particle-in-cell code *par-T* is presented. It is intended to model plasma wave phenomena in extended plasmas with low density inhomogeneities. The code uses a leapfrog scheme to push particles and a finite-difference time-domain method to update the electromagnetic field. The parallelization is based on a domain decomposition, ensuring both low communication cost and good load balancing. The implementation is based on Fortran 90/MPI and is therefore easily portable. Performance results on a Cray-T3E, a SUN HPC and a Beowulf-Cluster are presented.

Keywords: Particle-in-Cell, Kinetic Plasma, TRISTAN, Performance Evalutation, Beowulf

1 Introduction

Particle-in-cell (PIC) codes are a tool to investigate the kinetic behavior of collisionless plasmas. They trace the motion of a representative number of charged particles in the discretized self-consistent electromagnetic field.

In order to simulate kinetic processes as they occur in astrophysical plasmas, a wide range of spatial and temporal scales has to be resolved. The cell size is about given by the Debye length of the plasma, whereas the number of cells is determined by the longest wavelength of interest. Therefore the computational resources determine the physical problem which can be tackled.

In this paper, the parallel relativistic fully 3D electromagnetic particle-mesh code *par-T* (*par*allel-*Tristan*) is presented. It is based on the sequential code TRISTAN [2] and has been parallelized using the idea of the General Concurrent PIC code [3], [5]. The aim was to develop a tool to perform large scale plasma simulations on a variety of computing platforms. Due to the Fortran 90/MPI-based implementation, the code is highly portable. The performance on three different platforms is reported.

2 The PIC Algorithm

The relativistic motion of charged particles in an electromagnetic field is governed by the Newton-Lorentz law

T. Sørevik et al. (Eds.): PARA 2000, LNCS 1947, pp. 350–355, 2001.
© Springer-Verlag Berlin Heidelberg 2001

$$\frac{d\gamma v}{dt} = \frac{q}{m}\left(E + \frac{v}{c} \times B\right) \qquad\qquad \frac{dx}{dt} = v \qquad (1)$$

where x, v and q/m are position, velocity and charge to mass ratio of each particle, $\gamma = (1 - v^2/c^2)^{-1/2}$ and c is the velocity of light. The electrical field E and the magnetic inductivity B have to satisfy the full set of Maxwell equations

$$\nabla \cdot E = \rho \qquad (2) \qquad\qquad \nabla \cdot B = 0 \qquad (4)$$

$$\frac{\partial E}{\partial t} = c\nabla \times B - J \qquad (3) \qquad\qquad \frac{\partial B}{\partial t} = -c\nabla \times E \qquad (5)$$

where J denotes the current density. The equations are written in rationalized c.g.s. units to avoid factors 4π. To tackle this problem, the particles are located anywhere in the computational domain, whereas the field quantities E and B are discretized in a uniform cubic grid.

The particle-in-cell algorithm consists of two phases: In the *particle push* phase, the new particle position and velocities are determined according to Eq. (1). In the following *field solve* phase, the fields are updated, according to the particle motion.

In order to push the particles, the field quantities at the particle positions are determined by linear interpolation.

To solve the field equations, various techniques can be applied. Widely used are methos based on Fourier transforms of the whole grid. Due to the high non-locality of the Fourier transform, these methods are not well suited for a parallel implementation. In *par-T*, the approach of a rigorous charge conserving algorithm [4] is used: The current flux through a cell surface is determined from the charge carried across this surfaces due to the particle motion in one time step. In that case, the continuity equation

$$\frac{\partial \rho}{\partial t} = -\nabla \cdot J \qquad (6)$$

is satisfied, which implies that Eq. (2) remains satisfied provided it held initialy.

The field quantities are known at different positions in space: The E field components are defined on the cell edges, whereas the B field components are known on the cell surfaces. The components are staggered in a so-called Yee lattice [6]. In this setup, e.g. the B_z, E_x and E_y components reside on the same z-plane. This allows to compute the z-component of $\nabla \times E$ at the location of B_z by means of finite-differences. Analogously, all other curl operators can be computed. Providing the current flux at locations of the E field grid, the B and E fields can be updated by leap-frog-scheme, using only Eqs. (5) and (3).

All finite difference operators are time and space centered to guarantee second order accuracy. For the whole scheme, only local information is needed.

2.1   Parallel Implementation

The parallel implementation is based on the General Concurrent PIC (GCPIC) algorithm, which both minimizes interprocessor communication and optimizes load balancing: Each processor is assigned a subdomain and all the particles and grid points in it. To ensure that the field interpolation (gather) and current deposit (scatter) operations can be performed concurrently, additional guard cells at the subdomain boundaries are stored on each processor. Assuming low particle density inhomogeneities, a mesh which is evenly distributed among processors leads to good load balancing.

Both the particle push phase and the field update phase run entirely in parallel. Communication is needed after the particle push phase, when particles leaving a subdomain have to be transferred to the neighboring processor. Additionally, after each field update, the field guard cells have to be exchanged. Evidently, in all the communication phases there is only nearest neighbor communication.

Par-T is designed for rod-shaped geometries, which often occur in astrophysical problems (e.g. particle beams in magnetic flux-tubes). Therefore only domain decomposition in one spatial direction is supported. Overlapped with the communication along the extended dimension, the boundary conditions in the other two dimensions can be applied. The current implementation of *par-T* features periodic boundary conditions in all three dimensions.

The code is implemented using Fortran 90 and MPI. No platform-specific optimizations at source-code level were applied.

3   Performance Results

The aim of the parallelization is to both track large numbers of particles and to run each time step quickly. Therefore the scaling behavior of the algorithm was investigated, both for a fixed size problem as well as for a scaled problem size. The test case consists of two counter streaming electron beams in a background plasma of thermal ions. The mean beam velocity is $\pm 0.2c$ and the thermal velocity $0.05c$. The simulation run was carried out for 55 time steps which represents a physical time of $5 \cdot 2\pi/\omega_p$, where ω_p is the electron plasma frequency.

3.1   Test Systems

The experiment was run on three different systems: A Cray T3E-900, a Sun HPC 3500 and a Beowulf-Cluster.

The Cray T3E-900 system of the Edinburgh Parallel Computing Center (EPCC) consists of 344 Alpha EV5.6 processors running at 450 MHz. The local memory size varies between 64 or 128 MB. The interconnect speed varies between $0.8 - 3.8$ Gbit/s. For this experiments, at most 128 Processors were available.

The SUN HPC 3500 of EPCC features 8 400 MHz UltraSparc II processors and 8 GBytes of shared memory.

The Beowulf-Cluster of ETH Zürich consists of 192 nodes, each equipped with 2 Pentium III processors running at 500 MHz and 1 GBytes RAM. The interconnect is a 100 Mbit/s switched Ethernet between frames of 24 nodes and a 1 Gbit/s switched Ethernet between the eight frames. [1]

3.2   Fixed Problem Size

Figure 1 shows the scaling behavior of *par-T* for a fixed size problem of $16 \times 32 \times 512$ grid points with $3.1 \cdot 10^6$ particles (≈ 12 particles per cell).

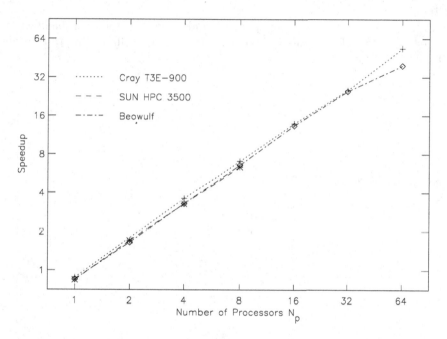

Fig. 1. Speedup for a fixed problem size ($16 \times 32 \times 512$ cells, $3.1 \cdot 10^6$ Particles) on the three test platforms.

The sequential code on the T3E-900 spends 92% of the time for the main computation phases, namely for particle pushing, current deposition and field update. The parallel code running on 1 processor spends 82% in the main computation routines, due to the additional work for particle and field exchange. Running on 64 processors, only 64% of the time is spent in the main computational phases. The grid-size per processor is then $16 \times 32 \times 8$ cells and only $4.9 \cdot 10^5$ particles per processor.

Let N_p be the number of processors, $T(N_p)$ the time it takes to run the problem on N_p processors and T_{seq} the running time of the sequential code. The

---

[1] In the meantime, the system size has increased to 251 Dual-CPU nodes.

parallel efficiency $\epsilon = T_{seq}/N_pT(N_p)$ for the fixed problem size on 8 processors is then 88% for the T3E-900, 79% for the Sun HPC 3500 and 82% for the Beowulf-Cluster. In case of 64 processors, the parallel efficiency is still 84% for the T3E-900, whereas it drops for the Beowulf-Cluster to 62%. This is due to the slower network if communication is performed among more than one frame.

3.3   Scaled Problem Size

Beside the reduced time to complete a given simulation, a parallel architecture can be used to tackle physical problems which would not fit on a single-processor system.

Figure 2 shows the time per particle and time step for a problem of size $64 \times 64 \times 16 * N_p$. The number of particles is ≈ 16 per cell, leading to a total of $1.0 \cdot 10^6 * N_p$ particles.

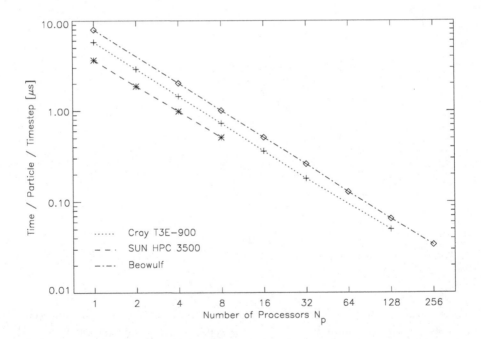

Fig. 2. Time spent per particle and time step for a variable problem size ($64 \times 64 \times 16 * N_p$, $1.0 \cdot 10^6 * N_p$ particles) on the three test platforms.

For the problem size of size $64 \times 64 \times 2096$ grid cells and $1.34 \cdot 10^8$ particles, the total time spent per particle and time step on 128 nodes of the T3E-900 is 50 ns. For the same problem size and the same number of processors, the Beowulf-Cluster needs 65 ns.

A major advantage of a Beowulf-Cluster is the vast amount of memory. The current configuration allows to tackle problems of size $64 \times 64 \times 80 * 256$ cells and $1.34 \cdot 10^9$ particles on 256 processors. The time per particle and time step is still 34 ns, resulting in an overall time of 46 s per time step.

4 Conclusions

The parallel relativistic fully 3D electromagnetic particle-mesh code *par-T* is presented. Particles are pushed using a relativistic leapfrog scheme and the electromagnetic field is updated using a charge conserving finite difference method. The parallelization is based on a domain decomposition among the processors and assigning them grid quantities and particles. Currently the code features periodic boundary conditions.

The code is written in Fortran 90 / MPI, simplifying porting among different platforms.

Performance measurements for the two-stream instability are presented. For large scale computations, most of the time is spent in the particle-push phase. Even on a Beowulf cluster with a slow network, the code scales well.

Times of 50 ns spent per particle and time step can be achieved on 128 nodes of a Cray-T3E for $64 \times 64 \times 2048$ grid cells and $134 \cdot 10^6$ particles.

Test simulations on $64 \times 64 \times 20480$ cells and $1.34 \cdot 10^9$ particles simulated were performed on 256 processors of a Beowulf-Cluster, using 34 ns per particle and time step.

Acknowledgments. Access to the Cray T3E System and the Sun Enterprise Server was granted by the Edinburgh Parallel Computing Center under the *Training and Research on Advanced Computing Systems* (TRACS)-Program. The Beowulf-Cluster is operated by the Institute of Theoretical Physics at ETH Zürich. The author thanks P. Arbenz (ETHZ) and C. Mulholland (EPCC) for many helpful discussions. His visit at EPCC was financially supported by the Swiss Federal Office for Education and Science.

References

1. Hockney, R.W., Eastwood, J.W.: Computer Simulation Using Particles. McGraw-Hill, New York (1981)
2. Buneman, O.: TRISTAN: The 3-D Electromagnetic Particle Code. Computer Space Plasma Physics, Eds. Matsumoto, H. and Omura Y., Terra Scientific Publishing Company, Tokyo (1993)
3. Liewer, P.C., Decyk, V.K.: A general concurrent algorithm for plasma particle-in-cell simulation codes. J. Comput. Phys. **85** (1989) 302–322
4. Villasenor, J., Buneman, O.: Rigorous charge conservation for local electromagnetic field solvers Comput. Phys. Commun. **69** (1992) 306–316
5. Wang, J., Liewer, P.C., Decyk, V.K.: 3D electromagnetic plasma particle simulations on a MIMD parallel computer. Comput. Phys. Commun. **87** (1995) 35–53
6. Yee, K.S.: Numerical solution of initial boundary value problems involving Maxwell's equations in isotropic media, IEEE Trans. Antennas Propagat. **14** (1966) 302–307

Evaluation of MPI's One-Sided Communication Mechanism for Short-Range Molecular Dynamics on the Origin2000

T. Matthey[1] and J. P. Hansen[2]

[1] Department of Informatics, University of Bergen
Høyteknologisenteret i Bergen, N-5020 Bergen, Norway
`Thierry.Matthey@ii.uib.no`
[2] Institute of Physics, University of Bergen
Allégaten 55, N-5007 Bergen, Norway
`JanPetter.Hansen@fi.uib.no`

Abstract. In this paper we evaluate the possibilities of one-sided communication, a new feature of the MPI-2 standard, on the Origin2000 for relatively short-range molecular dynamics (MD) simulations. Our algorithm is based on an asynchronous message-passing multi-cell approach using MPI as message-passing layer and the Leap-Frog/Verlet algorithm for the time integration. We compare one-sided with two different two-sided communication approaches for typical production runs ($10^5 - 10^9$ atoms) where we discuss the communication vs. computation time for increasing number of processes. We also show how the partitioning of the problem affects the different communication approaches. Using one-sided communication we achieved 10-70% better performance over two-sided.

1 Introduction

Molecular dynamics (MD) [1,2] has been known for several decades and successfully used in atomistic simulation models with a few thousands interacting particles. With the introduction of large scale parallel computers it became possible to study more realistically sized systems[3,4] and nowadays MD is an appreciated tool for a whole range of scientists to study dynamic properties of micro- or macro phenomena [5].

For computer scientists MD simulation is interesting in order to evaluate new parallelization and communication approaches, since it includes several challenging parallelization problems. MD simulations may also be used in benchmarking, to study the efficiency of different machines.

In this paper we evaluate the one-sided communication mechanism of MPI-2[6] for relatively short-range MD simulations. For our evaluation we use a shared memory machine, the Origin2000 with 128 CPU's. Our system contains approximately $10^5 - 10^7$ particles representing a general metal atom with a dense, regular distribution. The implementation is well optimized[7] and in C++. For testing and evaluation purposes we implemented the two-dimensional case, but

T. Sørevik et al. (Eds.): PARA 2000, LNCS 1947, pp. 356–365, 2001.

the design and the algorithms extend naturally to higher dimensions. MPI[1] was used as communication layer.

The paper is organized as follows: In Section 2 we review the basic physical principles for MD simulations, in Section 3 we explain the parallelization, distribution and load balancing of the work. In Section 4 we describe our three different communication approaches based on MPI and in Section 5 we give a short overview of the one-sided communication mechanism of MPI-2. In Section 6 we discuss the timing results and in Section 7 we close with a conclusion and prospects for future work.

2   Short-Range Molecular Dynamics

The MD method is based on the solution of Newton's equation of motion for N interacting particles. This general N-body problem first involves the calculation of $N(N-1)/2$ pairs of interactions to compute all forces. On a given particle at position r_j,

$$m_j \ddot{r}_j = F_j = \sum_{i=1, i \neq j}^{N} -\nabla u(r_{ij}).$$ (1)

Here m_j is the mass of particle j and r_{ij} is the distance between particle i and j. The complexity of the force calculation is governed by the potential function $u(r_{ij})$. To simplify the potential we define a finite range of interaction, which is a reasonable approximation of atomistic interactions. In our evaluation we use the general Lennard-Jones 6-12 (LJ) potential,

$$u(r_{ij}) = \begin{cases} 4\epsilon \left[\left(\frac{\alpha}{r_{ij}}\right)^{12} - \left(\frac{\beta}{r_{ij}}\right)^{6} \right] & : 0 < r_{ij} \leq r_{\text{cutoff}} \\ 0 & : r_{\text{cutoff}} < r_{ij} \end{cases}.$$ (2)

Where ϵ, α and β are LJ parameters with $\alpha \approx \beta$, often is $\alpha = \beta$, it is called σ. The potential $u(r_{ij})$ is cut-off at r_{cutoff}, i.e. no interactions are evaluated beyond this distance.

The number of interacting neighbors for each particle is determined by the interaction range r_{cutoff}, and the particle density ρ. Considering that the repulsive core of $u(r_{ij})$ limits the local density (the maximal number of particles inside the range), we can conclude that the number of interactions has an upper constant bound, i.e. does not depending on the number of particles, N. Once the forces for each particle are computed, the positions and velocities have to be updated. For our purpose we use the Leap-Frog/Verlet algorithm, which despite its low order of accuracy has excellent energy conservation properties[2].

3   The Multi-cell Algorithm and Distribution

The multi-cell algorithm[3] provides a means of organizing the spatial information for each particle into such a form that the particle's neighbors can be quickly

---
[1] MPT 1.4 from SGI.

located for the force calculation. Thus the number of interactions evaluated is
minimized. We briefly outline the main features of the computation for the two-
dimensional case, which extends directly to three dimensions.

The region is divided into rectangular, equal cells (Fig. 1) with dimensions at
least the r_{cutoff} distance. Particles are assigned to the cells geometrically according their positions. The computation of the force on a single particle involves
only the particles of the same cell and the neighboring cells. The evaluation of
forces for all particles in a cell consists of two steps. First all interactions between particles in the original cell (5) are computed. Then the forces between
neighboring cells (6,3,2,1) and the original cell are computed, by following an interaction path that describes how the interactions with neighboring cells should
occur. Following the path the interactions for the original and the visited cells are

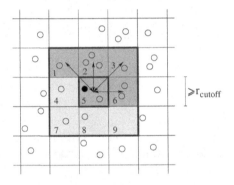

Fig. 1. Partition of cells, where cell nr. 5 is the origin and 1-4,6-9 the neighbor cells.
Cell nr. 6,3,2 and 1 describe the interaction path.

accumulated, e.g. using Newton's third law. The original cell (5) will accumulate
the interactions from its lower neighbors when they calculate their interactions.
To calculate all forces, this procedure has to be carried out for each cell. However, the defined path and the assumption that the dimensions of cells are at
least r_{cutoff}, guarantee that all interactions with a contribution will be taken into
acount and the total amount of computed interactions is of order $O(N)$.

With the cell structure now in place, the particles can be distributed. Each cell
with its particles is assigned to a particular process by a given partitioning algorithm. To study how partitioning affects the amount of communication and
communication conflicts (i.e. two or more processes request/need the same cell)
we introduced three different approaches to distribute the cells. The first partitioning (Fig. 2, *stripe*) uses a one-dimensional array partitioning simply assigning
an approximately equal stripe to each process. Fig. 3 illustrates a partitioning
(*metis*) minimizing the interface between processes for large systems. Our implementation uses METIS[8] – a family of programs for partitioning unstructured
graphs and hypergraphs. The last partitioning (Fig. 4, *rect*) is a two-dimensional
block partitioning trying to minimize the interface. Additionally, for the purpose

of testing the robustness of the communication approaches we used for some smaller systems a partitioning where we scattered the cells, trying to assign neighboring cells to as many as possible different processes.

Fig. 2. Stripe.   **Fig. 3.** METIS.   **Fig. 4.** Rectangle.

4   The Communication Approaches

Each process calculates the interactions in parallel, proceeding through all assigned (locally) – the owned – cells sequentially. As long as the neighboring cells are owned, the cell is independent from other processes. Once a process needs a neighboring cell owned by another process – a *shared* cell, we have to make sure that this cell with its particles is made available. Note that the passing of a cell may happen by either an explicit or implicit request. We have studied three communication approaches where one is based on MPI's one-sided communication mechanism and the other two on two-sided communication.

First we discuss the two-sided communication approaches which require a matching send for each posted receive. This criterion synchronize the involved processes and can in the worst case lead to a dead-lock. The first approach (*simple*) does an implicit passing depending strongly on the distribution algorithm, i.e. it works only for the *stripe* partitioning. Each process starts with the cells, which will be used (later) by another process. Once these cells are not needed any more for further computations they are sent asynchronously to the appropriate process. As soon as the receiver needs these cells it will receive them and continue to compute. The cells are sent back when the process has completed its computations. To simplify we assume that each process will request cells from only one process. *stripe* is the only approach which requests cells from at most one process.

The second approach (*dist*) passes cells by explicit requests. The owner of the cell will decide if it can give away the requested cell, i.e. the cell is not needed by the owner itself. The requesting process will hold the cell as long as the owner does not request it back, and the process has not finished all its computations. Thus the cell may be used for further interactions and passed back only if it is really needed. Further requests may occur for a cell, which actually is held by a third process. The owner has then to request back the cell and pass it to the requesting process. All this leads to a more complicated communication policy, but the method will handle all kind of partitionings including scattered distribution. Additionally we can show that the cells are coherent at any time.

The last approach (*one*) is based on MPI's one-sided communication mechanism. Before any cell can be passed we have to define the (memory-)*window*'s for the remote access (see next section) for *shared* cells, which will be needed by other processes. We assume that the allocated storage in each cell is large enough for the whole run, otherwise we have to redefine the *window* later. Before computing the forces each process will copy and make public the *shared* cells in its own *window*. Whenever a process needs a cell it will grab the cell from the according process *window*. Once the processes have computed all interactions they write their contribution of the *shared* cells for each process in a separate *window*. Then the owner of the *shared* cells will collect and accumulate all contributions scattered among all (neighboring) processes. The approach has a simple policy without any synchronization when passing or requesting cells. We elaborate on this communication mechanism next.

5   MPI's One-Sided Communication Mechanism

One-sided communication[6] or Remote Memory Access (RMA) extends the communication mechanisms of MPI by allowing one process to specify all communication parameters, both for the sending side and for the receiving side. This mode of communication suits applications with dynamically changing data access patterns, where the data distribution is fixed or slowly changing. For such applications, each process can access or update data at other processes where the remote process may not know which data in their memory need to be accessed or updated. RMA communication mechanisms also avoid the need for global communication or explicit polling.

Where regular send/receive communication requires matching operations by sender and receiver, RMA requires once the definition of the *window*[2] for the remote memory access or update. The definition of the *window* – a collective operation – specifies the memory that is made accessible to remote processes. Once the *window* is not any more needed or obsolete, it should be freed[3] and if necessary redefined.

The access[4] is similar to the execution of a send by the target process, and a matching receive by the origin process with the obvious difference that the receiver does not depend on the sender. The update[5] works similarly, except that the direction of data transfer is reversed. For a more sophisticated update than

---

[2] `MPI_WIN_CREATE()`.

[3] `MPI_WIN_FREE()`.

[4] `MPI_GET()`.

[5] `MPI_PUT()`.

copying data from the origin to the target process, a general accumulation function[6] is provided. RMA synchronization *fence*[7] – a collective operation – entails the completion of RMA's and the synchronization of RMA calls on the given *window*. Beside *fence*, there are several functions to implement finer and more subtle synchronization of RMA calls.

6   Results

To monitor the different communications approaches and distributions we performed a variety of 190 test runs with different lattice sizes (Table 1) and number of CPU's. All test runs were performed for 200 timesteps representing 1.01805×10^{-4} [ns]. The lattices are two-dimensional and consist of 89.9% particles representing an idealized, optimal packed metallic alloy with 10% of an additive atomic species and 0.1% holes. Table 1 shows the number of particles

Table 1. The three different test cases. L2-bound is the lower bound of CPU's required so that all particles fit into the L2-cache on each process.

Cases	Particles	Cells	L2-bound	Sequential [s]
small	91,800	82×82	1.93	193.4
medium	925,600	260×259	19.4	2300.0
large	9,218,300	820×819	193.4	24692.5

for each test case and the number of cells. L2-bound is the lower bound of CPU's required so that all particles fit into the L2-cache (4MB) on each process, assuming an equal distribution. Note that each process will required some additional memory for the communication buffers (*window*'s) containing the *shared* cells. The last column is the time of a sequential run. Periodic boundaries were applied in all dimensions and $r_{\text{cutoff}} \approx 3\sigma$, where each particle has approximately 43 interacting neighbors and approximately 14 particles per cell. The test runs for one to 32 CPU's were performed on a normally loaded machine, where for 64 or more CPU's the machine was empty.

The following tables and figures is a selection of the runs of interest. Tables 2-4 are a summary of the achieved speedups and the gain of *one* compared to *dist*. The bold numbers represent the best achieved speedup for each approach. Fig. 13, 5 and 6 are the corresponding graphs for *small*, *medium* and *large*. Figs. 7, 9 and 11 show the time spent in [s] for the communication, which includes the redistribution of particles moving from one process to another and idle cycles when a process have to wait to receive data. Figs. 8, 10 and 12 present the time spent for the computation of forces and computation of new positions and velocities (upper curves) relatively to the total run time. Additionally the update timings, which is the amount spent for the update of the cell structure, are shown.

---

[6] MPI_ACCUMULATE().

[7] MPI_WIN_FENCE().

Table 2. Speedups for the test case *small*, 91,800 particles.

CPU's	dist			one				simple	Cells/
	stripe	metis	rect	stripe	metis	rect	Gain	stripe	CPU
1	1.00	–	–	–	–	–	–	–	6,724
2	1.45	1.49	**1.62**	1.63	**2.00**	1.80	23%	1.85	3,137
4	3.15	3.31	**3.38**	**4.21**	3.79	3.65	25%	3.67	1,569
8	4.21	**5.64**	5.60	5.80	**7.71**	6.86	37%	6.47	841
16	8.15	**8.92**	8.28	10.93	11.88	**12.49**	40%	11.70	421
32	13.83	**14.58**	14.44	17.19	**21.09**	20.16	45%	16.35	211
64	18.43	22.98	**24.10**	30.87	**41.68**	38.59	73%	–	106
121	21.30	29.30	**32.41**	30.50	33.25	**33.52**	3%	–	56

Table 3. Speedups for the test case *medium*, 925,600 particles.

CPU's	dist			one				simple	Cells/
	stripe	metis	rect	stripe	metis	rect	Gain	stripe	CPU
1	1.00	–	–	–	–	–	–	–	67'346
2	**1.77**	1.67	1.66	1.69	**1.83**	1.75	3%	1.61	33,673
4	3.08	**3.31**	2.83	**3.43**	3.14	3.15	4%	3.34	16,837
8	5.74	5.71	**5.85**	6.24	6.29	**7.17**	23%	6.92	8,419
12	7.23	8.01	**8.09**	9.06	**9.27**	7.87	15%	–	5,613
16	**10.39**	8.50	9.66	10.79	11.74	**12.56**	21%	12.39	4,210
20	**12.05**	10.84	10.99	14.71	**15.07**	14.22	25%	–	3,368
24	13.26	12.55	**13.70**	13.70	**15.23**	15.08	11%	–	2,807
28	12.37	13.80	**14.07**	19.23	17.86	**23.66**	68%	–	2,406
32	15.71	18.17	**21.76**	21.72	24.54	**24.90**	14%	23.11	2,105
64	–	**42.42**	–	–	**62.15**	43.47	47%	44.50	1,053
121	–	–	**54.43**	–	–	**102.73**	89%	62.97	557

Table 4. Speedups for the test case *large*, 9,218,300 particles.

CPU's	dist			one				Cells/
	stripe	metis	rect	stripe	metis	rect	Gain	CPU
1	1.00	–	–	–	–	–	–	671,580
2	1.73	1.70	**1.76**	**1.93**	1.84	1.89	10%	335,790
4	3.27	**3.33**	2.80	3.41	3.30	**3.65**	10%	167,895
8	6.17	**6.69**	5.40	7.05	**7.37**	6.95	10%	83,948
16	11.31	**11.62**	10.28	**13.22**	13.12	13.20	14%	41,974
32	17.82	**19.96**	–	23.83	**24.84**	23.93	24%	20,987
64	–	**34.24**	–	–	**36.09**	–	5%	10,494
121	–	–	**60.25**	–	–	**66.22**	10%	5,551
128	–	–	–	–	–	**91.31**	–	5,247

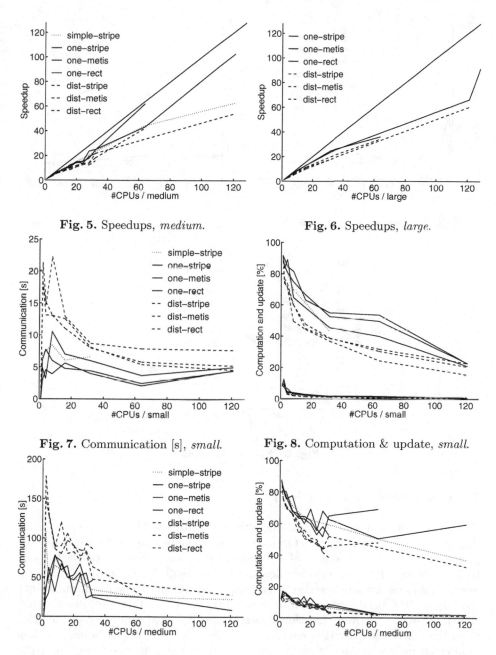

Fig. 5. Speedups, *medium*.

Fig. 6. Speedups, *large*.

Fig. 7. Communication [s], *small*.

Fig. 8. Computation & update, *small*.

Fig. 9. Communication [s], *medium*.

Fig. 10. Computation & update, *medium*.

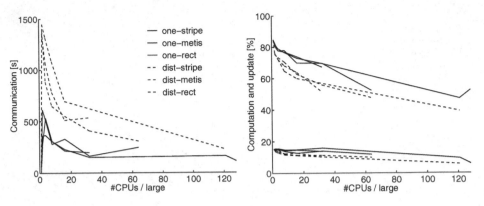

Fig. 11. Communication [s], *large.*   **Fig. 12.** Computation & update, *large.*

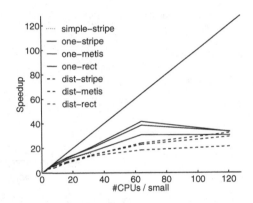

Fig. 13. Speedups, *small.*

7   Discussion and Outlook

From the figures it can be concluded that the use of MPI's one-sided communication mechanism on the Origin2000 is not only feasible but it improves performance as well. The gain of the one-sided communication approach – not to have to wait for a process sending data – is about 10% (Table 4). As soon as the domains get smaller and requesting conflicts occur more often, we get 20% up to 80% better performance. From Figs. 7, 9 and 11 we see clearly that the time spent for communication depends linearly on the number of particles regardless of the distribution approach: *one* compared with *dist* is more or less unaffected by the amount of communicated data since it uses non-blocking communication and does not need to wait when receiving data. Better performance, however, requires more memory; *one* will request more then twice the memory if the domains have less than 320 cells. Nevertheless we obtain a better performance, even around the L2-bound when only *dist* completely benefits from the L2-cache.

For the partitioning we did not find a clear "best choice": In some cases *rect* is slightly better than *metis* for large number of CPU's. METIS may produce a partitioning with non-connected domains and domains with irregular shape, which introduces more communication. Furthermore we noticed that *metis*'s irregular shape of domains reduces the conflicts (e.g. several processes requesting the same cell at the same time), especially for *dist. rect* may not generate an optimal load balance for small number of cells or for large numbers of CPU's with few, unequal factors (e.g. prime numbers). *stripe* is a competitive partitioning for small numbers of CPU's and scales better for huge number of cells.

With the cell algorithm we get an excellent local memory access pattern since the particles are stored in a dense array in each cell. This can be observed in Fig. 8 where the time spent for the update of the cells drops down from 10% for one CPU to 2% for four CPU's; we even get a super linear speedup. For *medium* (Fig. 10) we see that already for 16 CPU's parts of L2-cache can be reused for the next timestep, but it fits completely into L2-cache after 32 CPU's. *large* (Fig. 12) does not benefit from the L2-cache at all until 128 CPU's as expected (L2-bound in Table 1).

In future we will extend our systems to three dimensions and approve the timing number for this case. We here expect an even better improvement since the three-dimensional case involves more communication, which has been shown by [3]. The memory usage for *one* should be improved especially for the three-dimensional case in order to avoid running out of memory. Furthermore, we will consider dynamic load balancing to improve the performance for non-regular systems or systems where the local density is changing heavily.

Acknowledgment. The calculations are performed at Norwegian super-computing facilities in Bergen through a Norges Forskningsråd grant.

References

1. M. P. Allen and D. J. Tildesley, *Computer Simulation of Liquids*, Oxford Clarendon Press (1987).
2. D.C. Rapaport , *The Art of Molecular Dynamics Simulation*, Cambridge University Press, (1995).
3. D.M. Beazley, P.S. Lomdahl, *Message-Passing Multi-Cell Molecular Dynamics on the Connection Machine 5*, Theoretical Division and Advanced Computing Laboratory Los Alamos National Laboratory, Las Alamos, New Mexico 87545, (1993).
4. P.S. Lomdahl, P. Tamayo, N. Grøbech-Jensen, and D.M. Beazley, *50 GFLops Molecular Dynamics on the Connection Machine 5*, Theoretical Division and Advanced Computing Laboratory Los Alamos National Laboratory, Las Alamos, New Mexico 87545, (1993).
5. RS. J. Zhou, D. L. Preston, P. S. Lomdahl and D. M. Beazley, SCIENCE, **279**, 1525 (1998)
6. *MPI-2: Extensions to the Message-Passing Interface*, http://www.mpi-forum.org/docs/docs.html.
7. Igor Zacharov, *Origin Optimisation and Parallelisation Training*, notes, SGI, Bergen, Norway, (2000).
8. *METIS – Family of Multilevel Partitioning Algorithms*, http://www-users.cs.umn.edu/~karypis/metis/metis.html.

Ship Hull Hydrodynamic Analysis Using Distributed Shared Memory

João Silva* and Paulo Guedes

Instituto de Engenharia e Sistemas e Computadores
Rua Alves Redol, 9, Lisboa
Portugal
{joao.n.silva, paulo.guedes}@inesc.pt

Abstract. In this article we present the results obtained from the execution of a commercial Computational Fluid Dynamics program on a cluster of personal computers. The communication and data interchange between the nodes uses a Distributed Shared Memory (DSM) system called TreadMarks. The parallelized program was run on a 100Mbit/s Ethernet network connecting 4 personal computers. Using data similar to a real execution on a industrial environment, we achieved a maximum speedup of about 1.7 and measured an allmost constant communication time with various number of nodes. Comparing our DSM solution with a similar parallelization approach but using Message passing we observed that with less effort we can accomplish better performance with less communication if using a DSM system.

1   Introduction

With the increasing performance, personal computers have become a good computer platform choice for numerical simulation. Adding to this fact, the connecting networks are getting faster and with smaller latency times. All this leads to the development of systems that, using several computers connected by a fast Ethernet network, allow the development and deployment of parallel applications [1].

This kind of work has been centered on the use of some sort of message passing infra-structure to make data communication, PVM[2] or MPI[3]. In this paper another approach to data sharing between remote processes is used: Distributed Shared Memory (DSM).

We present the work done on the parallelization and distribution of a Fluid Dynamics Simulation program called Flash. This program simulates a towing tank experiment from which we get important results for the design of ships hulls. The system we use as the data sharing platform between machines is TreadMarks.

In the next section we present a description of the original application that was parallelized. In the following chapter we present the characteristics of a

---

* The presentation of this article was supported by a grant from the Portuguese Science and Technology Ministry

T. Sørevik et al. (Eds.): PARA 2000, LNCS 1947, pp. 366–372, 2001.
© Springer-Verlag Berlin Heidelberg 2001

general DSM system and of TreadMarks in particular. In the forth section we describe the how we parallelized and distributed the Flash application. In the last sections we present the results obtained and the conclusions.

2   Flash Application

The Flash application is part of the Flash project financed by the European Community under the Esprit Program. The goal of this project is to build a commercial ship design system to be used from the design of the ship hulls models to the hulls performance evaluation. This evaluation is based on the results of the towing tank simulation code.

One other goal of this project is to reduce simulation time of the hulls models. In the project three approaches where followed: changing computational algorithm, optimize code or parallelize the code. The two first approaches were successful, but computation time was still large.

To build a parallel platform, it was decided to use a cluster of workstations, as multiprocessors are rather expensive and are not usual at the premises of the target users of this application[4].

This simulation application solves the Navier–Stokes equations and, additionally, treats problems with free surfaces. A fractional semi-implicit step discretion of the incompressible Navier–Stokes equations is used and a finite element method is used for the spatial discretion.

This system stores the domain data in a list of Nodes belonging to the mesh and several lists of Elements. Each Node stores its coordinates and its degrees of freedom. Each kind of Element represents a type of physical element: fluid element, body elements and free surface.

At each iteration three equation systems are solved. In order to accomplish this, the matrices have to be assembled from data on the elements. The equation systems in the form of $A.x = b$ are solved, resulting x, a set of unknowns. After this, nodes are updated from x and new element values are obtained from the data stored in the nodes.

3   Distributed Shared Memory

Along with message passing, shared memory is the most used paradigm in parallel computing when using a single computer. With shared memory, all the threads of a process see the same memory space and all data transfer between threads is made through shared variables. The synchronization primitives used are those provided by the operating system and the memory consistency is provided by the underlying hardware.

If we wish to code applications using a shared memory paradigm, but the application is to run on several computer in a network we have to use a Distributed Shared Memory (DSM) system.

To develop our applications we have to use distributed memory allocation functions plus some synchronization primitives supplied by the DSM system.

When running a DSM application across the network, at each node the DSM system will intercept the local memory accesses and log that information. When a memory location is changed and it is then accessed by another computer, the DSM system has to provide a coherent view of the data changed. This coherent view is accomplished by fetching changed data from it's the last computer that has changed it.

The DSM system also provides synchronization objects that are used by the application to synchronize when access shared data.

Most DSM systems aggregate synchronization communication with coherence communication in order to reduce number of messages. This reduction in number of messages is accomplished but network traffic increases with the number of synchronization calls. With an high number of synchronization call, lots of redundant data is sent, with a better synchronization policie , only the data necessary to maintain distributed memory coherent is transmitted between nodes.

3.1   TreadMarks on Windows NT

TreadMarks[5] is a DSM system created at Rice University. The system implements the Lazy Release Consistency[6] to maintain the address space consistent in all processes.

This consistency protocols also reduces communication overhead between computer nodes when comparing with other protocols also used. When synchronization occurs, only information about data changes is sent between nodes and the data changed is only sent between nodes when it is needed. This may increase the time to access a remotely changed memory location but reduces communication time.

Another technique to reduce communication time is to send to remote nodes only the changed data in a page. This way, instead of sending the complete page were a change has occurred, only the changed data from that memory page is transmitted.

In the original version, TreadMarks uses the remote shell program (rsh) to launch the tasks on the remote nodes. In the Windows operating system this approach isn't the most suitable because remote shell daemon would have to be bought and not all daemons are fully compatible.

In order to reduce this problem, it was decided to eliminate the use of the rsh and instead of it produce a Windows service to install on all nodes that would launch the tasks. This service is a RPC server that has a known interface and waits in the computers for the master request to launch the slave process. The authentication and security is based on mechanisms provided by the Microsoft implementation of RPC.

4   Parallelization and Distribution

In the parallelization of the simulation code, a distributed shared memory programming model has been used. The underlying system used is TreadMarks.

The partition of the domain has been done by the use of the Metis[7] partitioning package. Only the fluid elements are partitioned among the computer nodes, all other elements are replicated among all computer nodes. Even though there is some replication in work and data, the partition of the fluid elements allows the distribution of most work between all machines.

The first parallelization experiment was to create all fluid elements in shared memory and give to each computer some elements. The nodes were also created in shared memory. This way all shared data was accessible by every computer participating in the computation.

The assembly of the matrices were made in parallel by all computer, but only one computer (the master) solved the system of equations. This computer then updated all nodes. From the new information on the nodes, each computer updated the new element state. This was a simplistic solution to the parallelization of the problem and the result from this experience was a decrease in performance. A lot of synchronization was required, thus increasing the communication between computers.

The second solution also uses the same element partition but there is no data (elements and nodes) created in shared memory, distributed shared memory is only used to passe information between computers.

In each step, the program updates the fluid elements in parallel. To solve the equations, a matrix is also constructed in parallel before the systems are solved. The assemblage of these matrixes needs communication between the slave nodes and the master. Each node assembles the correspondent part of the matrix to a shared memory location and after all computers have finished their partial assemblage, the master computer reads them and assembles the final matrix. The system of equations is then solved only on the master computer. After this, the result is written in another shared memory location and is read by all slave computers so that the local nodes and elements are updated.

This approach is similar to a message passing solution, there is no shared data (elements, nodes, ...) and synchronization occurs only where there is need for communication between computers.

Just for comparison, this same parallelization was also performed using PVM. The partition of the elements was the same as with TreadMarks, but to exchange information, instead of using shared memory, it was used message passing primitives.

5   Results

The execution environment is composed of a network of 4 personal computers connected by a 100 Mbit/s Ethernet network. The personal computers used are based on the Intel Pentium II processor with 128Mb of memory.

The test case was run with the data from the Series 60[8] ship hull that contains 15000 nodes and 85716 elements during 10000 time steps.

The results we obtained are the execution times (Figure 1) we got from the execution of the serial code and the TreadMarks parallel code.

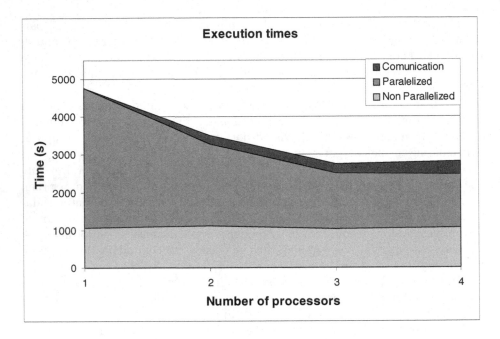

Fig. 1. Execution Times

Comparing TreadMarks with PVM, we got the values of data transmitted using each system (Figure 2)

In the Table 1 we present the real speedups accomplished with the parallelization proposed as well as the ideal speedup and the corrected speedup[9] (taking in account the non parallelized code).

Table 1. Speedups (TreadMarks)

	Number of processors			
	1	2	3	4
Ideal	1	2	3	4
Ideal corrected	1	1.63	2.07	2.38
Real DSM	1	1.36	1.72	1.68

6   Conclusion

With the work presented in this article we managed to produce a platform that allows the computer simulation of a towing tank experiment using several

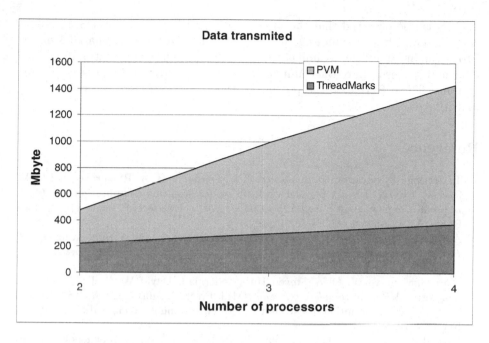

Fig. 2. Data transmitted (comparison TreadMarks and PVM)

personal computers in parallel. From the use of this new system we get exactly the same results as from the use of the serial version, but in a short period of time.

The use of several computers will not only allow the execution of more simulations on a period of time, but also allows the execution of finer resolution simulations (more elements and nodes) and longer simulations (more time steps), just by using computers that would be idle during that time.

The speedup accomplished in this experiment (1.68) doesn't come close to the ideal speedup, but is not far from the corrected speedup (about 2.38). The difference between the real and corrected speedups is due to the fact that we only partitioned the fluid elements but the other elements are replicated among all nodes and there are still some replicated work done by all computers.

We can see that the size of transfered data grows much less when using the DSM system than with the PVM proposal. This is due to the fact that in DSM systems, only the information updated by each node is transmitted to the other nodes without the programmer intervention.

By using DSM it is easier to obtain reduced communication than by using Message Passing, the programmer doesn't have to care about what data must be exchanged. To smaller data sets this is no longer true, there is a lot of false sharing (one page of memory contains data owned by several computers. The programmer must also be aware of synchronization, as redundant synchronization increases the data transfered between computers.

It was also observed that the programming API provided by DSM systems is more simple than the message passing API. The Distributed Shared Memory Programming interface is similar to the API's presented in multiprogrammed OS's and it doesn't have the communication routines present in message passing systems. These facts ease the porting of these applications to shared memory multiprocessors.

References

1. T. Sterling, D. Savarese, D. J. Becker, J. E. Dorband, U. A. Ranawake, and C. V. Packer. BEOWULF : A parallel workstation for scientific computation. In *International Conference on Parallel Processing, Vol.1: Architecture*, pages 11–14, Boca Raton, USA, August 1995. CRC Press.
2. A Geist, A Beguelin, J Dongarra, W Jiang, R Manchek, and V Sunderam. *PVM: Parallel Virtual Machine. A Users' Guide and Tutorial for Networked Parallel Computing*. Scientific and Engineering Computation. MIT Pres, 94.
3. Marc Snir, Steve W. Otto, Steven Huss-Lederman, David W. Walker, and Jack Dongarra. *MPI: the complete reference*. MIT Press, Cambridge, MA, USA, 1996.
4. G. J. Harvey. Description of end user design environment. Technical report, Flash Esprit Project, 1998.
5. Peter Keleher, Alan L. Cox, Sandhya Dwarkadas, and Willy Zwaenepoel. Tread marks: Distributed shared memory on standard workstations and operating system. In USENIX Association, editor, *Proceedings of the Winter 1994 USENIX Conference: January 17–21, 1994, San Francisco, California, USA*, pages 115–132, Berkeley, CA, USA, Winter 1994. USENIX.
6. P. Keleher, A. L. Cox, and W. Zwaenepoel. Lazy release consistency for software distributed shared memory. In *Proc. of the 19th Annual Int'l Symp. on Computer Architecture (ISCA'92)*, pages 13–21, May 1992.
7. George Karypis and Vipin Kumar. METIS, a software package for partitioning unstructured graphs, partitioning meshes, and computing fill-reducing orderings of sparce matrices. version 4.0. Technical report, University of Minnesota, Department of Computer Science / Army HPC Research Center, Minneapolis, MN 55455, September 1998.
8. F. Stern, J. Longo, Z.J. Zhang, and A. Subramani. Detailed bow-flow data and cfd for a series 60 cb=0.6 ship model for froude number 0.316. *Journal of Ship Research*, 1996.
9. David A. Patterson and John L. Hennessy. *Computer Organization and Design – The Hardware-Software Interface*. Morgan Kaufmann Publishers, 1994.

Domain Decomposition Solvers for Large Scale Industrial Finite Element Problems*

Petter E. Bjørstad[1], Jacko Koster[2], and Piotr Krzyżanowski[3]**

[1] Parallab, University of Bergen, 5020 Bergen, Norway. petter@ii.uib.no
[2] Parallab, University of Bergen, 5020 Bergen, Norway. jak@ii.uib.no
[3] Institute of Applied Mathematics, Warsaw University, Banacha 2, 02-097 Warszawa, Poland. przykry@mimuw.edu.pl

Abstract. The European research project PARASOL aimed to design and develop a public domain library of scalable sparse matrix solvers for distributed memory computers. Parallab was a partner in the project and developed a domain decomposition code for solving large scale finite element problems in a robust, yet efficient way. Although the PARASOL project finished in June 1999, Parallab has continued the development of the solver. In this paper, we report on the present status of the solver and show its performance on some challenging industrial problems.

1 Introduction

The domain decomposition solver DD is a distributed memory combined iterative/direct solver of large sparse systems of linear algebraic equations with a symmetric positive definite matrix. Though these matrices may be general in nature, DD has been developed primarily for solving finite element problems, such as solid elasticity and shell problems, even on disconnected subdomains.

The DD software package currently features two state-of-the-art domain decomposition techniques. The first technique is based on iterative substructuring with 1-level and 2-level Neumann-Neumann preconditioners [7,10,11,12,15]. The 2-level method features a variety of coarse spaces. Some of these coarse spaces are designed for specific classes of problems, but others are general purpose self-adaptive algorithms for which almost no additional problem information is required. The second technique supported by DD is to solve the whole partitioned system by using Additive Average preconditioners, see [3,4]. In this paper however, we restrict the discussion to the iterative substructuring algorithms and the Neumann-Neumann preconditioners.

---

* This work has been partially supported by the PARASOL project (EU ESPRIT IV LTR Project 20160).
** The work of this author has also been partially supported through the Polish Scientific Comittee research grant KBN 2P03A02116.

T. Sørevik et al. (Eds.): PARA 2000, LNCS 1947, pp. 373–383, 2001.

2   Iterative Substructuring in DD

Iterative substructuring algorithms are described in detail in [14]. In this section, we briefly mention the main features and the sources of parallelism that are exploited by the DD solver.

For iterative substructuring algorithms, the problem must be partitioned into nonoverlapping subdomains. The DD solver accepts a partitioning from the user or computes one internally (using METIS 4.0). DD accepts problems in assembled, substructured, and elemental format, but uses a common internal substructure data format. Let the physical domain be decomposed into N nonoverlapping subdomains i, $i = 1, \ldots, N$, and let the stiffness matrix $K^{(i)}$ and right-hand side $f^{(i)}$ for subdomain i be defined as:

$$
K^{(i)} := \begin{pmatrix} K_{II}^{(i)} & K_{IR}^{(i)} \\ K_{RI}^{(i)} & K_{RR}^{(i)} \end{pmatrix}, \qquad f^{(i)} := \begin{pmatrix} f_I^{(i)} \\ f_R^{(i)} \end{pmatrix}. \tag{1}
$$

Here, the matrix $K_{II}^{(i)}$ contains the contributions of the internal nodes, the matrix $K_{RR}^{(i)}$ the contributions of the interface nodes, and the off-diagonal matrix $K_{RI}^{(i)}$ contains the links between internal nodes and the interface nodes. $K_{IR}^{(i)}$ is the transpose of $K_{RI}^{(i)}$. The right-hand side vector $f^{(i)}$ is defined accordingly.

In a parallel environment, the subdomain matrices and vectors first need to be assigned to the available processors. DD allows a flexible mapping of the subdomains so that a good balance of the processor work load can be achieved. For example, if the number of subdomains is larger than the number of processors, more than one subdomain can be assigned to the same processor. This processor will treat these subdomains independently; that is, the subdomains are not merged. If the number of subdomains is smaller than the number of processors or if the subdomains differ widely in work and/or storage, more than one processor can be assigned to the same subdomain. These processors will perform the operations for this subdomain in parallel. In the remainder of this paper, we assume (unless stated otherwise) a one-to-one mapping of subdomains to processors.

Substructuring algorithms contain a preprocessing phase in which the internal unknowns in each substructure are eliminated. This is done by solving the N linear systems $K_{II}^{(i)} \cdot v_I^{(i)} = f_I^{(i)}$ for $v_I^{(i)}$ in parallel. DD uses the multifrontal direct sparse solver MUMPS for this. MUMPS is a software package that (like DD) was developed as part of the PARASOL project [1] and that is still being further extended (see elsewhere in this volume [2]).

The preprocessing phase also computes for each substructure i a right-hand side vector of the form $g_R^{(i)} = f_R^{(i)} - K_{RI}^{(i)} \cdot v_I^{(i)}$. Vector $g_R^{(i)}$ is defined for the interface variables of substructure i.

Finally, the preprocessing phase computes for each substructure i a local Schur complement matrix

$$
S_i := K_{RR}^{(i)} - K_{RI}^{(i)} \cdot [K_{II}^{(i)}]^{-1} \cdot K_{IR}^{(i)}. \tag{2}
$$

DD can construct these matrices in two ways. Either S_i is computed explicitly (using MUMPS) and available as a dense matrix, or the action of S_i on a vector is available indirectly through evaluation of the right-hand side of Equation (2). In the latter case, a matrix-vector product with matrix S_i requires three matrix-vector products and one forward/back substitution with the triangular factors of $K_{II}^{(i)}$.

After the preprocessing, a linear system remains to be solved that involves all the unknowns that lie on the interfaces between the subdomains. More precisely, DD solves the interface problem

$$S \cdot u_R = g_R , \quad \text{where} \quad S := \sum_{i=1}^{N} R_i S_i R_i^T , \quad g_R := \sum_{i=1}^{N} R_i g_R^{(i)} , \quad (3)$$

for interface vector u_R, by using a parallel Preconditioned Conjugate Gradient (PCG) iteration, see for example [9]. Here, R_i denotes the prolongation operator associated with subdomain i. In the sequel, a vector x_i denotes the interface vector x (that is defined for all interface variables) restricted to the interface variables of subdomain i, i.e., $x_i = R_i^T x$.

The interface matrix S in Equation (3) needs not be assembled. This provides a source of parallelism that can be exploited in the construction of the preconditioner and the PCG iteration. For example, a matrix-vector product $y = S \cdot x$ is computed as

$$y = \sum_{i=1}^{N} R_i \cdot (S_i x_i) , \qquad x_i = R_i^T \cdot x .$$

This requires N matrix-vector products $S_i x_i$, one for each subdomain, that can be computed in parallel, followed by one communication step in which the local contributions are summed.

After the interface solution vector u_R is computed, some postprocessing is still necessary to compute the internal solution vectors. Therefore, DD solves the N linear systems $K_{II}^{(i)} \cdot w_I^{(i)} = -K_{IR}^{(i)} \cdot R_i^T u_R$ for local vectors $w_I^{(i)}$ in parallel. For each system, this requires a matrix-vector product to compute the right-hand side and a forward/back substitution with the factors of $K_{II}^{(i)}$ that were computed during the preprocessing phase. Finally, the vectors $v_I^{(i)}$ and $w_I^{(i)}$ are summed to obtain the interior solution vectors $u_I^{(i)}$.

The pre- and postprocessing phases are 'embarassingly' parallel. The operations for a substructure do not depend on data from any of the other substructures that is not already available. Therefore, no communication is necessary between the processors. The PCG iteration is implemented in parallel and requires synchronization of the processors at each step of the iteration. The number of iterations that is needed strongly depends on the problem and the preconditioner that is used. In the next section, we describe the more important features of the Neumann-Neumann preconditioners that are available in DD.

3   The Neumann-Neumann Preconditioners

The DD solver provides 1-level and 2-level Neumann-Neumann preconditioners for the PCG iteration. We refer the reader to [14] for an introduction to Neumann-Neumann preconditioning and to [7,10,11,12] for some detailed analyses. The pseudo-code below shows how a preconditioned (local) interface vector $y_i = R_i^T y$ is obtained from a (global) interface residual vector x. The parts of the computation that are performed in parallel and/or require synchronization of the processors are indicated on the right. The matrices D_i are diagonal matrices and define a partition of unity, i.e., $\sum_{i=1}^{N} R_i D_i = I$.

> **foreach** subdomain $i = 1, \ldots, N$
>   /* apply 1-level Neumann-Neumann to x: */
> $y_i = D_i \cdot x_i = D_i \cdot R_i^T x$ parallel
> $x_i = S_i^{-1} \cdot y_i$ /* 'Neumann solve' */   parallel
> $z_i = D_i \cdot x_i$ parallel
> $y_i = R_i^T \cdot z = R_i^T \cdot \sum_{i=1}^{N} R_i z_i$ neigh-neigh comm.
>   **if** 2-level Neumann-Neumann **then**
>   /* balance the residual on the coarse level: */
> $w_i = R_i^T \cdot Sy = R_i^T \cdot \sum_{i=1}^{N} R_i \cdot (S_i y_i)$ parallel + neigh-neigh comm.
> $w_i = x_i - w_i$ parallel
> $t_i = \texttt{CoarseSolve}(w_i)$ gather-scatter comm.
> $y_i = t_i + y_i$ parallel
>   **end if**
> **end foreach**

The 1-level method preconditions an interface residual vector x by

$$y = \sum_{i=1}^{N} R_i D_i S_i^{-1} D_i R_i^T x \ .$$

This requires the solution to the Neumann problems $S_i x_i = y_i$ with Neumann matrix S_i. For this, DD factorizes the Neumann matrices using either MUMPS or LAPACK, depending on the sparsity of the matrix. However, the Neumann matrices can be singular. To avoid this, DD optionally regularizes the Neumann matrices by increasing the diagonal entries with a small constant α times the largest diagonal entry (see also [8]). DD then factorizes this modified matrix. We note that the regularization does not change the problem, but only the preconditioner. Alternatively, the direct solver MUMPS can handle linear systems with rank-deficient matrix and DD optionally makes use of this.

The 2-level Neumann-Neumann preconditioner requires the solution to a coarse problem (`CoarseSolve()` in the pseudo-code). Several options are available in DD to construct a coarse matrix that enables PCG to converge quickly for a range of problems. Mandel [12] showed that the coarse problem can be formulated in terms of N local coarse spaces $Z_i = \text{span}\{z_i^k : k = 1, \ldots, n_i\}$, one for each substructure i, where each local coarse space must satisfy $Ker\, S_i \subset Z_i$. The DD solver contains efficient coarse operators for problems for which the local coarse spaces Z_i (or rigid body motions) are known in advance. This is for example the case for diffusion equation and (2D and 3D) elasticity problems. However, DD also contains coarse operators that do not require such detailed *a priori* knowledge from the user about the local coarse spaces. This makes the Neumann-Neumann preconditioner quite flexible and it enables us to implement several variants of the preconditioner in a single framework.

A flexible strategy is to construct the local coarse spaces from selected eigenvectors of the local Schur operators [6]. This requires the computation of the eigenvectors that correspond to selected eigenvalues of a Neumann problem posed on each subdomain. Since the Schur matrices are assigned to different processors, these eigenanalyses can be performed in parallel. In general, larger eigenvector based coarse spaces are more expensive to construct but improve convergence of the PCG iteration. Hence, it is important to find a balance between the dimension of the coarse space and the number of iterations. Therefore, in [6] a new coarse operator is proposed that adjusts the number of computed eigenvectors adaptively to the number of small eigenvalues found (for each S_i individually). An eigenvalue λ is considered small if

$$\lambda \leq \Lambda := \theta \cdot \max_i \lambda_{min} K_{II}^{(i)} . \tag{4}$$

Here, θ is a user-defined parameter. The threshold Λ is obtained by first computing the smallest eigenvalue for each of the matrices $K_{II}^{(i)}$, followed by a reduction operation to get the largest. This reduction operation requires a synchronization of the processors. The smallest eigenvalues (and eigenvectors) for each S_i or Neumann matrix are then computed in blocks of k values until the largest computed eigenvalue exceeds Λ. We denote this coarse option by ADAPT(k).

If a good bound n_i for the dimension of $Ker\,S_i$ is known *a priori*, then it usually suffices to compute n_i eigenvectors. The DD solver contains a coarse space option that computes for each Neumann matrix a pre-defined number of vectors. This coarse option does not require synchronization of the processors. We denote this option by DETER(k), where $n_i = k$, for all $i = 1, \ldots, N$.

The coarse space constructors DETER(k) and ADAPT(k) were designed to make DD a general purpose and robust domain decomposition solver. Besides these eigenvector based coarse constructors, a variant of the cross-point method described in [11] is also implemented. This method was initially designed for plate and shell problems, but is also suitable for solid elasticity problems [13]. Another coarse option is available for direct substructuring (by defining the coarse problem as the complete interface problem), and an option in which the direct solver MUMPS computes a(n exact) null space for the Schur complement matrices. More details can be found in the forthcoming report [5].

After the vector sets Z_i are determined, the (sparse) coarse matrix can be constructed. The order of the coarse matrix is equal to the sum of the local coarse space dimensions ($\sum_{i=1}^{N} n_i$) and the entries in the matrix are of the form

$$(R_j D_j z_j^l)^T \cdot S \cdot R_i D_i z_i^k ,$$

where i, j are neighbouring subdomains or subdomains that have a common neighbour subdomain. The entries of the coarse matrix can be computed in parallel, but only after the local coarse vectors are exchanged between neighbouring subdomains. DD makes use of higher level BLAS routines where possible. After the entries of the coarse matrix are assembled, this matrix is factorized. This is done in parallel (provided the order of the matrix is large enough).

4 Experimental Results

We performed numerous experiments on various types of finite element problems. These include solid, plate, and shell elasticity problems (see [6]) and large scale industrial problems provided by partners in the PARASOL project. In this section, we illustrate the performance of the DD solver on three large scale problems. The experimental results were obtained on the 128-processor (195 MHz) SGI Origin 2000 at Parallab. More experimental results can be found in [5]. The DD software is written in Fortran 90 and uses MPI for message passing.

The tables in this section use the following notation. N denotes the number of subdomains, #procs denotes the number of processors, #iter denotes the number of PCG iterations, and 'coarse' denotes the dimension of the coarse space. Furthermore, 'pre' denotes the elapsed time (in seconds) for the preprocessing phase, which includes the factorization and elimination of the internal unknowns, the construction of the Schur complement matrices, and the construction of the 2-level preconditioner, 'CG' denotes the elapsed time for the PCG iteration, and 'total' denotes the total elapsed time for solution required by DD. (The times for the postprocessing are much smaller and not listed here.) DD 3.0 is the version of the code that is currently being developed. DD version 2.5 is the code that is part of the public domain PARASOL library that was released in August 1999. DD 2.5 uses for example only implicit Schur complement matrices.

The DD solver was given the partitioned stiffness matrices from Equation (1). Right-hand sides that were not available were generated randomly. No other information (e.g., the type of elements/unknowns, decomposition details, or geometry) was provided to the solver. The subdomains are mapped onto the processors in a cyclic way; i.e., subdomain i is assigned to processor $(i-1)$ **mod** #procs. Unless stated otherwise, all Schur complement matrices are computed explicitly and the following parameters are used. Regularization parameter $\alpha = 10^{-12}$. Threshold relaxation parameter $\theta = 0.005$ for the ADAPT coarse variant (see Equation (4)). The PCG iteration was stopped when the l_2-norm of the residual was reduced by at least six orders of magnitude.

MSC-BMW-crankshaft. The first test case comes from the automotive industry and is provided by MSC.Software (Munich, Germany). It is an elasticity problem with solid elements and contains 148,770 degrees of freedom. The problem was partitioned (by MSC.Software) into 2 and 4 subdomains. The 4-subdomain case contains two subdomains that consist of two disconnected pieces.

Table 1 shows the experimental results for the coarse space variants DETER (12) and ADAPT(6), for varying numbers of processors. For a 3D elasticity pro-

Fig. 1. The MSC-BMW-crankshaft test case decomposed into four subdomains. One of the disconnected subdomains is shown on the right.

Table 1. Results for the MSC-BMW-crankshaft test case.

DD version	N	#procs	DETER(12)						ADAPT(6)					
			coarse	#iter	times				coarse	#iter	times			
					pre	CG	total				pre	CG	total	
3.0	2	2	24	19	243	6.6	256		24	19	243	6.9	256	
3.0	4	4	48	22	123	13.8	140		54	20	120	8.9	131	
3.0	4	1	48	22	438	38.8	493		54	20	445	30.8	490	
3.0	4	2	48	22	224	27.8	262		54	20	241	20.4	269	
3.0	4	4	48	22	123	13.8	140		54	20	120	8.9	131	
3.0	4	8	48	22	96	12.3	112		54	20	99	11.6	114	
2.5	2	2	24	19	659	130	793		24	19	672	115	790	
2.5	4	4	48	22	353	113	468		54	20	404	88	496	

blem, the deficiency of a Neumann matrix is normally 6, but since some of the subdomains consist of two disconnected pieces, we use parameter $k = 12$ for the `DETER` option. The first set of two results (rows in the table) shows the default case where one subdomain is mapped onto one processor. For the 4-subdomain case, the `ADAPT(6)` option computes a slightly larger coarse problem (54 vs. 48). This results in a slightly smaller number of iterations (20 vs. 22). The speed-down in the elapsed time for the PCG iteration is not only due to the larger number of iterations for the 4-subdomain case. More precisely, the partitioning in two subdomains has an interface between the subdomains with 1533 degrees of freedom. For the partitioning into four subdomains however, the subdomains have local interfaces of 2073, 1845, 1461, and 651 degrees of freedom. These larger local interfaces increase the cost (wall-clock time) of the distributed matrix-vector product in the PCG iteration. We attribute the poor scaling of the PCG iteration also in part to inaccurate timings that we obtained at the time of experimentation (due to a heavy load on the machine).

The middle set of four results shows the performance of DD on the 4-subdomain case for 1, 2, 4, and 8 processors. For the run on eight processors, two processors are assigned to each subdomain; one processor (the slave) helps the other (the master) in the factorizations and forward/back substitutions that are performed by the direct solver MUMPS. We attribute the differences in the

times for the DETER(12) and the ADAPT(6) option mainly to variations in the
load of the machine at the time of experimentation. By comparing the first set of
results for DD version 3.0 with the results for version 2.5, we can see a significant
improvement in the performance of the solver over the past twelve months (even
though the number of iterations remains the same).

DNV tubular joint (MT1). The second test case that we consider was provided by Det Norske Veritas (Oslo, Norway). The problem comes from the modelling of a tubular joint and contains a mixture of solid, shell, and transitional
finite elements. The problem was partitioned by Det Norske Veritas into 58 subdomains of widely varying sizes and shapes. The problem contains 97,470 degrees
of freedom; 13,920 lie on the interfaces between the subdomains.

Table 2. Results for the 58-subdomain MT1 test case using the ADAPT(6) coarse
variant. [a] : the solver uses only implicit Schur matrices; [b] : the solver uses $\theta = 0.012$.

DD version	#procs	coarse	#iter	times		
				pre	CG	total
3.0	1	1548	33	217.4	78.4	305.5
3.0	2	1548	33	115.5	40.2	160.7
3.0	4	1548	33	69.6	23.0	95.5
3.0	8	1548	33	42.2	12.8	56.5
3.0	16	1524	35	26.6	7.7	35.3
3.0	1	1548	33	217.4	78.4	305.5
3.0	1	1338	37	198.9	71.5	278.2
3.0	16	1524	35	26.6	7.7	35.3
3.0 [a]	16	1548	33	88.5	27.0	116.9
3.0 [b]	16	1836	22	41.8	7.9	50.8
3.0 [a,b]	16	1836	22	103.3	23.0	127.6
2.5 [a]	16	1850	22			380.0

The first set of results in Table 2 shows the experimental results using the
ADAPT(6) coarse operator for various numbers of processors. The order of the
coarse matrix is quite large (> 1500). Therefore, we experimented with a variant
of the ADAPT coarse option in which we remove all the computed eigenvectors
from the local coarse spaces that correspond to eigenvalues that exceed the
threshold value Λ from Equation (4). This reduces (see the second set of results)
the dimension of the coarse space for the 1-processor run from 1548 to 1338. As
a result, the number of iterations goes up from 33 to 37, but the overall solution
time decreases. This shows that the coarse space must be well tuned in order to
preserve both the convergence and the performance of the DD solver.

We want to stress the importance of the ability to choose between explicit or
implicit Schur complement matrices, which is now implemented in DD version
3.0. DD 3.0 computes 27 Schur complement matrices explicitly; the other 31
matrices are represented implicitly. For some time it has been believed that
iterative substructuring algorithms should 'naturally' use implicit S_i. For the
MT1 problem however, the use of implicit Schur complement matrices leads to

a performance degradation of more than a factor of three (116.9 vs. 35.3 seconds total elapsed time).

Note however that the better performance of DD 3.0 is not only due to the use of a smaller coarse space and explicit Schur complement matrices. To illustrate this, we used DD 3.0 with (a) only implicit Schur complement matrices and (b) $\theta = 0.012$ in order to obtain a coarse space of dimension approximately 1850. As a result, the elapsed time for solution increases to 127.6 seconds. This is still significantly less than the 380 seconds required by DD 2.5. We attribute this to many algorithmic improvements in all parts of the code. These include for example the better use of the direct solver, faster construction of the coarse matrix, and reduced memory requirements. More details can be found in [5].

MSC-inline skater. The third test case was provided by MSC.Software and comes from the modelling of an inline skater. It contains 503,712 degrees of freedom and was partitioned (by MSC.Software) into 2, 8, and 16 subdomains. Some of the subdomains for the 8 and 16-subdomain case consist of disconnected parts. For example, the 16-subdomain test case contains eight subdomains that consist of two parts and two subdomains that consist of three parts.

Table 3. Results for the MSC-inline test case using the `ADAPT(6)` coarse variant.

DD version	N	#procs	coarse	#iter	times		
					pre	CG	total
3.0	2	2	12	25	671	9.2	699
3.0	8	8	108	39	247	57.0	309
3.0	16	16	228	37	140	43.6	188
3.0	16	1	228	37	1185	328.0	1556
3.0	16	2	228	37	621	198.5	844
3.0	16	4	228	37	333	107.1	453
3.0	16	8	228	37	190	62.7	260
3.0	16	16	228	37	140	43.6	188
3.0	28	1	300	32	918	217.2	1177
3.0	28	2	300	32	459	123.0	606
3.0	28	4	300	32	249	64.2	326
3.0	28	8	300	32	151	39.9	198
3.0	28	16	300	32	98	29.5	131
2.5	2	2		22	1571	861	2446
2.5	8	8		31	594	257	851
2.5	16	16		30	391	126	517

Table 3 shows the experimental results for the `ADAPT(6)` coarse variant. The first set of (three) results shows the default case where one subdomain is mapped onto one processor. The largest local interface for the 2-, 8-, and 16-subdomain partitioning contains 1413, 3273, and 2958 degrees of freedom, respectively. The second set of results shows the performance of DD on the 16-subdomain test case for 1, 2, 4, 8, and 16 processors. For the third set of results, we performed the same runs as for the second set, but now the subdomains that consist of disconnected parts are split and each part is treated as a separate subdomain. This increases the number of subdomains from 16 to 28. The size of the largest local interface decreases from 2958 to 1977. It also leads to an increase in the dimension of the coarse space. This is due to the ADAPT(6) option that computes a multiple of six eigenvectors for each subdomain. In general, some of these eigenvectors correspond to eigenvalues that exceed the threshold Λ from Equation (4). As a result of the larger coarse space, the number of iterations decreases from 37 to 32. If we use the earlier mentioned variant of the ADAPT(6) option that deletes the eigenvectors with eigenvalues that exceed Λ, the dimension of the coarse space remains the same; i.e., 150 without splitting and 152 with splitting of the subdomains. The number of iterations is 56 for both cases.

5   Concluding Remarks

The iterative substructuring technique in the DD solver in combination with the Neumann-Neumann preconditioners has proven capable of solving different kinds of problems in engineering and industry, in a robust, yet efficient way in a distributed memory environment. Iterative substructuring allows a high level of parallelism throughout the solution process. The new eigenvector based coarse spaces that have been designed for the 2-level preconditioner can be constructed efficiently in parallel and work well on various kinds of finite element discretizations. The DD solver uses the sparse direct solver MUMPS for the local and the coarse subproblems, because it is equipped with a range of functionalities that are usually not available in other sparse direct solvers and that have proven very useful in all phases of the solution process of our domain decomposition solver.

In our experiments with the large scale industrial problems, we observed that it is often beneficial to split large subdomains that consist of disconnected pieces. Furthermore, we also observed that it is usually more efficient to compute the Schur complement matrices explicitly than to use an implicit representation (given by the right-hand side of Equation (2)). Obviously, there is a balance in the general case. If the order of S_i is small compared to the order of the subdomain matrix $K^{(i)}$, then computing a matrix-vector product $S_i \cdot x$ with a small dense matrix S_i is cheaper than using forward/back substitution with the (large) factors of the matrix $K_{II}^{(i)}$. However, if the Schur complement matrix is relatively large, then constructing and using the dense matrix S_i is more expensive and an implicit representation of S_i is preferred. We have verified this with some 3D Poisson-type model problems (not shown in this paper).

We plan to continue the development of the DD solver. This will include algorithmic improvements that enhance the basic performance, tuning of the software, and the integration of new functionalities. We are currently looking at hybrid versions of the available coarse space variants.

The DD software is available in the public domain and up-to-date versions of the code can be obtained by contacting one of the authors.

References

1. P. R. Amestoy, I. S. Duff, J.-Y. L'Excellent, and J. Koster. A fully asynchronous multifrontal solver using distributed dynamic scheduling. Technical Report RAL-TR-1999-059, Rutherford Appleton Laboratory, Chilton, Didcot, England, 1999. To appear in *SIAM J. Matrix Anal. Appl.*
2. P. R. Amestoy, I. S. Duff, J.-Y. L'Excellent, and J. Koster. MUMPS: a general purpose distributed memory sparse solver. In *Proceedings of the PARA2000 Workshop on Applied Parallel Computing.* Springer-Verlag, 2000. Lecture Notes in Computer Science.
3. P. E. Bjørstad and M. Dryja. A coarse space formulation with good parallel properties for an additive Schwarz domain decomposition algorithm. Submitted to *Numerische Mathematik*, 1999.
4. P. E. Bjørstad, M. Dryja, and E. Vainikko. Robust additive Schwarz methods on unstructured grids. In P. E. Bjørstad, M. Espedal, and D. Keyes, editors, *Domain Decomposition Methods in Sciences and Engineering.* DDM.org, 1997. Ninth International Conference, Bergen, Norway.
5. P. E. Bjørstad, J. Koster, and P. Krzyżanowski. DD: a parallel domain decomposition solver for large scale industrial problems. User's Guide 3.0. Technical report, Parallab, University of Bergen, Norway, August 2000. In preparation.
6. P. E. Bjørstad and P. Krzyżanowski. An eigenvector based two-level Neumann-Neumann coarse space. Technical report, Warsaw University, Poland, May 2000.
7. Y.-H. De Roeck and P. Le Tallec. Analysis and test of a local domain decomposition preconditioner. In R. Glowinski, Y. Kuznetsov, G. Meurant, J. Périaux, and O. Widlund, editors, *Fourth International Symposium on Domain Decomposition Methods for Partial Differential Equations*, pages 112–128. SIAM, 1991.
8. M. Dryja and O. B. Widlund. Schwarz methods of Neumann-Neumann type for three-dimensional elliptic finite element problems. *Comm. Pure Appl. Math.*, 48(2):121–155, 1995.
9. G. H. Golub and C. F. Van Loan. *Matrix Computations.* Johns Hopkins University Press, Baltimore, MD, 3rd edition, 1996.
10. P. Le Tallec, J. Mandel, and M. Vidrascu. Balancing domain decomposition for plates. In *Domain Decomposition Methods in Scientific and Engineering Computing, University Park, 1993*, pages 515–524. Amer. Math. Soc., Providence, 1994.
11. P. Le Tallec, J. Mandel, and M. Vidrascu. A Neumann-Neumann domain decomposition algorithm for solving plate and shell problems. *SIAM J. Numer. Anal.*, 35(2):836–867, 1998.
12. J. Mandel. Balancing domain decomposition. *Comm. Numer. Methods Engrg.*, 9(3):233–241, 1993.
13. J. Mandel and P. Krzyżanowski. Robust Balancing Domain Decomposition. August 1999. Presentation at The Fifth US National Congress on Computational Mechanics, University of Colorado at Boulder, CO.
14. B. F. Smith, P. E. Bjørstad, and W. D. Gropp. *Domain Decomposition. Parallel Multilevel Methods for Elliptic Partial Differential Equations.* Cambridge University Press, Cambridge, 1996.
15. M. Vidrascu. Remarks on the implementation of the Generalised Neumann-Neumann algorithm. In C.-H. Lai, P. E. Bjørstad, M. Cross, and O. B. Widlund, editors, *Domain Decomposition Methods in Sciences and Engineering.* DDM.org, 1999. Eleventh International Conference, London, UK.

A High Parallel Procedure to Initialize the Output Weights of a Radial Basis Function or BP Neural Network

Rossella Cancelliere[1]

Department of Mathematics, University of Turin, v. C. Alberto 10,
10123 Torino, Italy
Cancelliere@dm.unito.it
http://www.springer.de/comp/lncs/index.html

Abstract. The training of a neural network can be made using many different procedures; they allow to find the weights that minimize the discrepancies between targets and actual outputs of the network. The optimal weights can be found either in a direct way or using iterative techniques; in both cases it's sometimes necessary (or simply useful) to evaluate the pseudo-inverse matrix of the projections of input examples into the function space created by the network. Every operation we have to perform to do this can however become difficult (and sometimes impossible) when the dimension of this matrix is very large, so we deal with a way to subdivide it and to obtain our aim by a high parallel algorithm.

1 Introduction

Neural networks have excited great interest during the last years thanks to many reasons as their robustess, flexibility and also thanks to their interdisciplinarity in fact researchers in this field naturally deal with subjects in numerical analysis, artificial intelligence and physics that are among the most advanced today.

These architectures face and resolve the well known problem of learning from examples: given a training set of N data couples $(x_i, d_i / x_i \in \Re^p, d_i \in \Re)$ i=1,.....N, we want that

1) if we have x_i as input to the network, the output be something "near" the desired answer d_i

2) the network has generalization properties too, that is it gives as output d_i even if the input is only something " near " x_i , for instance a noisy or distorted or incomplete version of x_i .

We can formalize the preceding requests, using some notions from numerical analysis, as a classic curve-fitting or approximation problem: we design the network to implement a function F that minimizes the standard error (distance) between the desired answer d_i and the actually obtained answer $F(x_i)$, so our aim is the minimization of the functional

$$E_S(F) = \frac{1}{2} \sum_{i=1}^{N} [d_i - F(x_i)]^2 \tag{1}$$

T. Sørevik et al. (Eds.): PARA 2000, LNCS 1947, pp. 384–390, 2001.

Dealing with RBF neural networks we add to this term the so called regularization term, as discussed in [6],:

$$E_C(F) = \frac{1}{2} \parallel PF \parallel^2 \tag{2}$$

where P is a linear (pseudo) differential operator called regularization operator.

We don't linger over the addition of this term here : the reader caring for more details on regularization theory applied to neural networks can find an extensive treatment in [5],, where he can find proofs of the relations used in this paper.

The neural network realizes a mapping F between the input and the output spaces that minimizes the global functional:

$$E = E_S(F) + E_C(F) = \frac{1}{2} \sum_{i=1}^{N} [d_i - F(x_i)]^2 + \frac{1}{2} \lambda \parallel PF \parallel^2 \tag{3}$$

($\lambda \geq 0$: regularization parameter).

In order to obtain this aim, the network constructs $F(x)$ as a linear superposition of Green functions of P (see [6]), whose analytical expression is determined by the properties of the P operator: for instance if P is invariant under rotations and translations, the Green functions depend only on the Euclidean norm of the vector $x - t_i$:

$$F(x) = \sum_{i=1}^{M} w_i G(\parallel x - t_i \parallel), \; M \ll N \tag{4}$$

The centers $t_i \in \Re^p$ have to be properly determined i.e. they can be M random vectors among the N data points x_i or they can form a completely different set.The neural network that realizes the expansion (4) has the architecture shown in Fig. 1.

It consists of three layers: the first layer is composed of input nodes whose number is equal to the dimension p of the input vector x .

The second layer is a hidden layer, composed of nonlinear units that are connected directly to all of the nodes in the input layer.

The weights that connect the input layer to the hidden one are not trainable and contain the informations about the coordinates of the M Green functions centers t_i (i.e. each of the p weights connected to the second unit of hidden layer contains one component of the vector t_2). The output of the i-th hidden unit is therefore $G(x - t_i)$. The output layer consists of a single linear unit, fully connected to the hidden layer; by linearity we mean that the output of the network is a linearly weighted sum of the output of hidden units.

Fig. 1 depicts the architecture of the network for a single output unit but it is easy to extend it to accomodate any number of output units.

The weights of the output layer are the unknown coefficients of the expansion (4), because the form of Green functions is determined by the P operator: so the next step is to determine the coefficients w_i and to obtain in this way the approximant surface $F(x)$.

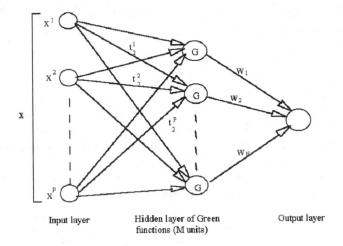

Input layer   Hidden layer of Green   Output layer
                     functions (M units)

Fig. 1. The regularization network

2   Towards the Applications

It is useful now to introduce a matrix notation for the quantities we deal with:

$$w = \begin{vmatrix} w_1 \\ \vdots \\ w_M \end{vmatrix}, \quad d = \begin{vmatrix} d_1 \\ \vdots \\ d_N \end{vmatrix}, \quad G = \begin{vmatrix} G\left(\|x_1 - t_1\|\right) & \cdots & G\left(\|x_1 - t_M\|\right) \\ \vdots & \ddots & \vdots \\ G\left(\|x_N - t_1\|\right) & \cdots & G\left(\|x_N - t_M\|\right) \end{vmatrix} \quad (5)$$

It can be shown [6] that the unknown vector w is a solution of the linear system:

$$(G^T G + \lambda G_0)w = G^T d \quad (6)$$

with

$$G_0 = \begin{vmatrix} G\left(\|t_1 - t_1\|\right) & \cdots & G\left(\|t_1 - t_M\|\right) \\ \vdots & \ddots & \vdots \\ G\left(\|t_M - t_1\|\right) & \cdots & G\left(\|t_M - t_M\|\right) \end{vmatrix} \quad (7)$$

When $\lambda = 0$ the vector w converges to the pseudoinverse (minimum-norm) solution to the overdetermined least-squares data-fitting problem, [1]:

$$w = G^+ d \quad (8)$$

where G^+ is the pseudoinverse of G , i.e.

$$G^+ = (G^T G)^{-1} G^T \quad (9)$$

In the case of a back-propagation neural network, whose architecture is shown in Fig. 2, request (1) leads to the well known method of gradient descent [7]; a

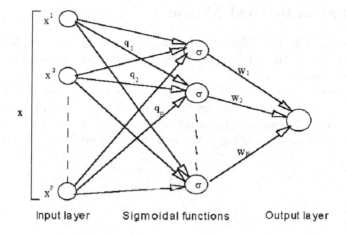

Input layer   Sigmoidal functions   Output layer

Fig. 2. A Sigmoidal Neural Network

new possibility we present here concerns the way to adapt the described theory to deal also with a BP neural network.

During the training phase all the weights that connect different layers of the network are changed according to the relations

$$\Delta w_i = -\eta^w \frac{\partial E}{\partial w_i} \quad \Delta q_i = -\eta^q \frac{\partial E}{\partial q_i} \tag{10}$$

The gradient terms are usually calculated for every weight of the network applying the "chain rule", an iterative procedure whose starting point is randomly chosen in the weight space.

For the output weights of the network shown in Fig. 2, in which there is one linear unit in the output layer (but it is possible to have more units), the starting point can alternately be chosen, in analogy with the RBF network, as the solution of the overdetermined system of equation:

$$F(x) = \sum_{i=1}^{M} w_i \sigma^i (q \cdot x) \tag{11}$$

so having to face a problem similar to that of RBF case.

One of the most used techniques to evaluate G^+ is the Singular Value Decomposition (S.V.D.) [3],[4]. We can find it implemented in efficient versions that however require to store in memory the entire matrix G and this can be sometimes a prohibitive task (in speech recognition applications sometimes $N \sim 10^6$,$M \sim 10^4$ and this makes practically impossible the first step that is the storage of G).

At this point we illustrate a method to overcome these difficulties.

3   A Method to Treat Matrices

The dimension N is determined by the number of examples to be learned while M is the number of units belonging to the hidden layer ; because it is $M < N$, G is a "long" matrix.

So the idea is to subdivide G in Q subsets each made by an arbitrary number a_i of rows $(i = 1, \cdots Q)$.[2] Then we built a set of new matrices named A_i each having the same dimension of G; in the matrix A_i the only set of rows different from zero is that corresponding to the i-th subset of matrix G, so we have

$$G = \sum_{i=1}^{Q} A_i \tag{12}$$

Substituting eq. (12) in eq. (8) we have

$$w = G^+ d = \left(\sum_{j=1}^{Q} A_j^T \sum_{i=1}^{Q} A_i \right)^{-1} \left(\sum_{j=1}^{Q} A_j^T \right) d = \left(\sum_{j,i=1}^{Q} A_j^T A_i \right)^{-1} \left(\sum_{j=1}^{Q} A_j^T \right) d \tag{13}$$

Now we want to emphasize that:

- every product $A_j^T A_i$ is a square M-by-M matrix
- because of the particular form of the matrices A_i , every product $A_j^T A_i$ with $i \neq j$ is the null M-by-M matrix
- if we define a set of reduced matrices $\hat{A}_i$ made only out of the rows different from zero of A_i, we have $A_i^T A_i \equiv \hat{A}_i^T \hat{A}_i$.

So, if we name H the matrix $G^T G$

$$H = \sum_{i=1}^{Q} \hat{A}_i^T \hat{A}_i \tag{14}$$

The first step necessary to obtain H and then w according to eq. (13) is therefore the iteration of the following procedure (for $i = 1, \cdots Q$):

1. store $\hat{A}_i$ (dim $\hat{A}_i = a_i \times M$)
2. compute $\hat{A}_i^T$ and therefore $\hat{A}_i^T \hat{A}_i$
3. add this product to a list that after Q steps will contain the whole matrix H .

In this way we only store in memory the matrices $\hat{A}_i$ one at a time; the quantities $\hat{A}_i^T \hat{A}_i$ can also be computed by different processing units, so realizing a high parallel procedure to perform this step.

Then in order to obtain w we invert H

$$w = H^{-1} \left(\sum_{j=1}^{Q} A_j^T \right) d \tag{15}$$

We evaluate $\left(\sum_{j=1}^{Q} A_j^T\right)$ using a technique analogous to that described before: since

$$H^{-1}\left(\sum_{j=1}^{Q} A_j^T\right) d = H^{-1}\sum_{j=1}^{Q} A_j^T d, \tag{16}$$

we introduce Q reduced vectors $\hat{d}_i$, each having the right dimension to be multiplied by $\hat{A}_j^T$.

Each vector $\hat{d}_i$ is just a piece of the vector d : more precisely, each $\hat{d}_i$ is made only by the rows $\left(\sum_{k=1}^{i-1} a_k\right) + 1$, $\left(\sum_{k=1}^{i-1} a_k\right) + 2$, $\cdots \left(\sum_{k=1}^{i-1} a_k\right) + a_i$ of the vector d.

In this way $A_j^T d \equiv \hat{A}_j^T \hat{d}_j$ and

$$w = H^{-1}\sum_{j=1}^{Q} \hat{A}_j^T \hat{d}_j \tag{17}$$

Iterating the following procedure $(j = 1, \cdots Q)$ we can now calculate w according to eq. (17):

1. store $\hat{A}_j$
2. compute $\hat{A}_j^T$ and therefore $\hat{A}_j^T \hat{d}_j$
3. add the term $\hat{A}_j^T \hat{d}_j$ to a list
4. multiply the obtained sum by H^{-1}

As described before, it is possible to use Q processing units, each performing one step to completely parallelize the procedure. The execution of the entire algorithm thus requires to store twice in memory the whole set of matrices $\hat{A}_j$, the first time in order to evaluate H^{-1} and the second time in order to obtain the solution w of the linear system (6). Thanks to the minor size of the matrices $\hat{A}_j$, every step is now feasible, allowing to find a solution that we could not find otherwise.

References

1. Broomhead, D.S., Lowe, D.: Multivariable functional interpolation and adaptive networks. Complex Systems **2** (1988) 321–355
2. Cancelliere, R.: Large database recognition tasks: a proposal. International Workshop on Neural Networks for Identification, Control, Robotics, and Signal/Image Processing, (1996)
3. Golub, G.H., Van Loan, C.F.: Matrix Computation 2nd edn. Baltimore, MD. John Hopkins University Press (1989)
4. Haykin S.: Adaptive Filter Theory. 2nd edn. Englewood Cliffs, NJ: Prentice-Hall (1991)
5. Haykin S.: Neural Networks, a Comprehensive Foundation. IEEE Computer Society Press (1994)

6. Poggio, T., Girosi, F.: Networks for approximation and learning. Proceedings of the IEEE **78**, n.9, (1990) 1481–1497
7. Rumelhart: Learning internal representation by error propagation. Parallel Distributed Processing (PDP): Exploration in the Microstructure of Cognition, MIT Press, Cambridge, Massachussetts Vol. 1, (1986) pp.318–362

High-Performance Computing
in Geomechanics
by a Parallel Finite Element Approach

Felicja Okulicka – Dluzewska
Faculty of Mathematics and Information Science
Warsaw University of Technology, Pl. Politechniki 1, 00-661 Warsaw, Poland
okulicka@mini.pw.edu.pl

Abstract. The parallelization of the finite element method (FEM) is considered. The parallel versions of the FEM package are developed on the base of the sequential version. The elasto-plastic behaviour of the geotechnical structure can be modeled, calculated and analyzed by the package. The Cray Fortran compiler directives are used for the parallelization of source code. The supercomputer SUN E10000 HPC is used for package test. As the engineering example of the geotechnical problem the dam building is modeled, calculated and analyzed. The different versions of parallel FEM package are compared by the reached speed-up.

1 Introduction

The parallelization of finite element algorithm is considered in the paper. Since the finite element method is widely used in the engineering problems the parallelization of the source code of the program, which leads to the considerable time reduction of the calculation process is very important from the practical point of view.

The FEM package HYDRO-GEO (Dłużewski 1997) oriented at hydro and geotechnical problems is used. The program has been developed at Warsaw University of Technology in the sequential form and next extended to allow the parallel calculations. The sequential version of the program has been the starting point for the step by step development of the parallel versions. The package is written in the Fortran language. The parallel versions of the finite element code are tested on the multiprocessor Cray 6400 and next the Sun E10000 HPC. The directives of Cray Fortran compiler are applied to create threads in the parallel versions.

The engineering problems like a dam building, filling of the reservoir and the settlement of the subsoil, can be modelled, calculated and analysed by means of finite element approach in parallel version. The calculation speed-up depends on the scale of the problem i.e. finite element mesh, nonlinearity of the problem, level of parallelization – how many subproblems are generated during the calculations.

Section 2 contains the short description of numerical procedure used in the finite element modelling of the the elasto-plastic behaviour of the structure in geomechanics and implemented in the HYDRO-GEO package. In the section 3 the parallelization methods of the source code by the Cray Fortan compiler directives, applied in the package, are presented. The parallel algorithm implemented in the package is described in section 4. Section 5 contains the calculation results for chosen example – the dam in Besko, Poland.

T. Sørevik et al. (Eds.): PARA 2000, LNCS 1947, pp. 391–398, 2001.
© Springer-Verlag Berlin Heidelberg 2001

2 Numerical Procedure

The virtual work principle, continuity equation with boundary conditions are the starting points for numerical formulation. The finite element method is applied to solve the initial boundary value problems. Several procedures stemming from the elasto-plastic modelling can be coupled with the incremental iterative procedures. The elasto-plastic soil behaviour is modeled by means of visco-plastic theory (Perzyna, 1966). The finite element formulation for the elasto-plastic material behaviour combines overlapping of the iterative processes.

The pseudo-viscoplastic algorithm for numerical modeling of elasto-plastic behaviour is used after Zienkiewicz and Cormeau (1974). The stability of the time marching scheme was proved by Cormeau (1975). The pseudo-viscous algorithm developed in finite element computer code is successfully applied to solve a number of boundary value problems, Dłużewski (1993). The visco-plastic procedure is extended to cover the geometrically non-linear problems by Kanchi et al (1978) and also developed for large strains in consolidation, Dłużewski (1997). The parallel version of pseudo-viscous procedure is developed herein for modelling elasto-plastic behavior in the multistage process.

The global set of the equilibrium equations takes the form as follows

$$ K \ \Delta u = \Delta F \tag{1} $$

where K is the stiffness array, Δu are the nodal displacements increments, ΔF is the nodal vector defined below

$$ \Delta F^i = \Delta F_L + \Delta R_I{}^i + \Delta R_{II}{}^i \tag{2} $$

ΔF_L is the load increment, ΔR_I^i is the vector of nodal forces due to pseudo-visco iteration, ΔR_I^i takes the following form

$$ \Delta R_I^i = \underset{V}{} B^T D \Delta \varepsilon_i^{vp} dv \tag{3} $$

which stems from the pseudo visco-plastic flow. In the above equation B is the geometrical array, D is the elastic material array and $\Delta \varepsilon_i^{vp}$ is the visco-plastic strain increment in the i-th iteration. ΔR_I^i stands for the nodal vector which results from the relaxation of the stresses. For each time step the iterative procedure is engaged to solve the material non-linear problem. The i-th indicates steps of the iterations. Both local and global criterions for terminating the iterative process are used. The iterations are continued until the calculated stresses are acceptable close to the yield surface, F < Tolerance at all checked points, where F is the value of the yield function. At the same time the global criterion for this procedure is defined at the final equilibrium equations. For two phase medium, the unbalanced nodal vector $\Delta R_{II}{}^i$ is calculated every iterative pseudo-time step.

$$\Delta R_{II}^{i} = P - \int_{V} B^{T}(\sigma' + mp)dv \qquad (4)$$

where P is a load vector, σ' are the effective stresses, m is zero, one vector, p is the pore pressure.

The square norm on the unbalanced nodal forces is used as the global criterion of equilibrium. The iterative process is continued until both, local and global criterions are fulfilled.

3 Parallel Machines and Parallel Programs

The parallel versions of the FEM package are run on the Cray 6400 and next on the Sun E1000 HPC machines, which are located at COI PW (Computing Centre of Warsaw University of Technology). These supercomputers are the shared memory parallel machines working as the multiusers systems. The Sun supercomputer has symmetric architecture with 12 processors 400MHz, 2GB RAM and 72GB of the disk memory. The highly specialised software, especially Cray Fortran which enables the parallelization of the sources during the compilation process, is used. Special directives (pragmas) which are inserted into the source code of the program control the creation of the threads.

3.1 Parallelization of the Source Code

The parallelization of the code by the compiler is done by splitting the loops into parts executed concurrently by separate processors. The compiler can parallelize certain kinds of loops, which include arrays. We can allow the compiler determine which loops can be parallelized or point them up by special pragmas inserted just before these loops. The parallelizer is integrated with the optimisation during the compilation process.

Two methods of parallelization of the FORTRAN sources code are possible: automatic and explicit. In the case of automatic parallelization of Fortran program by the compiler the loops only are executed concurrently. The compiler analyses the dependency between the data in the DO Fortran commands and divides its execution between the separate processors by creation of the set of threads. The compiler does not analyse the method. Automatic parallelizator splits DO loop which have no inter-iteration data dependencies. The compiler finds and parallizes any loop which meets the following criteria:

- the values of the array variables for each iteration of the loop do not depend on the values of array variables for any other iteration of the loop
- calculation within the loop does not conditionally change any pure scalar variable that is referenced after the loop termination
- calculation within the loop does not change a scalar variable across iteration (this is called loop-carried dependency).

The number of processors required or expected to be available during program execution should be set before the compilation process. This number should be smaller than the number of processors in the system. In other case i.e. if we compile our program splitting it for more processes than it is possible to execute concurrently by our parallel machine, the threads will wait in the queue for execution, what can degrade the performance of the program.

The second method is the explicit parallelism. The appropriate pragma should be inserted before each loop command we require to be split for the concurrent execution. In this method the knowledge on the implemented methods is necessary. So more in most cases same changes in the code are required because the same conditions as during automatic parallelization should be fulfilled. The result generally is better than in the automatic parallelization.

3.2 Debugging

UNIX commands allow debugging the time of calculation. The speedup of parallel calculation can be obtained for three times: user, system and real. The user time is the amount of time the CPU is active for a given process, excluding the operating system overhead. The system time is the amount of time when the operating system is active for the given thread. The real time is the difference between the time when the process begins and the time that it ends. It is a time the user is waiting for the execution of his program to complete.

4 FEM Algorithm Implementation

FEM algorithm can be divided into the three separate parts:
1. mesh generation, mesh optimalization and data input
2. stiffness matrix calculation, solving the global set of equations and analysis of strains and stresses
3. graphical and numerical presentation of the computed results

The HYDRO-GEO package is composed of the three main programs: preprocessor, processor for main numerical calculations and graphical postprocessor. These programs are running under the management program.

In the approach each part of the finite element code can be paralleled independently. In this paper the tests of the processor parallelization (with the stiffness matrix calculation, solver and analysis of strains and stresses) is presented. This part is the most time consuming and worth to be parallelized.

4.1 Parallelization of HYDRO-GEO Package

The parallel version has been developed step by step. The sequential package works with the auxiliary files for collecting data in the calculation process. First these files are replaced by COMMON statements to have the processor program working in the memory.

Next, the compilation with the option of automatic parallelization is done. The first parallel version of the processor is prepared.

In the second parallel version the group of source code of the procedure which calculate the values for the single elements is changed to allow the parallel execution. The critical sections connected with global variables are cancelled. The pragmas that control the explicit parallelization are inserted before loops, which should be split onto threads.

Replacing the solver procedure creates the latest version, which is tested, the third one. In the sequential version a slow front procedure is implemented. It needs a small amount of memory. It was replaced on the parallel Gaussian elimination with band matrix and parallel backward substitution procedures.

The processor parallel algorithm for the stages of structure building looks as follows:

```
Start
Data reading, initial computations
For each stage of the structure buiding and each
increment of  load do
    For each subset of the set of element do in
    parallel
        Calculate the local stiffness matrix, load
        vectors
        Calculate the initial stresses,
    End do in parallel
    Calculate the global set of equilibrium system
    First part of the solver (forward substitution)-
                                   parallel calculation
    For each plastic points do
        Second part of the solver (backward
        substitution)-parallel calculation
        For each subset of the set of element do in
        parallel
            Calculation of strains and stresses
        End do in parallel
    End do
End do
Stop
```

The loops where the values for each element are calculated are divided into threads, which are executed concurrently. The problem is tested for different number of threads. The Sun E10000 HPC has 12 processors, so 11 limit the number of threads. The threads are created for each element loop. The number of elements n is much bigger than the number of possible concurrent threads p so one thread calculates the values for n/p elements. All results are collected in the common memory.

5 Example of the Parallel Calculation

As the engineering example the dam construction is modelled, calculated and analysed by means of finite element approach in parallel version. The speed up depends on the scale of the problem i.e. finite element mesh, nonlinearly of the problem, level of parallelization – how many subproblems are generated during the calculations.

The Besko dam has been risen on the Carpathian flysch. The height of the dam is about 40m. The inclined schist layers are located in the subsoil. The parallel schist layers with various material parameters create the specific foundation typical for Polish dams in the south.

The dam is built from concrete. The clay-concrete screen of 0.8 m thickness is performed. The height of the screen is about 25m. The material zones are presented on Figure 1.

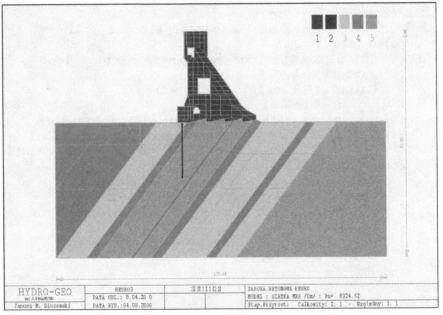

Fig. 1. Material zones.

The numerical modelling is done in three stages. In the first stage the initial stresses are introduced into the subsoil. In the second one, the heavy concrete dam is built. The special teeth are done between the subsoil and the dam body for better interaction between the dam and subsoil. The rising of the dam is done by adding the elements. In the third stage the loading causes by filling the reservoir is applied. The one of the calculated results – the stresses in vertical direction are presented on Figure 2.

The user time for different parallel versions of the package is compared to evaluate the performance. The system time is connected mostly with the synchronisation of the threads and the number of input/output commands, which is the same in all parallel versions.

The average speed up for sequential version in the memory comparing the sequential version of the HYDRO-GEO processor working with the auxiliary files is about 10.

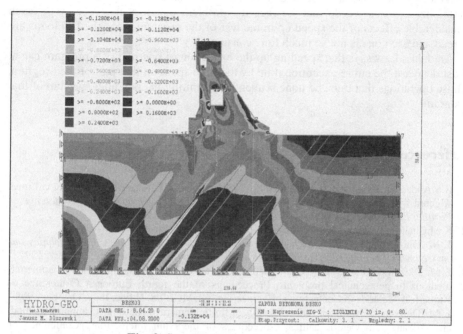

Fig. 2. Stresses in the vertical direction

In the case of automatic parallclization the speed up for user time comparing the sequential version working into memory is about 2 and does not change much when we change the number of processors. The supercomputer works as the multiuser system. The speedup depends on the number of processes that are working simultaneously.

The parallel versions are compared with the sequential version working in the memory because they take roots from it. The average speed up is following:

1. for the version of the processor obtained by the compilation with the option autoparallel the speed up is about 2
2. for the version of the processor obtained by the compilation with the option parallel (i.e. explicit and auto) the speed up is about 3,5
3. for the version of the processor obtained by the compilation with the option parallel (i.e. explicit and auto) and with band matrix parallel solver the speed up is about 5

Conclusions

The highest speed up obtained in the example is about 6, running the program on 12 processors supercomputer. Considering the parallelization of the finite element source code the first steps are obvious. First the solver and second the element procedures (calculating the stiffness matrix and nest calculating the strains and stresses) are parallelized. When we do that we face the problem that it is not so easy to reach considerable effects of the speed up in the rest of the program since the DO loops are spread and each one is not so much time consuming.

Amdahl's Law says that speeding up the execution of one part of a program can at most decrease the entire execution time by the time spent on that part of the program. Those operations that must be done sequentially limiting the potential speedup of that program.

References

1. B.S.Andersen, F.Gustavson, A.Karaivanov, J.Waśniewski, P.Y.Yalomov, LAWRA: Linear Algebra with Recursive Algorithms, *Proceedings of the Third International conference on Parallel Processing and Applied Mathematics*, Czestochowa 1999
2. K.M.Chandy, J.Misra, *Parallel Program Design*, Addison-Wesley 1989
3. J. M. Dłużewski, *HYDRO-GEO – finite element code for geotechnics, hydrotechnics and environmental engineering*, (in Polish), Warsaw University of Technology, Warsaw, 1997
4. J. M. Dłużewski, Large strain consolidation for elesto-plastic soils, Aplication of numerical methods to geotechnical problems, Proceedings of the fourth European Conference on Numerical Methods in Geotechnical Engineering, NUMGE'98, Udine, Italy, October 1998
5. J. M. Dłużewski, Nonlinear consolidation in finite element modelling, Computer methods and Advances in Geomechanics, Proceedings of the Ninh International Conference on Computer Methods and Advances in Geomechanics, Wuhan, China, November 1997
6. L.D.Fosdick, C.J.C.Schauble, E.R.Jessup, Computer Performance: A Tutorial, Draf: December 19, 1994, http://www//netlib/benchmark.

Author Index

Lecture Notes in Computer Science

For information about Vols. 1–1920
please contact your bookseller or Springer-Verlag

Vol. 1952: M.C. Monard, J. Simão Sichman (Eds.), Advances in Artificial Intelligence. Proceedings, 2000. XV, 498 pages. 2000. (Subseries LNAI).

Vol. 1953: G. Borgefors, I. Nyström, G. Sanniti di Baja (Eds.), Discrete Geometry for Computer Imagery. Proceedings, 2000. XI, 544 pages. 2000.

Vol. 1954: W.A. Hunt, Jr., S.D. Johnson (Eds.), Formal Methods in Computer-Aided Design. Proceedings, 2000. XI, 539 pages. 2000.

Vol. 1955: M. Parigot, A. Voronkov (Eds.), Logic for Programming and Automated Reasoning. Proceedings, 2000. XIII, 487 pages. 2000. (Subseries LNAI).

Vol. 1956: T. Coquand, P. Dybjer, B. Nordström, J. Smith (Eds.), Types for Proofs and Programs. Proceedings, 1999. VII, 195 pages. 2000.

Vol. 1957: P. Ciancarini, M. Wooldridge (Eds.), Agent-Oriented Software Engineering. Proceedings, 2000. X, 323 pages. 2001.

Vol. 1960: A. Ambler, S.B. Calo, G. Kar (Eds.), Services Management in Intelligent Networks. Proceedings, 2000. X, 259 pages. 2000.

Vol. 1961: J. He, M. Sato (Eds.), Advances in Computing Science – ASIAN 2000. Proceedings, 2000. X, 299 pages. 2000.

Vol. 1963: V. Hlaváč, K.G. Jeffery, J. Wiedermann (Eds.), SOFSEM 2000: Theory and Practice of Informatics. Proceedings, 2000. XI, 460 pages. 2000.

Vol. 1964: J. Malenfant, S. Moisan, A. Moreira (Eds.), Object-Oriented Technology. Proceedings, 2000. XI, 309 pages. 2000.

Vol. 1965: Ç. K. Koç, C. Paar (Eds.), Cryptographic Hardware and Embedded Systems – CHES 2000. Proceedings, 2000. XI, 355 pages. 2000.

Vol. 1966: S. Bhalla (Ed.), Databases in Networked Information Systems. Proceedings, 2000. VIII, 247 pages. 2000.

Vol. 1967: S. Arikawa, S. Morishita (Eds.), Discovery Science. Proceedings, 2000. XII, 332 pages. 2000. (Subseries LNAI).

Vol. 1968: H. Arimura, S. Jain, A. Sharma (Eds.), Algorithmic Learning Theory. Proceedings, 2000. XI, 335 pages. 2000. (Subseries LNAI).

Vol. 1969: D.T. Lee, S.-H. Teng (Eds.), Algorithms and Computation. Proceedings, 2000. XIV, 578 pages. 2000.

Vol. 1970: M. Valero, V.K. Prasanna, S. Vajapeyam (Eds.), High Performance Computing – HiPC 2000. Proceedings, 2000. XVIII, 568 pages. 2000.

Vol. 1971: R. Buyya, M. Baker (Eds.), Grid Computing – GRID 2000. Proceedings, 2000. XIV, 229 pages. 2000.

Vol. 1972: A. Omicini, R. Tolksdorf, F. Zambonelli (Eds.), Engineering Societies in the Agents World. Proceedings, 2000. IX, 143 pages. 2000. (Subseries LNAI).

Vol. 1973: J. Van den Bussche, V. Vianu (Eds.), Database Theory – ICDT 2001. Proceedings, 2001. X, 451 pages. 2000.

Vol. 1974: S. Kapoor, S. Prasad (Eds.), FST TCS 2000: Foundations of Software Technology and Theoretical Computer Science. Proceedings, 2000. XIII, 532 pages. 2000.

Vol. 1975: J. Pieprzyk, E. Okamoto, J. Seberry (Eds.), Information Security. Proceedings, 2000. X, 323 pages. 2000.

Vol. 1976: T. Okamoto (Ed.), Advances in Cryptology – ASIACRYPT 2000. Proceedings, 2000. XII, 630 pages. 2000.

Vol. 1977: B. Roy, E. Okamoto (Eds.), Progress in Cryptology – INDOCRYPT 2000. Proceedings, 2000. X, 295 pages. 2000.

Vol. 1978: B. Schneier (Ed.), Fast Software Encryption. Proceedings, 2000. VIII, 315 pages. 2001.

Vol. 1979: S. Moss, P. Davidsson (Eds.), Multi-Agent-Based Simulation. Proceedings, 2000. VIII, 267 pages. 2001. (Subseries LNAI).

Vol. 1983: K.S. Leung, L.-W. Chan, H. Meng (Eds.), Intelligent Data Engineering and Automated Learning – IDEAL 2000. Proceedings, 2000. XVI, 573 pages. 2000.

Vol. 1984: J. Marks (Ed.), Graph Drawing. Proceedings, 2001. XII, 419 pages. 2001.

Vol. 1987: K.-L. Tan, M.J. Franklin, J. C.-S. Lui (Eds.), Mobile Data Management. Proceedings, 2001. XIII, 289 pages. 2001.

Vol. 1989: M. Ajmone Marsan, A. Bianco (Eds.), Quality of Service in Multiservice IP Networks. Proceedings, 2001. XII, 440 pages. 2001.

Vol. 1991: F. Dignum, C. Sierra (Eds.), Agent Mediated Electronic Commerce. VIII, 241 pages. 2001. (Subseries LNAI).

Vol. 1992: K. Kim (Ed.). Public Key Cryptography. Proceedings, 2001. XI, 423 pages. 2001.

Vol. 1993: E. Zitzler, K. Deb, L. Thiele, C.A.Coello Coello, D. Corne (Eds.), Evolutionary Multi-Criterion Optimization. Proceedings, 2001. XIII, 712 pages. 2001.

Vol. 1995: M. Sloman, J. Lobo, E.C. Lupu (Eds.), Policies for Distributed Systems and Networks. Proceedings, 2001. X, 263 pages. 2001.

Vol. 1998: R. Klette, S. Peleg, G. Sommer (Eds.), Robot Vision. Proceedings, 2001. IX, 285 pages. 2001.

Vol. 2000: R. Wilhelm (Ed.), Informatics: 10 Years Back, 10 Years Ahead. IX, 369 pages. 2001.

Vol. 2003: F. Dignum, U. Cortés (Eds.), Agent Mediated Electronic Commerce III. XII, 193 pages. 2001. (Subseries LNAI).

Vol. 2004: A. Gelbukh (Ed.), Computational Linguistics and Intelligent Text Processing. Proceedings, 2001. XII, 528 pages. 2001.

Vol. 2006: R. Dunke, A. Abran (Eds.), New Approaches in Software Measurement. Proceedings, 2000. VIII, 245 pages. 2001.

Vol. 2009: H. Federrath (Ed.), Designing Privacy Enhancing Technologies. Proceedings, 2000. X, 231 pages. 2001.

Vol. 2010: A. Ferreira, H. Reichel (Eds.), STACS 2001. Proceedings, 2001. XV, 576 pages. 2001.

Vol. 2024: H. Kuchen, K. Ueda (Eds.), Functional and Logic Programming. Proceedings, 2001. X, 391 pages. 2001.